NEXT GENERATION TELECOMMUNICATIONS NETWORKS, SERVICES, AND MANAGEMENT

IEEE Press
445 Hoes Lane
Piscataway, NJ 08854

Technical Reviewer
Lakshmi Raman

Books in the IEEE Press Series on Network Management

Telecommunications Network Management Into the 21st Century,
Co-Editors Thomas Plevyak and Salah Aidarous, 1994

Telecommunications Network Management: Technologies and Implementations,
Co-Editors Thomas Plevyak and Salah Aidarous, 1997

Fundamentals of Telecommunications Network Management,
by Lakshmi Raman, 1999

Security for Telecommunications Management Network,
by Moshe Rozenblit, 2000

Integrated Telecommunications Management Solutions,
by Graham Chen and Quinzheng Kong, 2000

Managing IP Networks: Challenges and Opportunities,
Co-Editors Thomas Plevyak and the late Salah Aidarous, 2003

NEXT GENERATION TELECOMMUNICATIONS NETWORKS, SERVICES, AND MANAGEMENT

Edited by

THOMAS PLEVYAK

VELI SAHIN

IEEE Communications Society, *Sponsor*

Thomas Plevyak and Veli Sahin, *Series Editors*

IEEE PRESS

A JOHN WILEY & SONS, INC., PUBLICATION

Published by John Wiley & Sons, Inc., Hoboken, New Jersey.
Published simultaneously in Canada.

Library of Congress Cataloging-in-Publication Data:

Plevyak, Thomas.
Next generation telecommunications networks, services, and management / Thomas Plevyak, Veli Sahin.
p. cm.
ISBN 978-0-470-57528-4 (cloth)
1. Telecommunication systems–Forecasting. 2. Computer networks–Forecasting.
I. Sahin, Veli. II. Title.
TK5102.5.P59 2010
621.382–dc22

2009036488

Printed in the United States of America.

10 9 8 7 6 5 4 3 2 1

The Editors and Authors
Dedicate This Book to Their Families:
The Cornerstone of Successful Societies

CONTENTS

GUEST INTRODUCTIONS xv

EDITOR AND CONTRIBUTOR BIOGRAPHIES xix

CHAPTER 1 *CHANGES, OPPORTUNITIES, AND CHALLENGES* 1

Veli Sahin and Thomas Plevyak

1.1 Introduction 1
1.2 Scope 2
1.3 Changes, Opportunities, and Challenges 2
  1.3.1 Major Life Style Changes: Desktops, Laptops, and Now Handtops 2
  1.3.2 Major Network Infrastructure Changes 3
  1.3.3 Major Home Network (HN) Changes 4
  1.3.4 Major FCAPS Changes 4
  1.3.5 Major Regulatory Changes 5
  1.3.6 Service Aware Networks to Manage Expectations and Experiences 5
1.4 Major Management Challenges for a Value-Added Service: Triple Shift Service 7
1.5 The Grand Challenge: System Integration and Interoperability of Disjoined Islands 8
1.6 Some Examples of Management System Applications 10
  1.6.1 Event Correlation 10
  1.6.2 Hot Spot Identification and SMS Actions 11
  1.6.3 SLAs, Contracts, and Policy Management 12
    1.6.3.1 Service Assessment 12
    1.6.3.2 Contract Assessment 12
    1.6.3.3 Service and Contract Assurance 12
  1.6.4 SMS Integration with Planning and Engineering Systems 13
1.7 Overview of Book Organization and Chapters 13
1.8 References 14

CHAPTER 2 *MANAGEMENT OF TRIPLE/QUADRUPLE PLAY SERVICES FROM A TELECOM PERSPECTIVE* 15

Jean Craveur

2.1 Introduction 15
2.2 Context of Triple/Quadruple Play for Telecom Operators 15
2.3 The Economic, Service, and Commercial Challenges 18
  2.3.1 General Conditions 18
  2.3.2 Service Offer Requirements 19
2.4 The Technical Challenge 20

2.4.1 The Technical Tool Box **21**
2.4.1.1 Customer Equipment **21**
2.4.1.2 Access Line and Aggregation/Backhaul Networks **21**
2.4.1.3 Backbone Networks **22**
2.4.1.4 Control Platform **22**
2.4.1.5 Service Platform **22**
2.4.1.6 IS Equipment **22**
2.4.2 The Global Vision **23**
2.4.2.1 Vision for an Overall Architecture Supporting Triple and Quadruple Play **23**
2.4.3 Key Issues to Consider When Designing Network and IS Infrastructures for Triple and Quadruple Play **24**
2.4.3.1 Convergence and Mutualization **25**
2.4.3.2 Quality of Service (QoS) **25**
2.4.4 Customer Premises Equipment (CPE) and Home Network **26**
2.4.4.1 The Home Network Complexity **26**
2.4.4.2 Distribution of Functions between Network and IS Platforms and Residential Gateways **27**
2.4.4.3 The Home Network Paradox **27**
2.4.4.4 The Home Device and Applications **28**
2.4.5 Access Lines **28**
2.4.6 Access Networks, Aggregation, and Backhauling **29**
2.4.7 An Illustration of the Fixed Access Network Transformation from Internet Access Support to Triple Play Support **30**
2.4.8 Backbone Networks **31**
2.4.8.1 Content Delivery **32**
2.4.9 Service and Resource Control **33**
2.4.9.1 Core Control and Application Servers **33**
2.4.9.2 Service Platforms **33**
2.4.10 Information System **33**
2.4.10.1 A Renovated IS Architecture for Triple/Quadruple/Multiple Play Business **35**
2.4.10.2 The Customer Front-End **36**
2.4.10.3 The Aggregation Layer **37**
2.4.10.4 The Back-End **37**
2.4.10.5 Order Management and Delivery **39**
2.4.10.6 A Crucial Cooperation between IS, Network, and Service Platform **39**
2.5 The Operational Challenge **40**
2.5.1 Focus on the Service Management Center Function (SMC) **42**
2.5.2 IS Tools for the SMCs **43**
2.5.3 Operating IT and Service Platforms in Triple and Quadruple Play Contexts **44**
2.5.4 Roles and Responsibilities of the Different Functions **45**
2.5.5 New Skills in Operations **47**
2.6 The Customer Experience in Broadband Triple Play **47**
2.6.1 Definition of the Offerings **48**
2.6.2 Distribution Channels **49**
2.6.3 Relationship with the Local Operator **49**

2.6.4 The Customer Journey 49
2.7 The Organizational Challenge 51
2.8 Conclusions 51
2.9 Acknowledgments 52
2.10 References 52
2.11 Suggested Further Reading 52

CHAPTER 3 *MANAGEMENT OF TRIPLE/QUAD PLAY SERVICES FROM A CABLE PERSPECTIVE* 53

David Jacobs

3.1 Introduction 53
3.2 The HFC Network 55
3.2.1 HFC Planning and Inventory 55
3.2.2 HFC Network Maintenance 56
3.2.3 HFC Network Upgrades 56
3.3 Digital TV 57
3.3.1 Digital TV: Coding and Transmission of Analogue Information 58
3.3.2 Network Information Table (NIT) 62
3.3.3 DVB-SI Program Decoding 62
3.3.4 ATSC-PSIP Program Decoding 62
3.3.5 Conditional Access 63
3.3.6 Out-of-Band Channels 64
3.3.7 Digital Storage Media—Command and Control (DSM-CC) 64
3.3.8 Switched Digital Video 65
3.3.9 Enhanced TV/Interactive TV 67
3.3.9.1 Enhanced TV Binary Interchange Format 69
3.3.10 DOCSIS Set-Top Gateway 69
3.3.11 Digital TV Head-End 70
3.3.12 Integrated Receiver/Decoder or Set-Top Box 71
3.3.13 Point of Deployment Module/CableCard 72
3.4 Data over Cable Service Interface Specification (DOCSIS) 73
3.4.1 Physical Layer 74
3.4.2 Data Link Layer 76
3.4.2.1 Media Access Control (MAC) Sublayer 76
3.4.2.2 Link Layer Security 78
3.4.2.3 Logical Link Control (LLC) 79
3.4.3 Network Layer 79
3.4.4 Multicast Operation 80
3.4.5 Cable Modem Start-up 80
3.4.6 IP Detail Records 81
3.4.7 DOCSIS Evolution 82
3.5 Cable Telephony 83
3.5.1 Cable IP Telephony 84
3.5.1.1 Network Control Signaling PacketCable 1.0 and 1.5 85
3.5.1.2 Distributed Call Signaling 90
3.5.1.3 Embedded MTA Start-up 90
3.5.1.4 PacketCable 2.0 91
3.6 Wireless 96

3.7 Cable Futures 97
3.8 References 98

CHAPTER 4 *NEXT GENERATION TECHNOLOGIES, NETWORKS, AND SERVICES* 101

Bhumip Khasnabish
4.1 Introduction 101
4.2 Next Generation (NG) Technologies 102
4.2.1 Wireline NG Technologies 102
4.2.1.1 Fiber to the Premises (FTTP) 103
4.2.1.2 Long-Haul Managed Ethernet (over Optical Gears) 103
4.2.2 Wireless NG Technologies 104
4.2.2.1 Broadband Bluetooth and ZigBee 104
4.2.2.2 Personalized and Extended Wi-Fi 104
4.2.2.3 Mobile Worldwide Inter-operability for Microwave Access (M-WiMax) 105
4.2.2.4 Long Term Evolution (LTE) 106
4.2.2.5 Enhanced HSPA 106
4.2.2.6 Evolution Data Optimized (EVDO) and Ultra Mobile Broadband (UMB) 106
4.2.2.7 Mobile Ad Hoc Networking (MANET) and Wireless Mesh Networking (WMN) 106
4.2.2.8 Cognitive (and Software Defined) Radios and Their Interworking 107
4.2.3 Software and Server NG Technologies (Virtualization) 107
4.3 Next Generation Networks (NGNs) 108
4.3.1 Transport Stratum 108
4.3.2 Service Stratum 110
4.3.3 Management 110
4.3.3.1 Fault Management 110
4.3.3.2 Configuration Management 110
4.3.3.3 Accounting Management 111
4.3.3.4 Performance Management 111
4.3.3.5 Security Management 111
4.3.4 Application Functions 112
4.3.5 Other Networks: Third-Party Domains 112
4.3.6 End-User Functions: Customer Premises Devices and Home Networks 113
4.3.7 Internet Protocol (IP): The NGN Glue 113
4.3.7.1 Internet Protocol version 4 (IPv4) 113
4.3.7.2 Internet Protocol version 6 (IPv6) 114
4.3.7.3 Mobile Internet Protocol version 6 (MIPv6) 114
4.4 Next Generation Services 114
4.4.1 Software-Based Business Services 114
4.4.2 High-Definition (HD) Voices 115
4.4.3 Mobile and Managed Peer-to-Peer (M2P2P) Service 115
4.4.4 Wireless Charging of Hand-Held Device 115
4.4.5 Three-Dimensional Television (3D-TV) 116
4.4.6 Wearable, Body-Embedded Communications/Computing Including Personal and Body-Area Networks 116
4.4.7 Converged/Personalized/Interactive Multimedia Services 116

4.4.8 Grand-Separation for Pay-per-Use Service 117
4.4.9 Mobile Internet for Automotive and Transportation 117
4.4.10 Consumer- and Business-Oriented Apps Storefront 117
4.4.11 Evolved Social Networking Service (E-SNS) 118
4.4.12 NG Services Architectures 118
4.4.13 Application Plane's Requirements to Support NG Services 120
4.4.14 Transport Plane's Requirements to Support NG Services 120
4.5 Management of NG Services 121
4.5.1 IP- and Ethernet-Based NG Services 121
4.5.2 Performance Management of NG Services 122
4.5.3 Security Management of NG Services 123
4.5.4 Device Configuration and Management of NG Services 123
4.5.5 Billing, Charging, and Settlement of NG Services 124
4.5.6 Faults, Overloads, and Disaster Management of NG Services 124
4.6 Next Generation Society 124
4.6.1 NG Technology-Based Humane Services 125
4.6.2 Ethical and Moral Issues in Technology Usage 125
4.7 Conclusions and Future Works/Trends 126
4.8 References 127

CHAPTER 5 *IMS AND CONVERGENCE MANAGEMENT* 129

Keizo Kawakami, Kaoru Kenyoshi, and Toshiyuki Misu
5.1 IMS Architecture 129
5.1.1 Serving CSCF (S-CSCF) 130
5.1.2 Proxy CSCF (P-CSCF) 131
5.1.3 Interrogating CSCF (I-CSCF) 132
5.2 IMS Services 133
5.2.1 Push to Talk over Cellular (PoC) Service 133
5.2.1.1 Service Authentication 133
5.2.1.2 Floor Information Management 133
5.2.1.3 Message Duplication and Transmission in 1-to-n Communication 133
5.2.2 IMS-Based FMC Service 134
5.2.2.1 CSCF 134
5.2.2.2 PDG 134
5.2.3 IMS-Based IPTV Service 134
5.3 QoS Control and Authentication 135
5.3.1 QoS Control in NGN 135
5.3.2 RACS 136
5.3.2.1 Functions Provided by RACS 136
5.3.2.2 Function Blocks Comprising RACS 137
5.3.3 Authentication in NGN 138
5.3.4 NASS 138
5.4 Network and Service Management for NGN 139
5.4.1 Introduction 139
5.4.2 Network Management Operation Requirements 141
5.4.3 Service Management Operation Requirements 142
5.4.4 Service Enhancement Requirements 143
5.4.5 B2B Realization Requirements 143

5.4.6 Compliance with Legal Restrictions Requirements 144
5.5 IMS Advantages 144
5.5.1 Reduction of Maintenance and Operating Cost 144
5.5.1.1 Reduction of Time Required for Introducing New Services (Time to Market) 145
5.5.1.2 Cost Merits 145
5.5.2 Roles of SDP and Development and Introduction of New Services 145
5.5.2.1 Positioning of SDP in NGN 145
5.5.2.2 Features of SDP 146
5.5.2.3 Examples of Application Servers 146
5.5.2.4 API 149
5.5.3 Services Implemented on NGN 150
5.5.3.1 Push to X 150
5.5.3.2 IPTV 151
5.5.3.3 IPTV Architectures 151
5.5.3.4 Advantages of NGN (IMS-based) IPTV 152
5.6 References 153
5.7 Suggested Further Reading 153

CHAPTER 6 *NEXT GENERATION OSS ARCHITECTURE* 155

Steve Orobec
6.1 Introduction 155
6.2 Why Are Standards Important to OSS Architecture? 156
6.3 The TeleManagement Forum (TM Forum) for OSS Architecture 158
6.4 Other Standards Bodies 159
6.5 TM Forum's Enhanced Telecommunications Operations Map (eTOM) 159
6.5.1 Relationship to ITIL (Infrastructure Technology Information Library) 162
6.6 Information Framework 163
6.7 DMTF CIM (Distributed Task Force Management) 165
6.8 TIP (TM Forum's Interface Program) 166
6.9 NGOSS Contracts (aka Business Services) 167
6.10 MTOSI Case Study 170
6.10.1 Will Web Services and MTOSI Scale? 170
6.11 Representational State Transfer (REST)—A Silver Bullet? 176
6.12 Real Network Implementation of a Standard 177
6.13 Business Benefit 179
6.14 OSS Transition Strategies 181
6.15 ETSI TISPAN and 3GPP IMS 182
6.16 OSS Interaction with IMS and Subscriber Management (SuM) 183
6.17 NGN OSS Function/Information View Reference Model 187
6.18 Designing Technology-Neutral Architectures 189
6.19 UML and Domain Specific Languages (DSLs) 189
6.20 An Emerging Solution: The Domain Specific Language 192
6.21 From Model-Driven Architecture to Model-Driven Software Design 193
6.22 Other Standards Models (DMTF CIM, 3GPP, and TISPAN) 194
6.23 Putting Things Together: Business Services in Depth 195
6.24 Building a DSL-Based Solution 200
6.24.1 Problem Context 200
6.24.2 Proposed Initial Feature Content 200

6.24.2.1 Desired Inputs 200
6.24.2.2 Desired Outputs 201
6.24.3 Open-source Tool Environments 201
6.25 Final Thought 205
6.26 Bibliography 205

CHAPTER 7 *MANAGEMENT OF WIRELESS AD HOC AND SENSOR NETWORKS* 207

Mehmet Ulema
7.1 Introduction 207
7.2 Overview 208
7.2.1 Wireless Ad Hoc Networks 209
7.2.2 Wireless Sensor Networks 210
7.2.3 Wireless Ad Hoc Networks *vs.* Sensor Networks 211
7.2.4 Network Management Aspects and Framework 212
7.3 Functional and Physical Architectures 213
7.4 Logical Architectures 214
7.5 Information Architectures 216
7.5.1 Manager-Agent Communication Models 217
7.5.2 Management Interfaces and Protocols 223
7.5.3 Structure of Management Information and Models 223
7.5.4 Others 228
7.6 Summary and Conclusions 228
7.7 References 229

CHAPTER 8 *STRATEGIC STANDARDS DEVELOPMENT AND NEXT GENERATION MANAGEMENT STANDARDS* 231

Michael Fargano
8.1 Introduction 231
8.1.1 General Drivers for Standards 232
8.1.2 Management Standards History 232
8.2 General Standards Development Process 233
8.2.1 Key Attributes of Standards Development Process 234
8.2.2 General SDO/Forum Types and Interactions 235
8.2.3 General Standards Development and Coordination Framework 235
8.2.3.1 Project Execution and Cross-Organization Interactions and Handoff Points 238
8.3 Management SDO/Forum Categories 239
8.3.1 General Network/Service SDO/Forum 239
8.3.2 Specific Network/Service SDO/Forum 239
8.3.3 Information Technology SDO/Forum 239
8.3.4 Management-Standards Focused SDO/Forum 240
8.4 Principles, Frameworks, and Architecture in Management Standards 240
8.4.1 Principles and Concepts in Management Standards Development 240
8.4.2 Frameworks and Architecture 241
8.5 Strategic Framework for Management Standards Development 244
8.5.1 Strategic Questions for Standards Engagement Determination 244
8.5.2 Strategic Progression of Standards Work 245
8.5.3 Strategic Human Side of Standards Development 245

8.6 Sampling of NGN Management Standards Areas and SDO/Forums **245**
8.7 Summary and Conclusions **248**
8.7.1 Chapter Summary **248**
8.7.2 General Standards Development Process **248**
8.7.3 Management SDO/Forum Categories **248**
8.7.4 Principles, Frameworks, and Architecture in Management Standards **248**
8.7.4.1 Principles **248**
8.7.4.2 Frameworks and Architecture **249**
8.7.5 Strategic Framework for Management Standards Development **249**
8.7.5.1 Strategic Progression of Standards Work **249**
8.7.5.2 Strategic Human Side of Standards Development **249**
8.7.6 Key Lessons Learned for Strategic NGN Management Standards Development **250**
8.7.7 Challenges and Trends **250**
8.8 References **250**

CHAPTER 9 *FORECAST OF TELECOMMUNICATIONS NETWORKS AND SERVICES AND THEIR MANAGEMENT (WELL) INTO THE 21ST CENTURY* **253**

Roberto Saracco
9.1 Have We Reached the End of the Road? **254**
9.2 "Glocal" Innovation **257**
9.3 Digital Storage **259**
9.4 Processing **261**
9.5 Sensors **262**
9.6 Displays **263**
9.7 Statistical Data Analyses **265**
9.8 Autonomic Systems **267**
9.9 New Networking Paradigms **268**
9.10 Business Ecosystems **270**
9.11 Internet in 2020 **274**
9.12 Communication in 2020 (or Quite Sooner) **276**
9.13 References **280**

INDEX **281**

GUEST INTRODUCTION

Rapid progress in information and communications technology (ICT) induces improved and new telecommunications services and contributes greatly to society in general and to vendors and network and service providers. In addition to existing services such as telephony or leased line services, spread of the Internet, the Internet Protocol (IP) phone, and new communications services like IPTV are making great progress with the development of digital subscriber lines (DSL) and high-speed communications technologies like fiber to the home (FTTH). Furthermore, with the deployment of Next Generation Networks (NGNs), development of still newer services is anticipated. Construction of NGNs, in accordance with standards specified by international standardization organizations and feasibility studies and investigations, have begun in Japan and many countries around the world. The amount of information that a user can exchange has been expanding exponentially. Services can be used simultaneously (anywhere, anytime, and any device) and seamlessly with the development of broadband wireless access technology in NGN. Moreover, since service and application functions are separated and transport functions are independent from access technologies such as xDSL, FTTH, WiFi, WiMAX, Third Generation (3G), and Long Term Evolution (LTE), services of fixed and mobile communications are also unified. Furthermore, since the service and application functions consist of several common components, cooperation with third party applications becomes easier, resulting in practical use of various kinds of existing communications services (e.g., IT-based services and broadcasting services). Simultaneously, network reliability and security are also improving with the development of related technologies. In summary, NGN creates a new market by offering new services and rejuvenates markets such as career, enterprise, IT, and broadcasting businesses with new business models.

Maintaining the outstanding aspects of the existing network, NGN aims at larger scale, higher quality, and greater reliability. NGN is considered the biggest turning point in the history of communications. Although the present Internet provides services very conveniently for a user, the design of the Internet as a social infrastructure is inadequate. NGN can apply the technology of the Internet, can realize service level agreements (SLAs) and can provide mission-critical services. Users can choose high-price services for mission critical systems, medium-price services with high security, and low-price services as seen with the existing Internet. Wide-area client/server systems, which have high investment cost, were difficult to realize but will become realizable in NGN with the availability of super-mass storage systems. These allow integrated servers using the high-quality network services offered by NGN. As services spread for individual subscribers using NGN, IPTV, and voice over data, with development of NGN, a higher-definition video can be provided inexpensively.

Software as a Service (SaaS), using NGN will develop for business users. A reliable SaaS solution can be offered with security and SLA features that guarantee quality-of-service to each user of NGN. NGN will be ubiquitous. If information from rain sensors deployed all over a country is transmitted via NGN and processed and analyzed by a server, accurate weather forecasts will become reality. NGN will connect the medical systems of an area. If a doctor and residents can share medical information via the service of "virtual visits" by medical specialists in remote areas then we can offer medical consultation, medical checkup, etc. If a mobile IP network with an access speed of 100 Mbps is available, the distinction between mobile and fixed networks will diminish. NGN applications can be common to mobile networks and fixed networks. The wide area client/server system, which unifies mobile and fixed networks, will be completed by 2012. NEC Corporation has advanced communications and computers (C&C) as a concept, marrying communications and computers. NEC has been working on research and development of the future architectures realizing long-term C&C goals and views NGN as the field that realizes the philosophy of C&C.

This book aims at deepening the understanding of NGNs, services, service management technologies, Operations Support Systems (OSS), cable services, IP Multimedia System (IMS) and convergence services, ad hoc networks, sensor networks, etc. The book provides detailed explanations of latest technology trends. I am pleased and honored to provide the introduction to this book, which will promote your understanding and construction of NGN. I believe that an important benefit of NGN is further fullness to society and personal lives. I also believe that NGN further expands economic activities and can contribute to ecosystems by, for example, measuring climate change and global warming via efficient network deployment and management.

Botaro Hirosaki
Senior Executive Vice President and Board Member, NEC

GUEST INTRODUCTION

To say that we live in the information age is, of course, a cliché, and a 20-year-old cliché, at that. But the fact that it is a cliché doesn't make it any less true. Communications networks developed over the last two decades have profoundly changed how we carry out our everyday lives—how we exchange information, engage in commerce, form relationships, entertain ourselves, protect ourselves, create art, learn, and work. The convergence of communications and computing, long anticipated, is now a fact.

The "modern" communications industry is actually more than 130 years old. For almost all of that history, the industry's goal has been the reliable delivery of a particular kind of analog signal—first speech, then music, then video—over links and networks established for *only that signal.* It is only since the two-pronged emergence of the Internet and mobile telephone networks that we have been able to glimpse the splendid opportunities made possible by multimedia networks operating over a diversity of channels—wireless, wireline, and cable—delivering a wide array of content to an assortment of devices, including PCs, notebooks, TVs, mobile phones, and PDAs.

But as communications networks have become more complex and the services offered over those networks have become more diverse and numerous, the problem of managing networks has become profound. Different types of data mean different requirements in terms of latency, quality-of-service, and security. Different types of communications media mean significantly different operating environments in terms of delay, reliability, and bandwidth efficiency. Fortunately, the Telecommunications Management Network (TMN) model offers system designers a framework for interconnectivity across heterogeneous networks. It is an architecture that enables network management and provides a "handle" to engineers and computer scientists seeking to design products and services that will become part of the information infrastructure.

This book goes beyond the Network Management Layer (NML) of TMN to the Service Management Layer (SML) and business frameworks. As new services and "apps" are rolled out every day—new ways to use your smartphone or your home network that you have not yet envisioned—the challenge of managing those new capabilities, efficiently and securely, and their solutions, are addressed in this book. Its chapters describe some of the latest multimedia services offered by the telecom and cable industries and provide insight into how they are best managed. It looks ahead to IP-based next-generation telecommunications networks, services, and management, as well as ad hoc and sensor networks. This book offers a vision of how pervasive, heterogeneous, and converged multimedia networks will be deployed and managed well into the 21st century.

What role will academia play in this evolutionary (and, sometimes, revolutionary) process? It will be a fundamentally important role. Universities will continue to educate the designers, managers, and implementers of these networks and carry out the long-term, basic research that will help enable the next generation of networks. As teachers, we have the obligation to make sure that graduating electrical and computer engineers and computer scientists understand the fundamental properties of heterogeneous information networks. As researchers, we have the opportunity to use our tools—modeling, analysis, simulation—and our imaginations, to fashion better networks and to manage them more efficiently, securely, and robustly.

Thomas Fuja
Chair, Electrical Engineering, University of Notre Dame

Peter Kilpatrick
McCloskey Dean of Engineering, University of Notre Dame

EDITOR AND CONTRIBUTOR BIOGRAPHIES

Jean Craveur presently heads the France Telecom Group Transverse Mission and is in charge of preparing the group networks transformation from PSTN to NGN/IMS. He has previously steered, in France Telecom, the IT and Network overall architecture and strategy department and headed the R&D center on core network. He held several responsibilities in international telecommunication organizations: as a member of the International Experts Group, which wrote the first CCITT N°7 signaling specifications; as chairman of the CEPT and ETSI Subtechnical Committee, which issued the roaming architecture and signaling for the GSM system; as chairman of the "Network Group" of the European Cooperation on ISDN; and, finally, as vice chairman of the ETSI Technical Commitee on Signalling Protocol and Switching. He was also one of the vice presidents for Europe in the TINA Consortium. Jean Craveur has published several papers related to signaling and telecommunication networks in telecommunication reviews and presented in the International Switching Symposium (ISS). He graduated from Ecole Nationale Supérieure de l'Aéronautique et de l'Espace (SUP'AERO) and holds a masters in economic science from Université des Sciences Sociales in Toulouse and a diploma in automatic and complex systems from Ecole Nationale Supérieure de l'Aéronautique et de l'Espace.

Michael Fargano has broad telecommunications industry leadership responsibility and is the current Industry Standards Program Manager at Qwest Communications International. His career spans more than 25 years and is grounded in leadership in many successful telecommunications R&D projects, advanced systems architecture and engineering projects, and standards projects such as AIN, TMN, 3G wireless, NGN, emergency services, and security management at several well-known and respected telecommunications companies/departments such as Bell Labs, Bellcore, US WEST Advanced Technologies, and Qwest. In addition, he has been an adjunct instructor at several institutions and universities including Bell Labs, Stevens Tech, University of Denver, and University of Colorado, covering a wide variety of engineering topics including telecommunications network management and standards. He was chairman of several standards committees and is a sought-after leader in standards development, for which he was honored with several industry awards including the ANSI Meritorious Service Award and ATIS Leadership in Standards Development Award. He also holds several patents. He graduated in 1980 from a special simultaneous bachelor/master program in general engineering and electrical engineering at the Stevens Institute of Technology. He also holds an advanced business/technology management graduate certification from the University of Denver–Daniels College of Business, with a specialty in Strategic Program Management.

David Jacobs is chief technical officer in the Amdocs Broadband, Cable & Satellite Division with responsibility for driving Amdocs products strategy for Cable MSOs and Satellite operators who provide next generation services to residential and commercial customers. He joined Amdocs following the acquisition of Jacobs Rimell by Amdocs in April 2008. As co-founder and CTO of Jacobs Rimell, he was responsible for the company's technology and product direction, enabling it to become one of the leading providers of customer-centric fulfillment solutions for the cable industry. Previously, he spent 11 years with Reuters in a number of senior roles, culminating in the deployment of a global frame relay infrastructure and one of the world's first global IP extranets for the delivery of Reuters' information services. He holds a BSc in electrical and electronics engineering from Middlesex Polytechnic, London and a Full Tech City and Guilds Certificate in Telecommunications. He holds two U.S. patents and contributed to a third.

Keizo Kawakami is a project manager in the Network Management Systems Division of the Network Software Operations Unit of NEC. He joined NEC in 1989 and he has been engaged in software development of mobile, satellite, and fixed networks management systems for 15 years. He is now in charge of strategic planning and development of service and management solutions for mobile and fixed operators. He is a principal contact of the TeleManagement Forum (TMF) in NEC.

Kaoru Kenyoshi is a chief manager in the first Carrier Solutions Operations Unit of NEC. He joined NEC in 1984. He has been engaged in software development for ISDN switching systems and B-ISDN for 10 years. From 1995 to 2000, he worked as a manager in the planning division for switching systems and was in charge of strategic planning for services and products of switching systems for the carrier market. From 2000 to 2006, he worked as a general and chief manager in the sales and solution department for fixed and mobile networks. He is now in charge of promotion of NGN and IPTV solutions for fixed and mobile operators. He is very involved in standardization activities and is leader of the IPTV Network Architecture Sub-working Group and member of the Strategy Committee of TTC in Japan and one of the vice chairman of ITU-T SG11.

Bhumip Khasnabish, PhD, is a Senior Member of IEEE and a Distinguished Lecturer of the IEEE Communications Society (ComSoc). He is a Director in the Standards Development and Industry Relations Division of ZTE USA Inc. with responsibility to set direction, goal, and strategy of the Company for Next Generation Voice over IP (VoIP) and peer-to-peer (P2P) multimedia services. Previously, Bhumip was a Distinguished MTS of Verizon Network & Technology in Waltham, MA, USA. He is the founding chair of the recently created ATIS Next Generation Carrier Interconnect (NG-CI) Task Force. Bhumip also founded MSF Services Working Group, and led the world's first IMS-based IPTV Interop during GMI08. At Verizon, he focused on NGN and Carrier Interconnection projects related to delivering enhanced multimedia services. He also represented Verizon in the Standards activities of MSF and ATIS NG-CI. An Electrical Engineering graduate of the University of Waterloo and the University of Windsor (both in Ontario, Canada),

Bhumip previously worked at Bell-Northern Research (BNR) Ltd. in Ottawa, Ontario, Canada. While at BNR he initially designed, implemented, and then led the implementation of trunking and traffic management software modules for BNR's flagship Passport® multi-service switching product. Dr. Khasnabish has authored/co-authored numerous patents, books, chapters, technical reports, Industry Standards contributions, and articles for various international archival journals, magazines, and referenced conference proceedings. His recent book entitled, "*Implementing Voice over IP*" [ISBN: 0-471-21666-6] is currently in its second printing. Previously, he edited/co-edited "*Multimedia Communications Networks: Technologies and Services*" [ISBN-10: 0890069360, ISBN-13: 978-0890069363], and many Special Issues of *IEEE Network*, *IEEE Wireless Communications*, *IEEE Communications Magazines*, and the *Journal of Network and Systems Management (JNSM)*. He is also a member of the Board of Editors of the JNSM, and an adjunct faculty member of Brandeis University, Bentley University, and Northeastern University.

Toshiyuki Misu is a chief manager of the Network Software Operations Unit, NEC Corporation. Since he joined NEC, he has been engaged in software development of digital switching systems, Intelligent Networks, VoIP, IMS/NGN, and Service Delivery Platform. He is now in charge of NGN Service Promotion and Service API Standardization. From 1991 to 1992, he was a visiting researcher of CTR (Center for Telecommunications Research), Columbia University.

Steve Orobec is British Telecom's (BT) lead OSS standards manager and enterprise architect. The focus of his work is in the TM Forum, where he is the leader of the architecture harmonization team. He also leads the BT OSS team in ETSI TISPAN Working Group 8, collaborating with 3GPP to specify IMS/NGN management systems and solutions for integrating them into OSS. He has also represented BT at ITU Study Group 4 meetings. He has worked in all parts of the software lifecycle from validation and test, software development, solution design, and architecture during his 17 years at BT. He reports at director level and co-ordinates his activities to ensure that BT's requirements are represented in the TM Forum and that TM Forum standards are utilized in BT's OSS. He is currently responsible for developing an automated, standards-based OSS management solution that will reduce BT's OSS costs and increase agility. He holds a degree in physics and astrophysics from Leicester University.

Thomas Plevyak is a past president of the IEEE Communications Society (ComSoc). He has served as ComSoc's editor-in-chief of *IEEE Communications Magazine*, director of publications, and Member-at-Large of the Board of Governors. Mr. Plevyak is an IEEE Fellow for contributions to the field of Network Management. He is a Distinguished Member of Technical Staff in Verizon's Network & Technology organization, currently responsible for domestic and international wireline and wireless operations and network management standards. He holds a BS in engineering from the University of Notre Dame, an MS in engineering from the University of Connecticut, a certificate from the Bell Laboratories Communications Development Training (CDT) program and an MS in advanced management from Pace University. He is co-editor of *Telecommunications Network*

Management into the 21st Century, as well as a series of six books in the field of network management. He is the author of many technical publications and holds two U.S. patents.

Veli Sahin, Ph.D., is senior director of Business Development at NEC Corporation of America in Irving, Texas. Previously, he held management and leadership positions at Bell Laboratories, Bellcore, Samsung, and Marconi. He has been working in the area of Telecommunications Networks for over 25 years. His current interest includes Next Generation Networks (NGN), IP Multimedia Subsystems (IMS), Triple/Quad Play Services, IPTV Services, development of TMN-based management systems and wireline/wireless national and global information infrastructures for the 21st century. Dr. Sahin has over 100 internal and external publications, is co-author of an IEEE Press book chapter and co-editor of the IEEE Press Book Series on Network Management. He received an MS and PhD (multi-hop packet radio networks) in computer science and an MS in electrical engineering at Polytechnic University, Brooklyn, New York. He also received a BS in electronics engineering at Istanbul Technical University, Istanbul, Turkey. He received IEEE/IFIP The Salah Aidarous Memorial Award in 2008 for his contributions to IT and Telecommunications Network Management. He was general chair of the 1998 and 2002 Network Operations and Management Symposium, co-founder and first chair of the IEEE ComSoc Technical Committee on Information Infrastructure (from 1995 to 1998), and chair of the IEEE ComSoc Technical Committee on Network Operations and Management (from 1998 to 2000). Dr. Sahin was also a member of the editorial board and/or advisory board of several respected journals. He is currently MSF Board Member and also project leader for the NEC MSF and Verizon VIF Interoperability Testing (IOT) activities.

Roberto Saracco holds a bachelor's degree in computer science, a master's degree in math, and a postdoctoral degree in physics. He joined Telecom Italia in 1971, contributing to the development of the first SPC system in Italy. Through the years he worked on data transmission, switching, and network management. In the last 10 years he has worked on the economic side of telecommunications, creating and directing a research group at the Future Centre in Venice. Author of many papers and nine books in the field of telecommunications, with the last five on the topic of living and communicating in the next decade, he has worked on the foresight Panel of the European Commission, charged to imagine the Internet beyond 2020. He is currently director of Telecom Italia Future Centre, in Venice, and co-chair of the Edge-Core group of the Communications Future Program of MIT. He is a senior member of IEEE ComSoc, serving in many roles, including TC secretary, NM chair, and vice president of Membership Relations. He is currently ComSoc's director for Sister- and Related-Societies.

He received the Salah Aidarous Award in 2005 for his contribution to network management and the 2007 Donald McLellan Meritorious Service Award for his contribution to strengthening the Communications Society presence worldwide.

Mehmet Ulema is a professor at the Computer Information Systems Department at Manhattan College, New York. Previously, he held management and technical positions in Daewoo Telecom, Bellcore (now Telcordia), AT&T Bell Laboratories, and Hazeltine Corporations. He has numerous publications in various international conferences and journals. He holds two patents. He gave a number of talks and tutorials on Network management and wireless networks. He is on the editorial board of the IEEE Transactions on Network and Service Management, the ACM Wireless Network Journal, and the Springer Journal of Network and Services Management. He is an active Senior Member of IEEE. He served as the chair and co-founder of the IEEE Communications Society's Information Infrastructure Technical Committee. Previously he served as the chair of the Radio Communications Technical Committee. He is involved in a number of major IEEE conferences as technical program chair (Globecom 2009, ICC 2006, CCNC 2004, NOMS 2002, ISCC 200). He was a general chair of NOMS 2008. He received MS & Ph.D. in Computer Science at Polytechnic University, Brooklyn, New York. U.S.A. He also received BS & MS degrees at Istanbul Technical University, Turkey.

CHAPTER 1

CHANGES, OPPORTUNITIES, AND CHALLENGES

Veli Sahin and
Thomas Plevyak

1.1 INTRODUCTION

Never have telecommunications operations and network management been so important. Never has it been more important to move away from practices that date back to the very beginning of the telecommunications industry. Building and connecting systems internally at low cost, on an as-needed basis, and adding software for supporting new networks and services without an overall architectural design will not be cost effective for the future. Defining operations and network management requirements at the 11th hour for new technologies, networks, and services deployments must also change. Planning and deployment of all aspects of telecommunications leading to Next Generation Networks (NGN) and services must be done in unison to achieve effective and timely results.

The need for new approaches can be seen everywhere in the global telecommunications industry. Competition in telecommunications can turn players into victims if functional and cost-effective operations and network management requirements are not deployed quickly. Technology advancements in this field have been enormous. Operations and network management technologies make new approaches a reality in designing NGN services.

The point of departure for architected network management systems will be NGN and services. Points of departure can't be expected to initially play out with incrementally lowest cost. There can be initial added costs, but the operations and network management setting put in place will make the next network and service less costly, with more rapid implementation than would otherwise have been the case. Telecommunications network and service providers will find themselves on a fully competitive playing field.

Next Generation Telecommunications Networks, Services, and Management, Edited by Thomas Plevyak and Veli Sahin

1.2 SCOPE

This book discusses NGN architectures, technologies, and services introduced in the last decade, such as Triple Play / IPTV [1] and services that are expected to become increasingly deployed in the coming decade such as Time Shift TV (TSTV), network Private Video Recording (nPVR), multi-screen video services, triple-shift services, location- and presence-based services, blended and converged services, etc.

In addition, this book also focuses on the Service Management Layer (SML) of the Telecommunications Management Network (TMN) [2]. In the past 30-plus years, the global industry spent considerable time and resources developing Element Management Systems (EMSs), Network Management Systems (NMSs), and Business Management Systems (BMSs). Changes in life style (expectations, viewing habits, calling habits, shopping habits, etc.), technologies, and the competitive business environment are now moving the industry to pay attention to Service Management Systems (SMSs).

Internet access, cellphones, laptops, and DVRs are integral to our lives today. How many of us can live a day without them? Daily personal and business lives are completely dependent on telecommunications services. End-to-end management of those services and Quality-of-Service (QoS) management and identification and management of Quality-of-Experience (QoE) metrics are very important to improve standards of living and increase productivity. Examples of QoE are quality-of-picture, channel switching time, easy use of user interface/programming guide, request response time, etc.

This is the seventh book in the IEEE Network Management Series. It follows the same approach as the first book in the series, *Telecommunications Network Management into the 21st Century,* published in 1994 [2], and the second, *Telecommunications Network Management Technologies and Implementations*, published in 1998 [3]. It is an orchestrated set of original chapters, written expressly for the book by a team of global subject experts. This is a technical reference book and graduate textbook.

1.3 CHANGES, OPPORTUNITIES, AND CHALLENGES

This section briefly discusses major changes and how service providers (SPs) and SMS vendors use this as an opportunity to develop solutions that address expectations of their customers. SPs work to offer new services such as IPTV, multi-screen, triple-shift, blended and converged services, etc. Vendors and SPs work to provide new SMS applications to manage those new services.

Today's users want to communicate, watch, shop and make payments, etc. anytime, anywhere, and with any device. This is a major paradigm shift and has major impact in designing NGNs and services as well as management systems.

1.3.1 Major Life Style Changes: Desktops, Laptops, and Now Handtops

We all know how personal computers (PCs) have changed our lives during the last two decades. First, we started with desktop PCs and then started using more and more laptop PCs, especially in last 10 years or so. Laptops allow us to carry our PC with us anywhere we go and use it. With wireless and mobile Internet access, users access the Internet anywhere and anytime. We can send and receive e-mails and exchange files at any time, from anywhere. Voice applications allow us to call and talk with anyone in the world who has a PC or a phone. PC-to-PC calls are free and PC-to-phone calls cost less than traditional calls.

Many of our traditional daily habits have been changing too—watching, calling, shopping, making payments, and many more. These changes affect the way we do business in many industries.

It wasn't so long ago that we watched a movie, a video, or a program just using the TV and made phone calls using only wireline phones. Today, we also use PCs to watch programs and wireless phones to make a great many of our phone calls. In more and more families and businesses, wireline phones are used for special cases (conference calls, interviews, other business calls, etc.). Increasingly, people do not have wireline phones. They use their cell phones. They watch TV programs using their laptops and/or "handtops." Handtops are mini personal computers such as iPhones and BlackBerry phones. Even though we refer to them as phones, they are small laptops, used to access the Internet, send/receive e-mails, make phone calls, etc. Millions use the Internet to shop, pay their bills online and manage their bank accounts. As a result, security management (SM) has risen to become a first priority concern.

In the future, user-generated content (UGC) will play a major role in designing NGNs, service, and management systems.

1.3.2 Major Network Infrastructure Changes

The first major network infrastructure change was to shift from time-division multiplexing (TDM) to statistical multiplexing. NGNs are now based on packet switching technologies rather than TDM. Internet Protocol (IP) became the winner. Today, NGNs are becoming IP-based packet-switched networks, end-to-end, including backbone, metro, and access networks. This is important because it caused a paradigm shift in Fault, Configuration, Accounting, Performance, and Security (FCAPS) operations and network management applications and in SP concerns, which we will discuss later in this section.

The second major change is the use of more and more wireless and mobile technologies in NGNs. Billions of cellphones are in use worldwide, and the number will continue to grow. The concept of telecommunicating (via phone or e-mail) and Internet access at any time and any place has become a reality.

The third major change is just starting and will be rapidly taking place in the next few years. This change is IP Multimedia Subsystems (IMS)-based signaling and control to replace traditional signaling systems. IMS will provide an end-to-end platform to offer most new services and, therefore, will eliminate current silos. With IMS signaling and control, many advanced location- and presence-based services will become a cost-effective reality. IMS is also expected to solve the problem of rapid introduction of new services at less cost. Details of IMS can be found in Chapter 5.

Finally, development and deployment of Service Delivery Platforms (SDPs) with open Application Programming Interfaces (APIs) for third-party application development will have major affects in introducing next-generation advanced services quickly and in more cost effective ways.

1.3.3 Major Home Network (HN) Changes

Residential customer premises networks, also called Home Networks (HNs), are now becoming extensions of SPs' networks.

Home connectivity is evolving from narrowband to broadband. SPs have deployed the technology needed to offer larger bandwidth with cable, xDSL, or fiber technologies. The Internet has been a major driver for evolution to broadband, creating a new experience for customers and offering new services, such as fast Internet browsing, video-on-demand (VoD), online shopping and banking, and digital video recording (DVR), while providing broadband connectivity among many devices at home such as PCs, TVs, Set Top Box (STBs), DVRs, residential gateways (RGs) / home gateways (HGs), game consoles, etc.

The main drivers for home networking that exist today are as follows:

1. As media become increasingly digital in nature (online music and video, digital photos etc.), consumers want to share content and listen to or display it on other, more consumer-friendly devices such as TVs, etc. This requires customers to connect their digital content storage devices (e.g., PCs, MP3 players, private video recorders (PVRs), and digital cameras/camcorders) to their entertainment systems over a home network.
2. More and more customers want to use digital voice and video. This is due mainly to the attractive price using triple play services. These new voice and video services should be capable of being received on a range of mobile consumer devices (laptops, mobile phones, etc.).
3. Devices such as laptops that are WiFi-enabled are encouraging consumers to access the Internet, work, and/or watch videos wherever it is convenient in the home.

Management and control of home networks have become a strategic challenge for SPs all over the world. Problems in home networks affect QoS and customers' experiences. Therefore, all SPs have been developing strategies to provide RGs / HGs as part of their triple play services.

1.3.4 Major FCAPS Changes

As stated previously, FCAPS stands for Fault Management (FM), Configuration Management (CM), Accounting Management (AM), Performance Management (PM) and Security Management (SM). Readers who are not familiar with basic FCAPS functions should read the FCAPS sections in [2] or brief further details in Chapter 4.

In the past, when networks were based on circuit switching, FM was a first-priority application, followed by CM, AM, PM, and SM. PM and SM functions were considered to have least priority in circuit switched/TDM networks. FCAPS has been used for a long time, perhaps implying order of importance. Technically speaking, for packet-based networks, PM applications are now more important than FM. We are going through a transition period. When subscribers start using delay- and quality-sensitive services such as voice over IP (VoIP), IPTV, and VoD, SPs will pay more attention to PM-based applications.

QoS can suffer even if there is no failure in the network due to congestion and/or over-utilized resources such as Central Processor Units (CPUs), buffers, bandwidth, etc., in packet-based networks. Congestion and over-utilization of resources will result in delays, packet loss, and jitter, which greatly affect QoS and customers' experience, such as snowy screen, unsynchronized voice and picture, longer time to receive a requested video, etc. All of these impairments can be detected and corrected in advance by using PM and SM systems using trend analysis, data correlation, and SLA management (proactive management).

We might want to rethink FCAPS priorities. Security Management is, now, arguably the highest priority. The amount of confidential data that is transmitted, collected, and stored is very large and must be protected. SM needs to take its place as the number one concern followed by PM as opposed to FM and, in turn CM. The order now is probably SPxxx, not FCAPS.

1.3.5 Major Regulatory Changes

In this book, we will not discuss regulatory/legal changes even though they greatly affect the types of services offered (e.g., network- vs. home-based video recording), security, copyright, wireless spectrum allocation, content distribution and usage, etc.

1.3.6 Service Aware Networks to Manage Expectations and Experiences

NGNs and Management Systems (MSs) must be aware of traffic generated by each service as well as which subscriber generated that traffic. In some cases, this is done to manage customers' expectations and legally satisfy their Service Layer Agreement (SLA). NGNs must treat traffic generated by each service separately while transporting them through the networks. Networks must have the ability to use different priorities and policies for traffic generated by different services such as VoIP, Video, e-mail, file transfer, etc. Furthermore, they must also assign different priorities for traffic generated by the same service depending on who owns the traffic. Traffic belonging to a residential VoIP service does not have the same priority as VoIP traffic from a large business customer who pays more and signs an SLA.

Networks and SMS must also have the ability to handle special cases, e.g., during national disasters (earthquake, war, terrorist attacks, etc.). Some services will get higher priority than all other services based on predefined policies and rules. Examples of these services are Government Emergency Telecommunications Services (GETS) in the USA. Note that GETS and 911 calls have higher priority than many other services even during normal circumstances.

Figure 1-1 is an NGN example, capable of offering triple play services such as voice (VoIP), video (IPTV, interactive gaming, video conference, distance

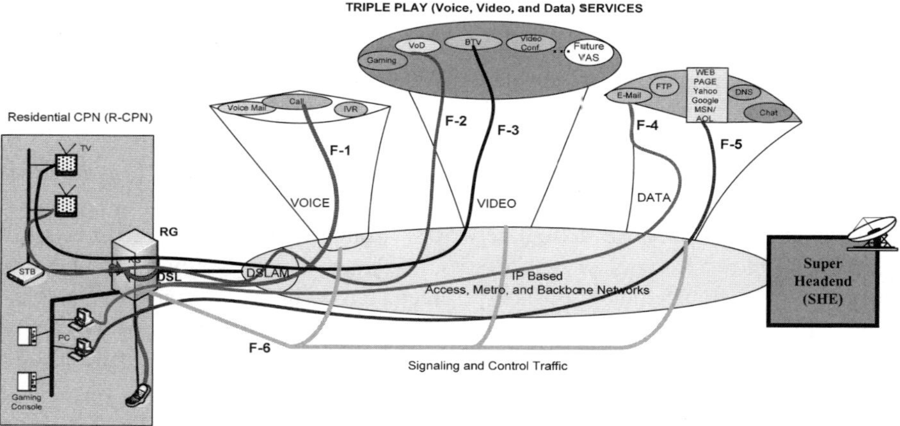

Figure 1-1. Example of triple play services and flows in Service Aware Networks

learning, etc.), and data (e-mail, File Transfer, Web page, DNS, chat, etc.). The architecture shows an example of metro and backbone networks common for most of the SPs. The biggest difference may exist in the last mile and access networks. A customer can have one, two, or all three services.

For example, the residential customer shown in Figure 1-1 uses ADSL in the loop and has VoIP, video, and high-speed internet access (HSIA) services. Traffic (or flow) for each service shares the limited bandwidth in the loop. In Figure 1-1, the residential customer generates six different flows using all three services at the same time. These are:

1. VoIP flow (F1)
2. VoD flow (F2)
3. IPTV flow (F3), also called broadcast TV flow (BTV)
4. E-mail flow (F4)
5. Web page flow (F5), and
6. Signaling and control flow (F6)

Also shown in Figure 1-1, triple play services will have shared resources such as the access, backbone, and parts of the Customer Premise Network (CPN). This will depend on each SP's network architecture. Customers have different expectations for each service. They can tolerate some delay for their data services but not for VoIP. VoIP is a real-time application, very sensitive to delay. It is important that SPs consider accumulated effects of impairments on the network to increase customer's satisfaction. For example, QoS requirements vary with each service. Voice services are stringent on latency and jitter, whereas data services are not. So each service has its own characteristics and it is not appropriate to treat them as the same. SPs must correlate and aggregate their customer's experiences across all services. Each service may be managed by different management systems but there is a need to manage a customer's expectation from one centralized place. This common system could be a Service Management System (SMS) interworking with other management systems, especially PM systems/applications. When video and VoIP services are widely deployed, there is an important need for information exchange and sharing among SPs. This area might be an important application for the SMS.

Customers/subscribers and SPs have related, but different, experiences for the same event. For example, let's assume that for the F1 VoIP call, the customer is having a bad quality call, hearing clicks, echo, low volume, etc. Customers do not care why the call quality is not good, but SPs do. It is the responsibility of an SP to find out:

1. Where is the problem? Is it in the CPN or the SP's network?
2. If the problem is in the SP's network, where is it? Is it in the access network, backbone network, media gateways, or servers?
3. Why does this call have bad quality? Is it because of delay, packet loss, jitter, error, over-utilized resources, etc?

Therefore, SMS vendors need to know customers' and SPs' expectations as well as experiences so their solution can solve the problems. Problems get more difficult when traffic flows across multiple SP networks.

1.4 MAJOR MANAGEMENT CHALLENGES FOR A VALUE-ADDED SERVICE: TRIPLE SHIFT SERVICE

As illustrated in Figure 1-2, a triple shift service means that customers can access and control (pause, play, rewind, fast forward, etc.) service contents at anytime (time shift), at any place (place shift), and using any device (device shift). In other words, customers do not have time, place, and device limitations to access, control, and replay any content. Accessing the information, regardless of time, place and device has already been happening. Our e-mails and voice mails (messages) follow us no matter where we go.

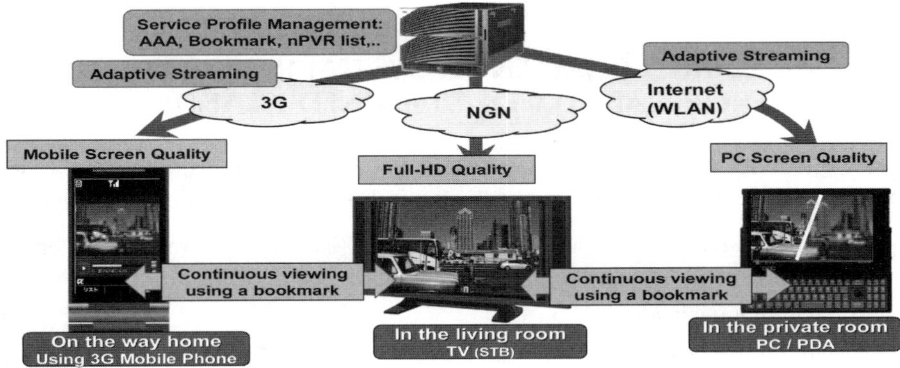

Figure 1-2. A triple (time, place, and device) shift service architecture

As an example, a subscriber starts watching a program on her mobile phone while in a taxi going to the airport. She pushes a button just before boarding the airplane and asks the system to record the program so it can be watched later. After a few hours of flight and while driving home, the system can be asked to start the program at the point where it was stopped. On arriving home, the program can be watched on HDTV.

Since several hours passed between stopping program and the replay, it is a time shift service. It is also a place shift service because the subscriber started watching in one place and finished watching in another place or places. Finally, it is also a device shift service because the program watching started on a mobile phone and ended on HDTV.

The triple shift service we just described creates many challenges for an SP. Let's briefly describe a few major tasks an SP must accomplish in order to provide triple shift services:

- The SP must check to see that the customer is authorized to use the requested services by BML applications.
- The SP checks to see that they can legally record the content and store it in their network by BML applications.
- When the customer requests a replay, the SP must make sure that content is suitable for the device that is being used. Each device requires different encoding because of size of screen, resolution requirements, type of service required (e.g., if it is TV then is it HD or SD?), available bandwidth, etc., by SML applications.
- Streaming and content distribution based on device, QoS, and SLA requirements by SML applications.
- While a customer is watching a program, SMSs must monitor the service to make sure SLA requirements are not violated.
- End-to-end PM functions at SML and NML monitor the whole network for impairments (packet loss, congestion, jitter, buffer overflow, etc.) so the SP can detect problems in advance and correct them before they affect QoS and QoE metrics [proactive management]).

1.5 THE GRAND CHALLENGE: SYSTEM INTEGRATION AND INTEROPERABILITY OF DISJOINED ISLANDS

Since divestures (1984 in the United States), large SP networks all over the world became collections of islands, such as shown in Figure 1-3:

- Merger and acquisition (M&A) islands (island of formerly different companies)
- Technology islands
- Cultural islands
- Regulatory islands

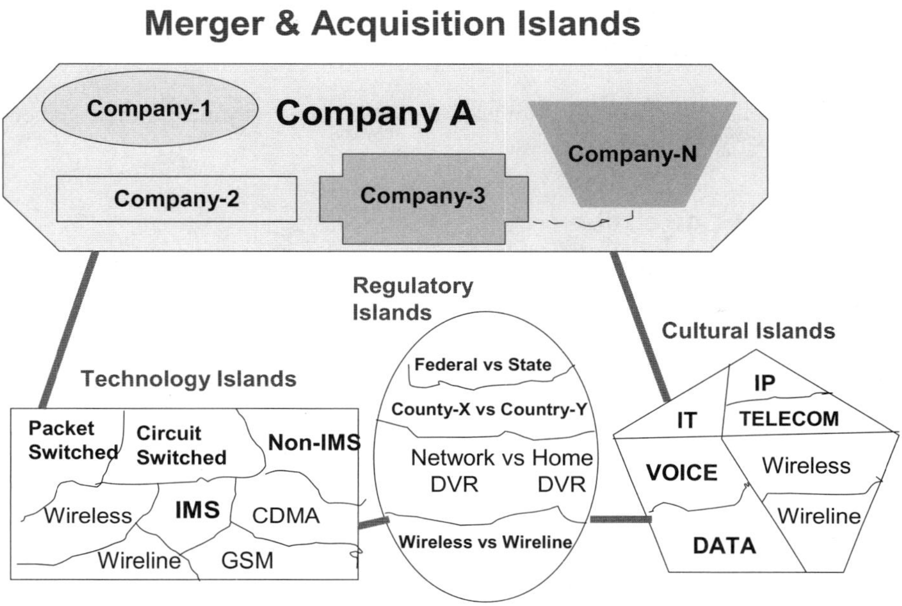

Figure 1-3. System integration and interoperability of disjoined islands

The grand challenge for SPs is to have end-to-end views and end-to-end management of services. With respect to networks and services management, SPs' networks may consist of disconnected islands.

Some of today's largest SPs, and even some smaller SPs, are the result of M&As.

For each case of an M&A, the former companies had their own networks, management systems, organizations, and cultures. They may also be using different architectures, technologies, standards, and products from many different equipment vendors.

The second important class of islands is the technology island, such as:

- Circuit switched networks and packet switched networks
- ATM networks and non-ATM networks
- CDMA networks and GMS networks
- Wireless and wireline networks
- IMS and non-IMS networks, etc.

Third, and perhaps the most important class of islands, is culture. In our professional backgrounds (e.g., Telecom, IT, IP, voice, data, etc.), we worked in different cultures. M&As can fail due to cultural differences.

These islands must be connected and/or interwork with each other in order to provide end-to-end services, meet QoS requirements, and satisfy customers' expectations and experiences.

1.6 SOME EXAMPLES OF MANAGEMENT SYSTEM APPLICATIONS

This section will briefly discuss some of the SMS and Performance Management System (PMS) applications.

1.6.1 Event Correlation

When an SMS detects a threshold crossing in a network that affects QoS, SMS has the capability to notify affected customers and the service provider. For example, if the voice router in the Business Customer Premises Network (B-CPN) shown in Figure 1-4 below exceeds a threshold, SMS can notify the enterprise customer via e-mail informing them that a) the voice router is having a problem, b) the problem is being worked on, and c) the estimated time to repair.

If SMSs can access contact information (e.g., e-mail addresses) of all the users served by the router in question, it can send e-mail to the complete list of users. If the number of affected customers is very large (e.g., several thousands) then SMSs can notify these users during non-peak hours when there is less traffic in the network, such as between midnight and 4:00 AM. Customers are informed proactively, which will result in a decline in new trouble tickets and a reduction in churn.

If an IP backbone network router, connected to a CPN, detects a threshold crossing violation (e.g., large delays or packet loss), the SMS can inform the business customer and all phone users (end users for that customer) by e-mail of the violation, as discussed above. If the CPN network has an alternate path to another router in the backbone and the system has the capability to re-route the

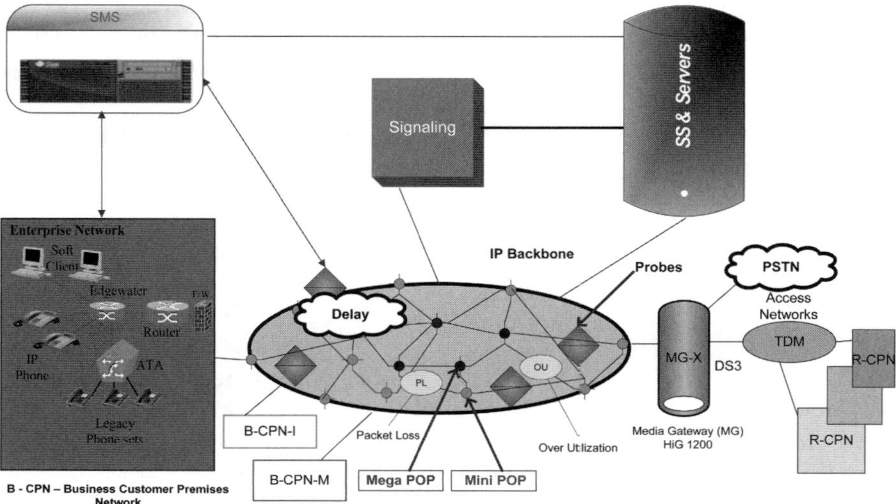

Figure 1-4. Event correlation and hot spots in a network

traffic using the back up router, the SMS will only notify the business customers about the violation.

Similarly, when a media gateway such as MG-X, shown in Figure 1-4, detects a threshold violation (e.g., a DS3 port threshold violation), SMS will inform all customers served by that DS3 port. It is important to understand that an SMS does not need to know which DS0 time slot is used and by whom. All that is needed is the list of affected customers and their phone numbers.

What an SMS does is similar to what cable TV companies do today when there is a problem in their network. They do not inform individual customers but send a broadcast message to all affected customers. In other words, a threshold violation is correlated to a group of customers, not to a specific customer in the case of MG threshold violation. On the other hand, a threshold violation in B-CPN can be correlated to the business customer and all of its employees (affected customers). So, the correlation event is dependent on network architecture and its location.

1.6.2 Hot Spot Identification and SMS Actions

With respect to SMS applications, a hot spot means that a part (a sub-network) of a network is not operating according to specified key performance indicators (KPIs) such as delay, packet loss, utilization, availability, jitter, etc., as shown in Figure 1-4. A sub-network can be as small as just a single node (e.g., a router, MG, IP PBX, an application server, a softswitch, etc.) or a single resource (e.g., a CPU, port, buffer, trunk, etc.). It is important to know that we can get bad QoS even though there is no infrastructure failure in the network such as cable cut, broken or burned equipment, CPU failure, etc., due to packet loss, packet delay, congestions, and over-utilization of physical resources (e.g., CPU, buffer, bandwidth, etc.) Therefore, a fault management system does not receive any alarms and it assumes that every thing is okay. However, performance management system SML applications can detect those impairments in advance before any SLA requirements were violated and inform other appropriate OSSs / BSSs to take the necessary actions. This is why PM now is more important than FM for packet based networks [4].

What an SMS will recommend to the appropriate OSSs depends on:

- Frequency of violation
- Duration of violation
- Location of violation (an MG, a backbone router, a softswitch, an application server, etc.)
- Time and day of violation

The effect of each KPI on service quality changes from service to service (VoIP, video, HSIA, data, etc.). VoIP QoS KPI dependencies are availability, delay (E2E delay up to 150 ms is acceptable), packet loss, utilization, and jitter.

VoIP service is more sensitive to delay than packet loss. Up to a few percentage points of packet loss does not affect quality and it is acceptable. If packet loss is evenly distributed, up to 5% packet loss is also acceptable.

1.6.3 SLAs, Contracts, and Policy Management

An SMS has a suite of tools that allow the SP to assess and report on service delivery, SLA contractual performance, service connectivity (including service layering), service reporting, and topographic information, in real-time through an SMS Web-based graphical user interface (GUI). Also, through service assurance, the service provider can automatically manage and correct service problems as they occur, creating a service operations center (SOC).

SOCs increase SPs' efficiency and reduce operational costs. By correlating customers, infrastructure, and service information, SOCs enable SPs to make smart decisions about subscriber services and infrastructure and manage service and contract incidents with the proper priority.

1.6.3.1 Service Assessment Service assessment monitors real-time service quality in the resource infrastructure and assesses it against specific SP-defined quality objectives. An SP can measure overall service quality against policy thresholds for specific services, or service segments, within their infrastructure. If policy thresholds are violated, service assessment initiates automatic actions that alert staff to service problems in the resource infrastructure. The Web-based GUI provides real-time graphical displays of policy violations, with a point-and-click drill down from top layer service violations to root cause resource violations, and specific KPI values and thresholds. Individually tailored service quality policies can automatically be applied to a service and assessed, based on the time of day, day of the week, legal holidays, and special events.

1.6.3.2 Contract Assessment Contract assessment monitors real-time customer committed service quality for end user services specified in customer SLA contracts. Contract assessment automatically measures the real-time performance of customers' delivered services against specific customer service quality commitments created with contract builder. If an SLA contract's thresholds are in jeopardy of violation, contract assessment initiates automatic actions that alert staff of specific customer service problems, so that prioritized corrective actions can be taken before SLAs are violated. By converting raw infrastructure performance data into business knowledge about end-to-end customer service quality, contract assessment gives a service provider the means to understand how well they are meeting customer expectations and commitments. The resulting knowledge can be used to assure the service for the customer.

1.6.3.3 Service and Contract Assurance Service and contract assurance enables the automatic restoration of service when service objectives, or SLA contract commitments, are violated. Service and contract assurance extend the capabilities of service designer actions to include connectivity to third-party provisioning and service activation platforms. When a resource infrastructure or customer service objective is missed, service and contract assurance requests third-party systems to re-route or re-provision resources, so that service is restored to normal operation.

1.6.4 SMS Integration with Planning and Engineering Systems

Network planning and engineering systems need historical data for expansion and reengineering of network. Therefore, it is very important that they have access to accurate historical data on important parameters such as utilization, traffic, delay, packet loss, call statistics, etc. Since an SMS collects and stores all that information, it is very important for SPs' planning and engineering systems to access an SMS database. Historical data will be very useful in modifying and improving routing algorithms, flow control algorithms, closing a point-of-presence (POP) that does not generate enough revenue, adding a new POP for growth, capacity / bandwidth planning, etc.

In addition to providing planning and engineering data to SPs' OSSs, an SMS will also make short- and long-term recommendations to the other OSSs to implement requested action or actions (re-routing, re-provisioning of resources, etc.), based on frequency and duration of threshold violations.

1.7 OVERVIEW OF BOOK ORGANIZATION AND CHAPTERS

This is an orchestrated set of chapters, written exclusively for the book by a team of subject experts from around the globe. As a technical reference book, users will find definitions and descriptions of every aspect of next generation telecommunications networks and services and their management. As a graduate textbook, students will have information that strikes at the center of where the telecommunications industry is going over the next 15 years and beyond.

In this chapter, the co-editors discuss changes, opportunities, and challenges in the field of next generation telecommunications networks, services and management and summarize the book. Chapter 2 and Chapter 3 address the nearly boundless arena of triple and quad-play services that have been deployed in the past three to five years and their management, from a Telecom and cable point-of-view, respectively. These services will migrate into more advanced next generation IP-based services such as IMS-based IPTV, triple shift, multi-screen, blended/converged services, social networking, shared video services, interactive advertisement with instant purchasing, etc. Chapter 4 goes into specific definitions of next generation technologies, networks, and services. Architectures are described. Importantly, for the purpose of this book, Fault, Configuration, Accounting, Performance and Security (FCAPS) requirements are addressed. Chapter 4 brings into clear focus the next generation point-of-departure for operations and network management.

Convergence is a key word in the telecommunications industry. Chapter 5 addresses convergence and an important convergence vehicle, IP Multimedia Subsystem (IMS), and associated management requirements. Chapter 6 is fundamental to steering the right course to the future. It defines next generation operations and network management architecture. This is the key for timeliness and functional and cost effectiveness. Ad hoc wireless and sensor networks and their management is the key to home networking. Chapter 7 defines these technologies, networks, and services opportunities. Chapter 8 approaches next generation operations and network management standards from a strategic perspective. This chapter offers users and students the information needed to understand the global standards landscape of forums and their scope and processes. Perhaps most importantly, Chapter 8 instructs users and students how to engage in next generation operations and network management standards. It concludes with specific information on current next generation operations and network management standards, existing and/or under development. Chapter 9 forecasts the future in this field. It is for reading enjoyment. One thing is clear: the future will be rich with opportunities for the global telecommunications industry.

1.8 REFERENCES

1. IPTV High Level Architecture Standard (ATIS-0800007), April 11, 2007. Washington, DC: Alliance for Telecommunications Industry Standards.
2. Aidarous S, Plevyak T, eds. 1994. *Telecommunications Network Management into the 21st Century: Techniques, Standards, Technologies, and Applications*. Piscataway, NJ: IEEE Press.
3. Aidarous S, Plevyak T, eds. 1998. *Telecommunications Network Management: Technologies and Implementations*. Piscataway, NJ: IEEE Press.
4. ITU-T Recommendation Y.1271: Framework(s) on network requirements and capabilities to support emergency communications over evolving circuit-switched and packet-switched networks, October 2004. Geneva, Switzerland: International Telecommunication Union Telecommunication Standardization Bureau.

CHAPTER 2

MANAGEMENT OF TRIPLE/QUADRUPLE PLAY SERVICES FROM A TELECOM PERSPECTIVE

Jean Craveur

2.1 INTRODUCTION

Managing broadband triple/quadruple play today represents a number of challenges for telecommunications operators at commercial, technical, and operational levels. To better understand such challenges, it is worthwhile to give some historical perspective and to explain today's telecommunications context where historical frontiers between technical domains and commercial domains have started to vanish.

2.2 CONTEXT OF TRIPLE/QUADRUPLE PLAY FOR TELECOM OPERATORS

Telecom operators are living a technological and business mutation. Their networks are entering a true second life (see ref. 1, paragraph 2.9) after a gestation period where digitalization of information, IP transport protocols, Web technologies, mobile communications, broadband, and a number of other technologies have caused that second birth.

Telecom networks today represent a prodigious tool for entertainment, communications, management, etc. at the disposal of customers who have access to a melting pot of services and content. Even more, these physical networks are becoming the foundation on which people are, more and more, building their human and social networks.

To say the least, users are to be considered at the core of the networks. From passive end-points, they became permanent active components of layered and meshed networks and sources of information—user generated content (UGC)—transferred or accessed worldwide.

Next Generation Telecommunications Networks, Services, and Management, Edited by Thomas Plevyak and Veli Sahin

In conjunction with a change in the rules of the game and in regulations, these technology and usage revolutions have also led to a deep transformation in the value chain. Competition, online service and content providers (newly created or coming from other sectors such as audio-visual), the apparent free access and use of services (a habit coming from the Internet generation), and advertising dispatched over the networks, etc., have created new businesses and business models. This context has incited operators (but also actors like suppliers and others) to explore new territories at the boundary of their core business in order to follow the value, the end customer.

At the technical level, historical bottlenecks vanish. Broadband on both fixed and mobile subscriber connection is now implemented. Initially focused to get Internet access (in Europe), broadband became multi-service. This presented the opportunity to connect homes more easily to a number of new applications via new equipment installed on customer premises. An example of such equipment is the residential gateway (RGW), which provides common entrance to the broadband communications world and a new and important service sale point, closer to the customer.

This has pushed network operators to study and set-up multimedia broadband infrastructures designed on the basis of a completely new framework. End-to-end digitalized information transport is now in place using packet techniques instead of circuit techniques. IP has become the universal and common transport protocol for any type of digitalized information. Adoption of new architecture principles, like separation of transport and control functions in the Next Generation Network (NGN) and common control for mobile and fixed services in IP Multimedia Subsystem (IMS), are enabling control and transport of data flows of any nature and origin. This includes the more stringent ones, i.e., those coming from conversational or real-time TV services. One can notice new access characteristics: the increasing symmetry of user flows on fixed services, from xDSL over copper to optics, and on mobile services, from Universal Mobile Telecommunications Services (UMTS) to 4th Generation (4G) access as well as widespread implementation of always-on connected user equipment.

This technical revolution provides a great opportunity for Telecom operators to share network infrastructures between fixed, mobile, Internet, and content services. It provides the opportunity for separation from legacy networks (PSTN, X25, PDH), thus contributing to medium term cost savings and complexity reduction, even though mass migrations from legacy to new technologies may be costly, painful, and risky.

Triple play is voice, Internet, and TV services access. Quadruple play adds mobile services. Tomorrow there will be multiple play services. These are made possible through a single generic broadband access. This is *the* challenge. The attempt is not new. Remember that the first step toward a fully integrated digital world was called Integrated Services Digital Network (ISDN). That was 25 years ago! Since then, technology has evolved tremendously. One part of the Holy Grail for customers is within reach.

"The value of networks is very much linked to their bandwidth, to the nodes, and to the content which they transport. New services are the oxygen of our networks" said Didier Lombard, Chairman and CEO of France Telecom Orange. Beside proposing higher access throughputs at home, on the move, and at the office, Telecom operators also have a fundamental imperative: to bring a continuous flow of innovation into their networks, services, and IS. This will lead, for instance, to enhancements in content offers (HDTV, 3DTV, mobile TV, etc.) and the daily operation of services, such as health and security. It will support the development of UGC and social networks to insure better experiences on existing services (VoIP, TV, VoD, etc.) and provide, for a given service, continuity and fluidity abilities on different devices (multi-screen strategy) and access (fixed and mobile).

This profusion of technologies and usage has resulted in a tremendous amount of complexity. The need to simplify has become more than evident. Like in Sisyphus's work, this is a continuous effort. This is the reason why convergence, mutualization, architectural efforts and so on, are essential tools to obtain simplicity for service usage as well as service and network operation.

In summary, Telecom operators have a number of challenges to face.

The commercial challenge is that historical business models, based mainly on voice transport, are no longer sustainable. In a world of abundance, protecting a viable business model by driving a broadband-everywhere strategy, while taking advantage of assets and traditional strength, is an essential issue. This includes the use of capabilities such as billing (useful for billing third-party services), business intelligence (profiling, localization, and so on) based on knowledge of their customers, Quality-of-Service (QoS), and customer experience. Historical know-how and lessons are important, pulled from dozens of years of real-time applications delivered to millions of customers.

The technical challenge is to select the best-of-breed of new technologies whose arrival rate has never been so rapid. The technical challenge is to maintain agility and secure robustness and scalability for new innovative services in a complete IP-based world of transformation.

Agility means the ability to evolve service platforms and IT to support faster service rollouts. To secure robustness and scalability means the ability for network and IT architecture and design and implementation to face the growth of traffic and number of customers generated by new services. And last, but not least, to improve customer experience.

Triple play/quadruple play is currently under deployment in conjunction with a "broadband everywhere" strategy in fixed and mobile domains (FTTx, HSPA, LTE, WIMAX, etc.). This is going to have deep consequences all along the technical/network chain, from the customer premises (home network), network access, backhaul and aggregation, transport backbone, service and network control, service platform, and finally to IT.

But IT and network infrastructure life is, and always has been, a world of anticipation. Anticipating "beyond triple play" by providing future architectures and mechanisms is also part of the technical challenge, e.g., enhanced content delivery networks (CDNs), Internet, satellite and mobile TV, services on fiber, enhanced home networking, network storage, and so on.

The technical challenge cannot be successfully achieved if network and IT operations challenges, e.g., new operations models and processes, are not addressed and achieved. The key differentiator will be the ability to ensure, day-after-day, the QoS and competitive cost expected by customers. This will have the ability to hide (from customers) the overall complexity. A number of quality problems with triple and quadruple play exist, such as dropped VoIP calls, bad audio or video quality, long IPTV channel zapping delay, and others. In the end, what matters is the quality of experience, the quality as perceived by the customer. This challenge should be pursued while keeping operating expenses (OPEX) under control. This is critical in triple play operation and is valid for service provision, network operation, after sale processes, etc.

The existing network and IT architecture, methods of operation, delivery process (commercial and technical aspects), and operational structure need to be adapted to better fit with the characteristics of new services and business challenges. This implies transformation programs by operators that will not be cut over in one night when millions of connected customers have to be looked after and satisfied in real time. Seeking end-to-end service assurance solutions is key to getting a permanent, end-to-end, quality solution for all services. It will be important to be able to identify and locate the root of quality degradation. Finally, it will be necessary to perform efficient monitoring and trouble-shooting.

2.3 THE ECONOMIC, SERVICE, AND COMMERCIAL CHALLENGES

2.3.1 General Conditions

Finding new business opportunities in the face of decreasing revenue due to competition and business-disruptive technologies is a necessity for Telecom operators in today's world. Triple play is a tool to drive the business of broadband to mass market.

Business growth will be achieved via a customer-centric approach, i.e., the customer is no longer separate from the technical structure of the service(s). This implies an integrated approach regarding the service offer (fixed/mobile/Internet) toward a world of full service convergence. Cost reduction will impose a mutualized and integrated network and IT approach via convergent, common, and reusable enablers. It will also require a revisit to all processes including commercial processes (marketing, sale, after sale, etc.).

Multiple play strategy may lead to the definition of new services or the enhancement of existing services. For instance, even for voice services it is possible to look for added-value and stimulate usage differentiation from classic telephony. This will be built on the presence/availability of the voice line in the home toward comfort and high quality sound, e.g., high-definition (HD) voice and the mixing of voice and data (implying new devices). Even if the new information paradigm is to offer "good enough" services instead of "ensured high quality" for everybody, everywhere and at any time, one thing should never be forgotten—users are not only looking at the level of functionality for a given price, they are also considering the level of trust they might have for a new service. Just look at the number of people still keeping their public switched telephone network (PSTN) line in parallel with their VoIP line.

Another element to consider is the economic investment network operators will have to face in new broadband technologies. Extension of access network coverage requires huge amounts of investment. PSTN infrastructure has been historically financed publicly and the amortized copper access infrastructures (whose maintenance remains expensive) have been able to carry abundant data bandwidths, enabling the development of a significant part of the Internet user base. Global System for Mobile (GSM) infrastructures benefited from an initial up-lift linked to terminating call prices higher than PSTN terminating call. The situation is completely different concerning fiber access networks. These are new infrastructures, built largely from scratch. Network operators need to find new sources of revenue able to provide reasonable return-on-investment (ROI).

Successful entry into the new business paradigm and creation of new sources of revenue are essential but are not the only necessary condition. Definition of a clear and fair regulatory framework is also necessary. Another factor comes from customer requests in a communications world that is more and more open and, therefore, more and more perceived as invasive. In such an open communications world, protection of personal data (various identities like names and addresses, family or profession, location or history of purchases) is already an inescapable customer requirement. Trust in the operator's ability to assure this will be a key.

Although this book does not address regulatory/legal issues, regulation regime should also be considered carefully when defining triple play offers and designing their implementation. Some operators could be requested to build a wholesale offer for every retail offer implemented. This situation, currently encountered in Europe with converged fixed and mobile offers, needs to be anticipated during the design phase since it adds new functions to the overall architecture and impacts network and IS mechanisms.

2.3.2 Service Offer Requirements

There are many options that can be taken at the marketing and commercial levels. Triple play services characteristics, i.e., Internet access, throughput, TV broadcast, video-on-demand (VoD), VoIP first line, VoIP second line, and so on, and the potential penetration expected are essential features in the design of network and IT architecture and operation.

In order to properly design the network and IT infrastructures to support targeted triple play services, one must take into account factors like technical eligibility of customer line, deployment pace, encoding techniques, service nature, and QoS for the particular service access.

In a situation where most of the triple play offers are accessed today on copper, service eligibility of the access lines is mainly based on the technical capability of the line to carry the requested throughput. That capability is dependent on copper quality, line length, and copper cable diameter. Encoding technique is also a parameter that could influence eligibility of the line to certain services. With MPEG2 video, in countries like France, only 60% of the lines are eligible for the service. With MPEG4, the percentage increases.

Service deployment speed is an element to take into account in the network and IT architecture and engineering in order to prevent over-dimensioning and anticipated expenses, on the one hand, and blocking points in traffic or customer number handling by equipment, on the other hand.

It is important to know from the outset the true nature of the triple play services sold. For instance, when accessing TV service, it is crucial to know if the access is restricted to one program or if simultaneous access to several programs is part of the service. Simultaneous access to two TV MPEG2 encoded programs will result in lower coverage thus impacting the business plan. When moving to MPEG4, there is the option to choose increased coverage or provision of simultaneous access to two programs on the same line. Other examples are VoD offer penetration, which has a strong impact on the VoD servers dimensioning (VoD pumps) and the personal video recorder (PVR) location, either provided by the network or by the set-top box (STB).

Additional parameters like peak bandwidth requested, service level agreement (SLA) and QoS level, mean time to recovery (MTTR) requirements, Nomadism envisaged or not, typical customer density average, customer distance and trends in bandwidth evolution, will be the determinant regarding, for instance, aggregation network evolution. Concerning traffic scalability, the service level expectations can vary from one operator to another. There is, therefore, no uniform service scalability rules to be applied for each service.

Several future milestones in service development or technology evolution need to be anticipated because they can have a strong impact on the chosen path of evolution. Examples are High-Speed Downlink Packet Access (HSDPA) which will multiply 3G backhauling scalability needs by 5, 3G radio access network (RAN) traffic transported on Gigabit Ethernet IP (GE IP) aggregation which could disturb existing traffic flows, MPEG4 broadcasting, TDM (time division multiplexing) end-of-life, etc.

In the content domain, the type of relation contracted with content providers structures billing function and equipment management arrangements. Questions like which partner is responsible for customer relations, how is revenue calculated and shared, who is responsible for STB management, etc., should be clearly answered in order to complete the technical design of the solution.

2.4 THE TECHNICAL CHALLENGE

To implement triple play and quadruple play services, with different technical characteristics and requirements, in infrastructures which have been built a long time ago to deal initially with a single service (voice on copper), is an interesting challenge. The objective of accessing more than one service across a single access is not new. It is well known that analog circuit connection access to the PSTN has supported voice and data (fax, minitel, tele-detection applications, and so on). One can also point out that ISDN is a tremendous standardization and implementation effort that benefited from digitalization of information supporting both circuit and data connections over unique copper access. But in all these attempts, throughputs were limited to dozens of kb/s, restricting the extent of the service offering.

Broadband fixed and mobile technologies are today providing exceptional opportunities for Telecom operators to transform their business and their infrastructures. These services require closer network and IT, bringing together fixed and mobile infrastructures. The target is new innovative services and cost savings through common service enablers. A renovated IT infrastructure has to be built on best-in-class customer relationship management (CRM) serving fixed, mobile, and Internet, and on e-care mechanisms that allow customers to interact directly with the platforms to configure and provision their services. This renovated IT infrastructure will support a converged, access-independent service infrastructure based on the NGN principle of separation between transport and control and using IMS as common service and resource control. In order to allow internal and external actors, and even customers, to develop services by blending communications and open content-rich libraries, application programming interfaces (APIs) must be developed and sold. This service infrastructure, enhanced by a network storage offering, enabling customers to store their personal production, accessible and shared from everywhere, will be delivered over a broadband access and a backbone infrastructure supporting advanced IP routing technologies.

2.4.1 The Technical Tool Box

The technical tool box making triple play or quadruple play happen consists of equipment, such as CPE, network (access, aggregation, backbone, control), service platforms, and Information Services (IS), organized in a true working chain (none is to be considered as fully independent from each other).

2.4.1.1 Customer Equipment The role devoted to end terminals is now very much extended compared to the historic situation. This contributes substantially to enrichment of service offers like voice (fixed and mobile), WebTV, broadcast TV, VoD, digital content, tele-actions, interactive video services, and so on. It applies to fixed and mobile handsets, PCs, TV sets, STBs, PLT (power line transmission) equipment, modems, residential gateways, wired, and wireless LANs. One of the key questions for operators is the functional distribution between customer's and operator's infrastructure equipment. More and more "intelligence" is now located in handsets and in home networks competing with functions located in networks.

2.4.1.2 Access Line and Aggregation/Backhaul Networks To provide broadband (several Mbit/s) on copper lines, Telecom operators need to cope with xDSL's constraints (throughput dependent on copper line length and quality). To reach higher throughputs between homes and networks requires the use of optical fiber, alone or combined with copper and very high bitrate digital subscriber line (VDSL) techniques. When copper lines do not have the right characteristics to carry the requested bandwidth, satellite access can also be used to extend the number of customers eligible for broadband. Fiber accesses benefit from symmetrical throughput. Increased bandwidth demand, with different traffic characteristics, impacts access, backhaul, and aggregation parts of the network in terms of bandwidth. But the main issue, essential for broadband service evolution, is the access network architecture. This includes the transformation of digital subscriber line access multiplexer (DSLAM) in multi-service access nodes (MSANs), the move towards an access using a non-specialized virtual circuit (VC), the move from ATM techniques towards Gigabit Ethernet techniques in aggregation and the reallocation of broadband remote access server (BRAS) functions to the MSAN.

2.4.1.3 Backbone Networks This covers transmission equipment, IP routers, internetworking gateways, etc. The inherent characteristics of IP protocol and networks impose various implemention techniques, aiming to increase transport QoS such as multi-protocol label switching (MPLS), virtual private network (VPN), Diffserv, etc., in order to be able to carry the most demanding services like voice, real time video and TV. In the field of video content distribution different content distribution network (CDN) techniques and architectures have been developped for cost and QoS reasons.

2.4.1.4 Control Platform Consequences of the NGN principle of separation of user-to-user information transport from control are that these platforms host the functions that control the services and resources to be set up, used, and released. One can notice that they are more and more based on the same computing technology and techniques than those used by Service Platforms. In the framework of IMS (IP Multimedia Subsystem), these control functions are more standardized than before, since a quite detailed functional architecture and a set of interfaces have been specified by 3GPP and TISPAN Standards Development Organizations (SDOs). IMS-controlled services give assurance to customers that the service offered remains fully controlled by operators.

2.4.1.5 Service Platform These platforms host the service software functions that are interacting with the IS and with the network control elements. They have the same operational real-time constraints as the network elements (NEs) of the control and transport levels, i.e., service continuity, QoS, real-time reaction, and so on. They are the place where a number of innovative computing features are implemented, aiming at cost reduction, enhanced security, and better time-to-market (shared software between services, reusable developed software components, virtualization, etc.).

2.4.1.6 IS Equipment Some of the IS equipment participates in the service offer by interacting in real time with network or customer equipment. But the overall architecture of IS is a subject in itself and needs a deep transformation to serve triple play and quadruple play. IS evolution is no longer separated from network and service platform evolution and vice versa (see Section 2.4.10).

All these elements need to be synchronized when working and reacting to each other in real time. To get fully robust and scalable systems is a huge enterprise. At the conception phase, this is the job of architects, and in complex architectures required by triple play and quadruple play, this is in particular the job of one category of architects, the overall architects. There is a need to get network architects and IS architects working together to define a truly global enterprise architecture encompassing all the elements of the chain.

When overall design and development of all pieces have been achieved, there is one last thing to do, that is to check that what has been conceived and developed is working together as planned, not only for the simple call but for massive calls and in abnormal conditions of traffic and operation. This is the place to conduct end-to-end stress testing to check the overall behavior of the complete chain and its ability to meet expectation of the customer perception as it was specified at the origin of the cycle.

2.4.2 The Global Vision

In the past, the main service issues were more separate and independent than with triple play and quadruple play, i.e., call access, call and traffic concentration, circuit switching and call routing from point to point, call supervision, and release. Today, with triple play there are new correlated actions to perform, i.e., service type requested analysis, mobility or nomadism and location update provision, appropriate required resources to set-up depending on the service request and the terminal used, security measures (IP open world), content provision (with partners), other services and content blending if required, usage monitoring, and so on. In addition, new requirements are to be taken into account at the service conception phase, i.e., convergence and mutualization objectives, maximum latency value (sensitive to voice and real-time video), differentiated qualities demands, APIs provision for external partners, security requirements, regulated wholesale constraints, etc.

Network operators have to elaborate global architectures combining agility and robustness. To do so and to take into account the world of complex interactions, an overall architecture is required. This will be the job of architects who have to think and act globally.

2.4.2.1 Vision for an Overall Architecture Supporting Triple and Quadruple Play Figure 2-1 illustrates what could be the long-term vision of the Network and Information System to provide triple, quadruple, and multiple play. This vision is based on the following assumptions:

1. Devices / home network are an inherent part of the service infrastructure.
2. MSAN has become the unique access point (access technology agnostic).
3. A common control access technology takes care of the services and associated resources.
4. Open interfaces for customized services (APIs) are used by third-party developers.
5. Packet transport is based on IP enriched with MPLS and Diffserv. mechanisms.
6. A single layer for service platforms exists (access technology agnostic).
7. A single IS for all applications exists (access technology agnostic).

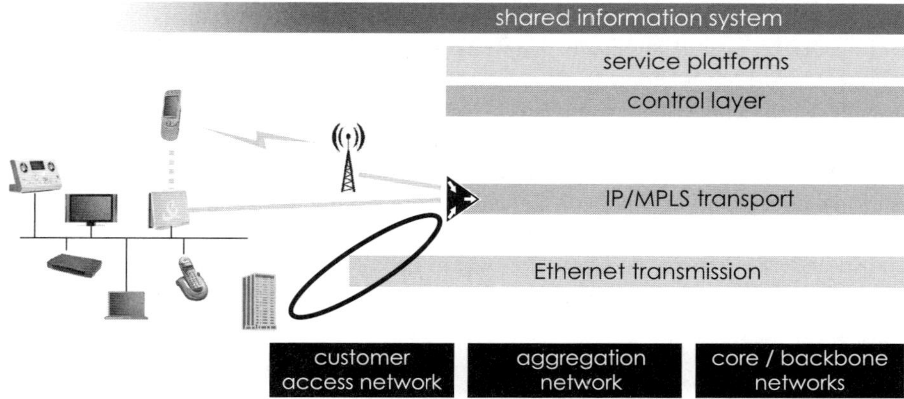

Figure 2-1. Multiple play global target architecture vision

2.4.3 Key Issues to Consider When Designing Network and IS Infrastructures for Triple and Quadruple Play

There are a number of points that require particular attention when designing network and IS infrastructures for triple and quadruple play services.

In the field of traffic, there is a necessity to design network architecture and network elements to take care of traffic blocking points related to access to broadcast programs. This point has consequences in terms of DSLAM architectures so that all customers get access to all programs without any restriction, even if all customers are willing to access the same program at the same time. In addition, DSLAM backhauling should be able to bring all the programs simultaneously to the same DSLAM. The traffic required for bundles of programs is, in general, much higher than that required to transport voice traffic or Internet access traffic for a smaller number of customers.

In the field of the required QoS for broadcast television, there are some peculiarities:

1. Permanence of service should be ensured during periods that are outside traditional working hours (peak hours for TV are between 8.00 PM and 10.00 PM in France).
2. Image quality can be maintained without any pixelization thanks to technology like forward error correction (FEC) on the access line. This reduces noise effect thanks to architecture arrangements like video dedicated ATM/VP (virtual path) on copper access to guarantee adequate throughput for video services. New fully IP based mechanisms are under development that will replace the ATM based VP.
3. When TV channel selection is done in the network, maintaining zapping time delay requires an Internet Group Management Protocol (IGMP) interaction between STB and DSLAM. The time delay of that interaction should remain small compared to synchronization time.

4. VoIP service quality is a critical issue, especially if Network operators are intending to offer VoIP as a 1st line. The issue is the competition between VoIP flow and Internet flows on the ADSL (Asymmetric Digital Subscriber Line) upstream low bit rate. One solution could be to choose a dedicated virtual circuit (VC) for Voice, but the drawback is the access management and the ATM layer to implement and to manage.

2.4.3.1 Convergence and Mutualization In an industry where fixed cost is important, there is a motivation to aggregate traffic, to group elements in order to decrease the average cost per unit (economy of scale through multiplexing). For example, in transmission systems, one 2.5 Gb/s SDH ring is far less costly than four 622 MBt/s rings. This logic, also found in IP networks where the granularity of the optical transmission links between IP routers is either 2.5 or 10 Gbit/s wavelength, has as a consequence, that any additional demand arriving at the IP backbone network is generally marginal in terms of cost per bit transported.

But mutualizing requires a number of careful studies. In particular, there should be some assurance that forecasted demand will fill up the shared infrastructure, otherwise the cost of the new offer may increase dramatically. Another element to consider is the fact that mutualizing means concentrating different traffic demands on a limited number of equipments or systems. Failure of one of them may have an effect on several service offers. In such a condition, reliability of the network is a very first priority via redundancies (n + p redundancy) to guarantee the required unavailability objectives. Load sharing or take over mechanisms need to be specified. The unavailability objectives should be explicitly specified at the very beginning of the service and network architecture design. Calculation of the projected unavailability should take care of equipment failures but also the unavailability resulting from software updates and reloading time delay.

In the field of service and control platforms, the type of redundancy to adopt will depend on the type of service and data manipulated. Indeed, data required for service handling should be accessed by the backup platform.

2.4.3.2 Quality of Service (QoS) Relevant QoS parameters for service offers should be well identified at the very beginning and then followed up. They are parameters linked to service continuity, session call set-up, and transport quality.

QoS Parameters Linked to Service Continuity Service continuity objectives are expressed via service unavailability objectives for one customer or a group of customers. In triple play and quadruple play, the broadband access unavailability issue is crucial. The customer experience with telephony and broadcast TV is rather good today. However, broadband access via a number of services has, by construction, more complex architecture, more equipment, and a higher unavailability than telephony access or aerial TV. One particular point to look after is the unavailability of DSLAMs. One can imagine reducing that unavailability by duplicating line couplers and by new software release loading without interrupting the equipment.

QoS Session Call Set-up Quality Parameters These parameters are only pertinent for services requiring session establishment. One can separate failures of session attempts due to protocol interworking from failures due to blocking by lack of resources, e.g., in transmission resulting from the activation access control admission mechanisms. This last case can be solved by dimensioning. The first is linked to protocol conception and development in the different equipments.

QoS Transport Quality Parameters International standards cover, generally, the basic parameters for transmission networks. However, the issue is not fully covered because the bit rate of more complex copper lines supporting broadband may depend on various factors like line attenuation linked to the length of the line, diaphonic disturbances, impulse noise, etc. These variations result in unacceptable pixelization in video signals. Possibilities exist to improve the situation by using redundancy of the video signal allowing FEC.

In IP networks, one important parameter is the packet loss rate. IP packet loss may result from temporary router overload or from failures of equipment that require a certain time for the network to reconfigure the routing tables. The impact of such packet loss is very much dependant on the type of service. For non-real-time data transport, the end-to-end retransmission mechanisms may be sufficient to deliver data with delay. But short interruptions are not acceptable for real-time services like VoIP or Video IP where packet loss can make voice unintelligible and can freeze video images. Therefore, it is important to decrease the convergence time of the routing algorithms to reduce the effect of temporary overloads via over dimensioning of links and nodes and implemention of prioritization for real-time packet flows (i.e., Diffserv mechanisms).

In the audio domain, voice quality of VoIP again becomes an important issue. Besides the IP transport mentioned above, the quality of speech coding is at stake. Packet encoding of voice signals introduces time delay and transcoding effects deepen the voice quality. Network operator architects should try, to the extent possible, to obviate network transcoding that is another drawback in traffic concentration on transcoding equipment. However, in the absence of international or regional regulation in this area, transcodings are today inevitable at network boundaries.

2.4.4 Customer Premises Equipment (CPE) and Home Network

As already mentioned, CPE and home network are the place for a number of important changes as far as communications, entertainment, and remote management are concerned. These new capabilities of hand sets and home networks have resulted in a number of interactions with functions located in networks or IS. When dealing with various service types, this has introduced a lot of complexity in terms of customer usage. It is therefore a first priority from an operator perspective to simplify the digital home experience through development of "plug and play" equipment.

2.4.4.1 The Home Network Complexity From single play to multiplay, the number of connected devices is going to increase (set top box, connected TV, infrastructure devices such as PLC plugs and wifi extender, PC, laptop, UMA phones, SIP phones, game consoles, etc.). All these devices belong to the same home LAN whose complexity is likely to also increase. These devices get access to various classes-of-service, locally and remotely, like voice, video, TV, Internet, tele-detection applications, and so on. Inside cabling is also a difficult issue. Usually, traditional telephony cabling existing in homes is not sufficient to transport high bit rate. Therefore operators may have to propose to their customers PLT (power line transmission) or wifi technologies in order for them to have TV sets or PCs located far from the STB location.

2.4.4.2 Distribution of Functions between Network and IS Platforms and Residential Gateways In classical networks, customer equipment was limited to network termination and quite simple hand sets. Interactions were rather limited, with a small number of network nodes (i.e., analog telephone hand sets interfacing the local switch in classical telephony and the DSL modem interfacing DSLAM).

Example of Additional Function in CPE More sophisticated functions are implemented such as analog to VoIP conversion functions to allow analog terminals to be connected to the VoIP supporting network. In classic telephony, the end of numbering for the called number, belonging to an open numbering plan, was performed by the local switching center. The situation with VoIP, accessed through residential gateways (RGWs), can be notably different and performed by either the RGW (by time out, which is simpler but increases the post dialing delay, or by received digit number analysis, based on data sent by the network via a specific protocol) or the service platform. In this case, digits are received in separate messages that multiply the signaling messages.

With the installation of RGWs, there are a number of decisions to be made that should balance the pros and cons of allocating the function to the RGW or the service platform.

One important point to keep in mind is the fact that millions of RGWs are disseminated and require remote reliable and fast software upgrade mechanisms. Another factor to consider is linked to the data stored in and used by the RGW, e.g., physical customer access configuration, some service subscription data, etc. The experience shows that the greater the number of functions implemented in RGWs, the more the customer calls arrive to the after-sale service. This results in significantly higer OPEX costs. These facts will certainly determine the limit of functions and data installed in RGWs.

2.4.4.3 The Home Network Paradox On one hand, from the function interaction point-of-view, operators need to consider the home network as the last meter/yard of their network. On the other hand, they cannot control it totally since, for instance, deployed wireless technologies (e.g., wifi, PLT) within the customer environment, have no guarantee as far as bandwidth, delay, and characteristics are concerned. Moreover, within the home network, there is cohabitation between devices managed and not managed by the operator. Therefore, the delicate question of the limits of responsibility is raised. In PSTN it was clear. But with multi-play, the question is, where is the internal ending interface? The first approach could be to include all the devices provided by the operator (RGW, STB, WIFI extender, PLT plugs, etc.). Then how far must the operator (alone or with partners) manage connected PCs, connected TVs, and so on?

2.4.4.4 The Home Device and Applications Devices and applications cover the communications, entertainment, and home management domains (in a mass market context). Self-installation could be an operator's strategy but the end-user may need to help set up his/her home LAN and services and maintain the right QoS. Assistance may be given through telephone, training, or operator's installation (with or without partners). On their side, operator technicians on the ground need tools for home network infrastructure installation and performance monitoring for PLC, wifi, cabling, and to set up end-to-end services. Finally, operator hotlines will need diagnosis and monitoring tools to remotely manage the home network.

2.4.5 Access Lines

For copper lines, network operators have to decide which of the xDSL techniques to use (ADSL 2, reach ADSL, VDSL, etc.) in order to make the concerned copper lines eligible, following economic analysis, to the targeted services and coverage.

Copper lines, being sensitive to noise impairments, especially for high bit rates, the implementation of dynamic line management (DLM) on DSL lines is recognized to give better conditions to offer triple play (see Figure 2-2 for an over-

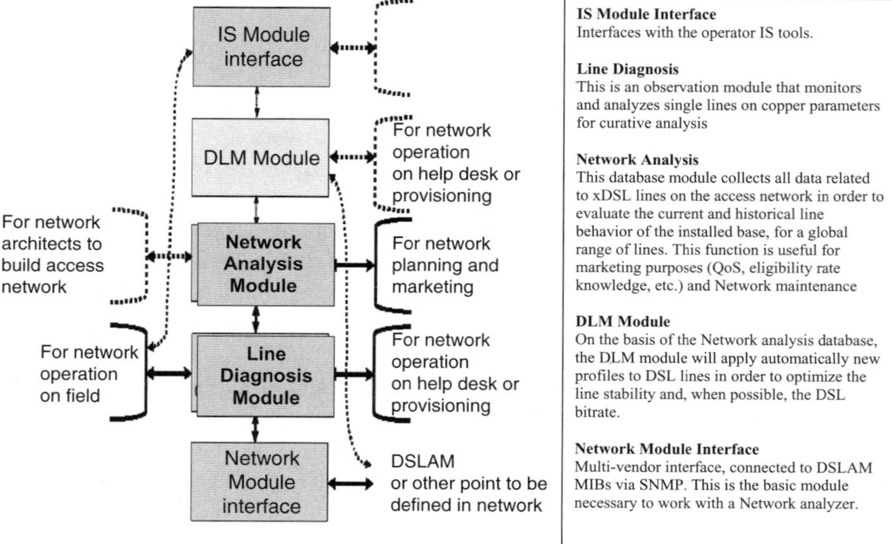

Figure 2-2. Dynamic line management for DSL lines

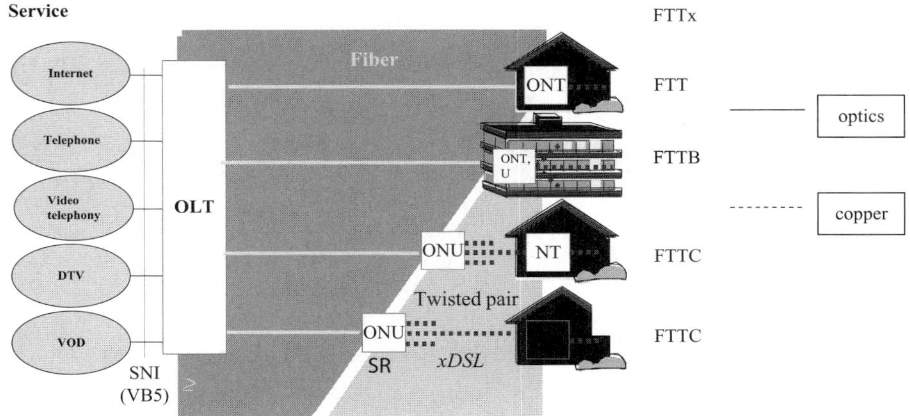

Figure 2-3. The different FTTx options. FTTB: fiber to the building, FTTCab: fiber to the cabinet, FTTH/O: fiber to the home/office, FTTCu: fiber to the Curb.

view of possible DLM implementation). By improving QoS of broadband services and the rate of eligibility for services of the installed copper lines, DLM generates some cash-in and saves after-sale internal OPEX (reduction of calls to the hotline and the number of interventions on the ground).

For optical lines, choice of fiber-to-the-home (FTTH) using point-to-point or GPON techniques, or fiber-to-the-building, plus VDSL (FTTB/VDSL), depends significantly on the considered geographical areas in a region or country. Figure 2-3 illustrates the different FTTx options to be chosen. Operators may have to consider not only investment in their own optical fiber network but also, use of possible wholesale offers (regulated or not) and possible associations to share investment. The main criteria to consider are fiber and duct availability, housing type in targeted area and environmental and regulatory constraints.

Today, operators' strategy is certainly pushing copper to its full potential and is leveraging satellite availability for TV broadcast. In the field of optics, Network operators are choosing either FTTH (GPON-based or point-to-point) or mixed optics and VDSL on copper (FTTB, FTTC). But massive deployment is very much linked to the regulatory regime and on acceptable return on investment (ROI).

2.4.6 Access Networks, Aggregation, and Backhauling

Converged fixed-mobile backhaul networks are the target for cost savings to anticipate traffic growth. Dramatic increase in traffic requires adoption of new technology that provides a breakthrough to manage traffic growth. For cost reasons, Ethernet technology is ramping up in aggregation networks (DSLAM with Gigabit Ethernet network interface). Concerning access line connecting nodes, DSLAM moves towards true multiservice access nodes (MSAN) with IP routing capabilities, supporting fiber and DSL, residential, business, wholesale, and mobile backhaul.

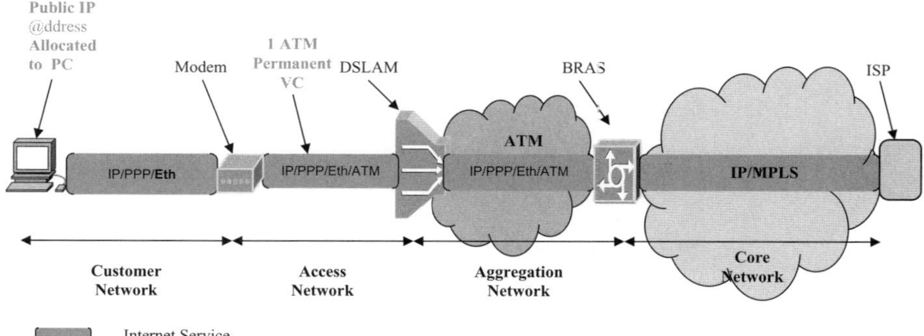

Figure 2-4. An example of an initial Internet deployment architecture (in France)

From a functional perspective, the evolution to a targeted architecture needs to cope with different issues like the unbundled-local loop (ULL), the choice of protocol to be used for Internet service e.g., point-to-point protocol (PPP) versus Dynamic Host Configuration Protocol (DHCP) based architecture, which is mainly dependent on marketing issues like volume requirements per offer and per user, and so on. Network operators should also make choices related to DSLAM/MSAN functions, virtual circuits (VCs) organization, aggregation, and so on.

2.4.7 An Illustration of the Fixed Access Network Transformation from Internet Access Support to Triple Play Support

From the first deployment of Internet to the full range of triple play and quadruple play services, access architecture has been transformed continuously. When broadband was first introduced on copper line, it was with the intention to boost Internet access. Access network architecture was quite simple at that time. Figure 2-4 illustrates an initial Internet deployment architecture.

Since the beginning of broadband Internet, the architecture was enriched progressively to cater to VoIP, video, and TV services, making the overall picture more complex (Figure 2-5).

In particular, this has lead to implement of new equipment at the customer premises, like residential gateways and set-top boxes, to cope with modem functions, voice and video signal transcoding, etc. In order to protect the different flows, some operators have chosen separated access paths for the various services (e.g., one VC for Internet, one for voice, and one for TV). Aggregation technology has moved from ATM to Gigabit Ethernet technologies (mainly for economic reasons in front of the huge increase of audio visual traffic, etc.). DSLAM technology has also evolved toward full Gigabit Ethernet (GEth) technologies. New service and control platforms have been developed for voice, TV, and video.

The increase in real-time or on-demand content consumption will probably lead to adoption of a more distributed and cost-optimised architecture using Content Delivery Network (CDN) technologies, Multi channel broadcast protocols (including for internet TV) and Peer-to-Peer dialogues techniques.

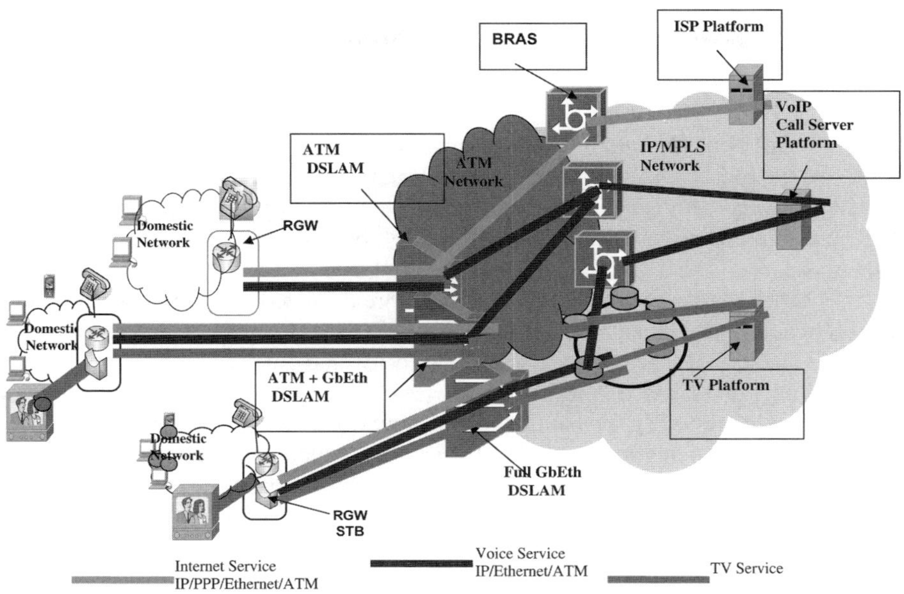

Figure 2-5. Broadband fixed access enrichment to deliver triple play services

These distributed and cost-optimized architectures are progressively moving the first IP routing point to the lowest possible in the network (e.g., DSLAM), whereas it is not yet the case in current networks, where such point is located in the BRAS (to access Internet). New principles for multiservice target architecture currently under intensive thinking refer to "full routed" mode box using mono virtual circuit (VC), end-to-end QoS based on Recommendation IEEE 802.1P and on IP priority mechanisms, access sharing simplification (i.e., mono VC and mono IPv4 or IPv6 @ddress, extension of IP to MSAN and use of DHCP for all services via a mutualised DHCP Server). Besides more simplicity, the targeted expectations with such new principles are reduced interactions between network and service and, thanks to residential gateway operating in full routed mode, accessibillity to all services from all terminals in the home network.

2.4.8 Backbone Networks

Without speaking about transmission networks where the issues are not specific to triple play or quadruple play (except the increased volume of traffic generated by video-oriented applications), what is at stake is the capability of IP networks to become the universal transport network. Key issues turn around the ability of IP transport network to be able not only to absorb the huge increase of traffic coming from Internet access, peer-to-peer, and other multi-play operations, but to satisfy the QoS requirements of voice, TV, and other real-time video in terms of packets lost and latency in normal and abnormal conditions.

When offering triple play and quadruple play on a large-scale basis, operators have to implement techniques like multi-protocol label switching (MPLS), Diffserv, etc., in order to fulfill the above requirements.

When operators are both fixed and mobile service providers, they may consider moving towards converged fixed-mobile core networks for economy of scale in support of increased traffic and for converged services.

2.4.8.1 Content Delivery In the domain of video content distribution using IP transport, the location of the content servers needs to be carefully studied. different content delivery network (CDN) architectures can be implemented using unicast and multicast techniques. Figure 2-6 illustrates the different options an operator can choose.

In centralized architecture, all the servers are localized at the network termination (NT)/network head location. Transmission of content is performed via a unicast flow, per client, from the source point to the residential gateway.

Caching and CDN technology improve VoD (and Web TV distribution) in the core network by reducing the bandwidth consumption over the network, although it has no effect on the aggregation part. Reduced server load and reduced latency are also benefits to expect. Cache and CDN servers can be centralized (e.g., at the NT level) for servers with low audience programs or they can be located at point-of-presence (POP) level for servers with the most popular programs. Unicast flow distributes contents from centralized or decentralized servers to the residential gateway. Figure 2-6 illustrates a simple caching technique decided locally with no content owner cooperation and a more sophisticated CDN technique where, after discussion with content owners, the content replication plan in cache servers is predetermined.

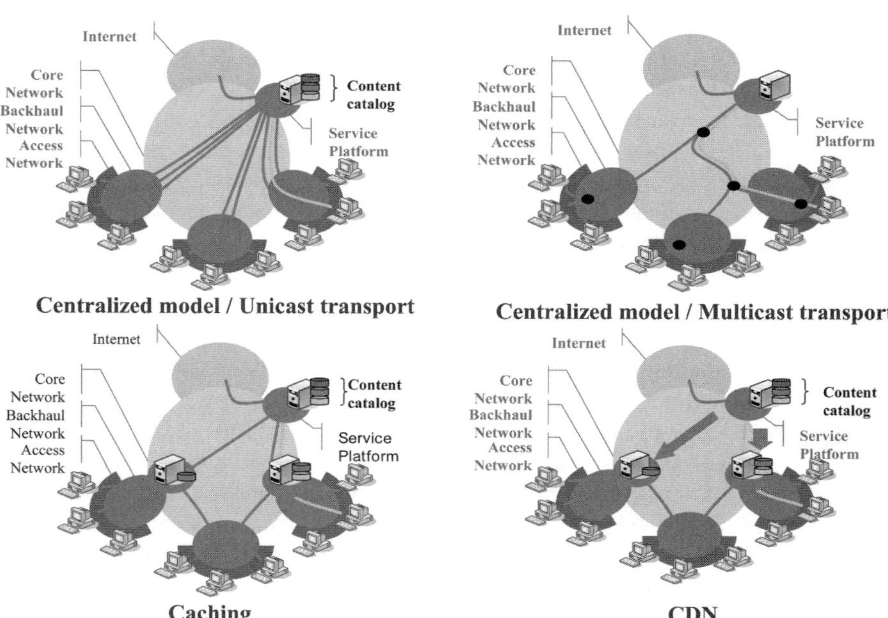

Figure 2-6. Content delivery architecture options

2.4.9 Service and Resource Control

2.4.9.1 Core Control and Application Servers Service and resource control based on IMS standards is a way for network operators to guaranty that QoS and security of services sold are well under their control. In the overall IMS architecture one can point out the importance of application servers (AS), among other important functions like home subscriber servers (HSS), where the service customer data are stored. These application servers contain the service logic. An important issue is the structure of this application layer. In other words, what is necessary is to define application servers' granularity and how these ASs cooperate for a given call (under normal and abnormal conditions, in case of failure or congestion).

2.4.9.2 Service Platforms Service Platforms cover a large range of applications. In the overall architecture, their role is to give flexibility in service offer evolution. One can say that it is, in many ways, the fulfillment, in NGN, of the Intelligent Network ambition in the public switched telephone network (PSTN) and public land mobile network (PLMN). Service platforms are considered by operators as a key tool for providing quickly innovated services. They are felt to be an important differentiation factor. This can be illustrated by three examples, the IPTV platform, central device remote management, and IMS.

In the IPTV platform domain, the evolution is to provide enhanced service offerings by providing a features-enriching interactive-content customer experience, in particular through the framework of three-screen strategy and by providing enhanced TV services to fiber access customers. In parallel with that service offer enhancement, scalability and robustness should be ensured through a multi-platform approach.

In the central remote device management domain, what is at stake is to make an agile delivery of new services with consistent configuration processes on both mobile and Internet ecosystems. This will build efficient customer care for home networks, including quadruple play configurations, and get accurate knowledge of device behavior.

In the IMS domain, the ongoing evolution is the implementation of the IMS technique on a larger scale to corporate customers (e.g., IMS-based mobile IP Centrex service), to mobile customers with the Rich Communication Suite (RCS) and to PSTN customers when PSTN will be shut down.

More globally, the aim is to build the foundation for new and enhanced services (e.g., video-sharing, multi-line) and for a third-party service development open ecosystem.

2.4.10 Information System

Telco's Information System has always been key to meeting business objectives by supporting the ability to deliver innovative offers on time (time-to-market) and to enable new business models.

Concerning the first point, it is more than ever the case that quick introduction of new and innovative offers like bundles, multi-play, etc., provide a means to stay competitive and become a full convergence operator. This requires efficient and flexible ordering, billing, and service delivery.

Concerning the second point, the imperative move toward new business models, replacing or completing the traditional ones, requires IS architecture to be agile, open, and secure. Optimizing their market presence through partnerships (Telco 2.0 model) leads to more and more interaction and interoperability with third parties (content suppliers, audience, partners, distributors, MVNOs, wholesale, etc.). Customer management, therefore, has to be extended to prospects as well as Internet users. New business models like audience, content, etc., drive revenue types far from Telco's traditional revenue streams. Last but not least, multiple external (including customers) accesses to IS will increase the requirement for security.

In addition to these considerations, there are two additional recent and important aspects to take into account, which will also contribute to quick time delivery and the development of new business models.

First, an online-driven customer relationship is becoming key today, where customers/end-users can manage their own services (self-service). This is a way to reduce OPEX (automated customer processes, simplified and consistent customer journey across all sale channels) and get higher quality, higher efficiency, and higher customer satisfaction via quick time to market. This will result in better customer loyalty, knowing that customer experience and customer satisfaction are the two success pillars to reduce churn. This new approach, increasing customer interactions, also provides useful additional customer knowledge to prepare new pertinent offers and to increase sales.

Second, leveraging of this customer knowledge asset is an important potential advantage that needs to be better exploited by developing a 360° view, not a segmented view of customers. Customers should be seen from different angles, such as contract holder, end-user, family, communities, etc. This will allow customer personalization (e.g., business intelligence to push offers adapted to customer needs and habits) and will give opportunity to monetize the audience. This strongly requires unified repositories (mainly customer base and installed base) and unified customer relationship management (CRM) in order to develop narrow segmentation and proactive campaign management. This might lead to enhanced customer knowledge, up-selling and cross-selling, and, finally, to making the most of all customer, prospect, and Internet user interactions.

The current and historic Telcos situation is characterized by a complex legacy IS supporting complex processes and a vertical silo structure (built by product line). This brings IS into the critical path for delivering new offers. This situation has even led to adoption of interim solutions that are very costly to maintain and operate and that have insufficient self-service-oriented customer experiences.

Because it contributes to business process efficiency (automation, online, data quality, shared processes, etc.), IS is an important contributing factor to the overall operator's performance (cost reduction). Therefore, an in-depth transformation of current IS needs to be undertaken to become agile while staying robust in the new world of multiple play. This implies development of best-in-class online driven and customer-centric IS and implementation of well-performing processes and improved real-time service delivery.

2.4.10.1 A Renovated IS Architecture for Triple/Quadruple/Multiple Play Business To offer triple/quadruple and, tomorrow, multiple play requires complete rethinking of the IS since it needs to be "online" operation-based and self-service minded. Service platforms and IS are part of the "new services" and are embedded with them (e.g., the marketing specification for the new triple play offers should include processes and IS to support them). This renovation should also consider, henceforth, that a part of IS has become an active element of real-time oriented services (e.g., e-care). Therefore, the IS architecture should be designed to support these "online driven" customer services.

To reach that goal, a standardized architecture is required, decoupling customer front-end and back-end, opened to new business models and with provision of IS entry points to partners.

Decoupling customer front-end and back-end minimizes cross-domain flows and reduces the impacts of frequent change requests on a given functional block. It also gives better organization to ensure semantic coherence and data integrity between the different blocks.

As illustrated in Figure 2-7, a renovated IS functional architecture can be split into main blocks such as customer front-end, customer back-end (encompassing customer platform, service platform and service, configuration and activation) and aggregation layer. The mediation bus ensures data exchanges during the execution of the services: passive mediation (post-event) and active mediation (e.g., credit information).

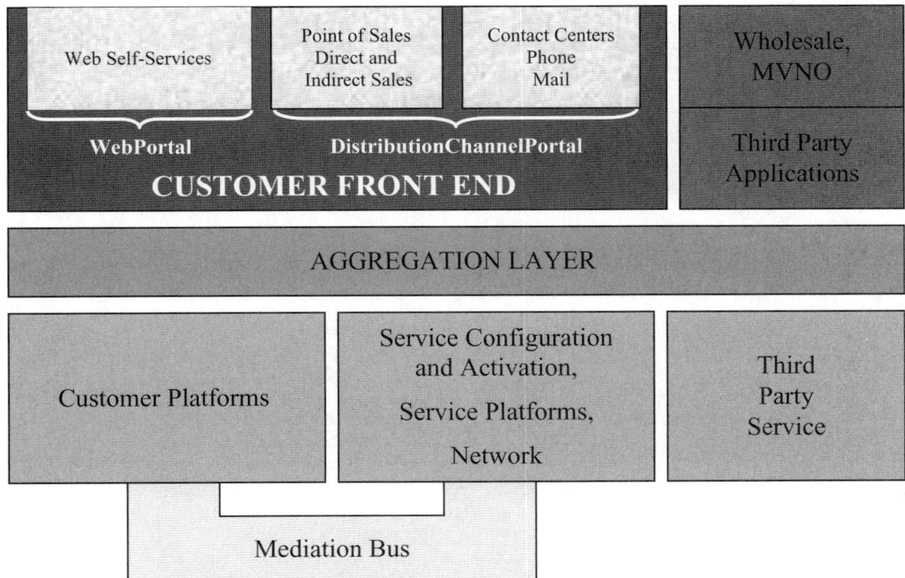

Figure 2-7. A renovated IS architecture framework

2.4.10.2 The Customer Front-End The customer front-end as illustrated in Figure 2-8 manages the user relationship through all channels and all the customer interactions like self-service as well as points-of-contact in direct and indirect distribution. Its scope covers the order capture process, including real-time platform responses to availability requests, the corresponding resource assignment done, the sale story board, and the intelligence of the interaction process. Front-end also covers order configuration (which can make use of a rule engine) and the offer catalogue (commercial presentation).

One of the main challenges of customer front-end evolution is to enhance the quality of the user interface (rich interface technology) by providing a simple, intuitive (and coherent across the channels) customer journey, with dialogue tailored to the customer profile and sales channel. Another challenge is to boost online operation (e.g., presentation of offers, up-selling, ordering, delivery follow-up, support, online payment, etc.) with the aim of having the most requests processed in real-time. Capacity to link properly with back-end services is key to reach such challenges.

The customer front-end, which has to implement the "online customer centric" vision adopted for multiple play services, can be divided into a portal layer, a presentation layer, and a business layer.

The portal layer aims to perform single authentication and identity management. It involves the content management system (e.g., to manage content and layout of the Web pages). It plays the role of a single syndication platform to collect back-end data. The presentation layer function is to adapt presentation to a particular channel, a particular device, and a particular user. The business layer is the place where common business logic has to be defined for all channels in order to allow inter-channel activity. This business logic contains the business rules, the access right management, etc. The business layer is the point to invoke services to the back-end.

It should be noted that the front-end, which does not store persistent data, accesses the customer platform or the service platform for service requests and/or repository requests. Tools required for troubleshooting, for advanced test and diagnosis, and for advanced sales are also part of the front-end.

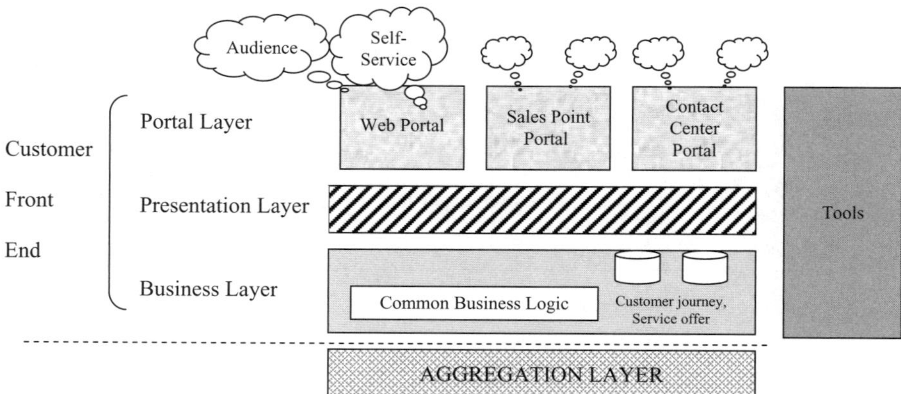

Figure 2-8. The customer front-end functional block

2.4.10.3 The Aggregation Layer The purpose of the aggregation layer is to clearly separate front-end and back-end. These blocks have a different evolution pace and the information exhanged between them needs to be organized. In fact, the aggregation layer hides back-end complexity from front-end. Furthermore, it limits the impact of back-end technology evolution. The aggregation layer executes customer front-end requests by invoking standardized services from the customer platforms from the service plaforms and from the service configuration and activation functions. When receiving a request (from front-end to back-end or from the back-end function to another back-end function), the aggregation layer role is also to translate it, if appropriate, before routing. Indeed, the customer front-end does not need to know the location of the destination function addressed. Finally, the aggregation layer controls execution of requests and can eventually aggregate results obtained. It includes a process rule engine. It is the entry point to partners.

2.4.10.4 The Back-End The back-end contains the customer platform as illustrated in Figure 2-9, the service configuration and activation functions, the service and Network platforms. Customer and service platforms will be accessed through standardized APIs.

The customer platform and their associated repositories are built to answer business needs. The customer platform includes order management, billing, commercial repositories (customer data, commercial catalogue and commercial installed

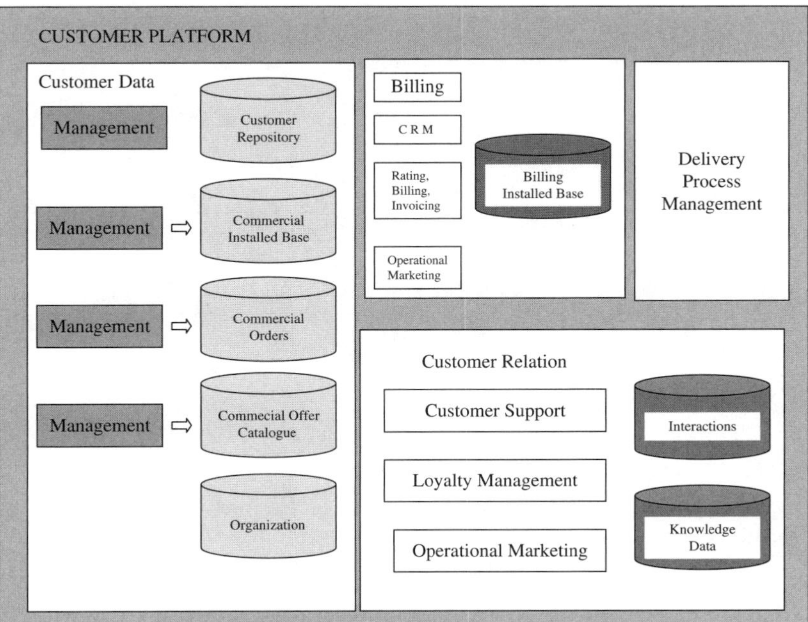

Figure 2-9. The customer platform functional block

configurations) and the delivery process manager. A new "360° CRM" includes a new customer data model, a configurator, the installed products per customer, etc. It exposes standard services to the aggregation layer.

Concerning billing, the new architecture should be able to support the most complex situations online or in a batch mode.

The aggregation layer receives the customer front-end request and invokes customer platform services through standardized language such as "get customer," "set customer," "get installed base," "update installed base," etc.

The customer data function contains the converged view of customer (360°), orders, and the updated situation of the installed base. It is the single reference for the IS. It is the place where repositories offer services to other functions. The customer relation function contains CRM tools and holds the interaction data since there is no data stored in the front-end. Bi-directional links with a business intelligence plane has to be implemented. The DPM function manages the delivery process.

Besides the customer platforms, the other main block (Figure 2-10) of the back-end deals with service and network platforms and their associated configuration and authentication functions. This block provides functions to manage configuration, activation, and administration of network platforms, added value service platforms, and content platforms.

The interface between the aggregation layer, the service, content, and network platforms, and the service configuration and authentication (SCA) function is unique

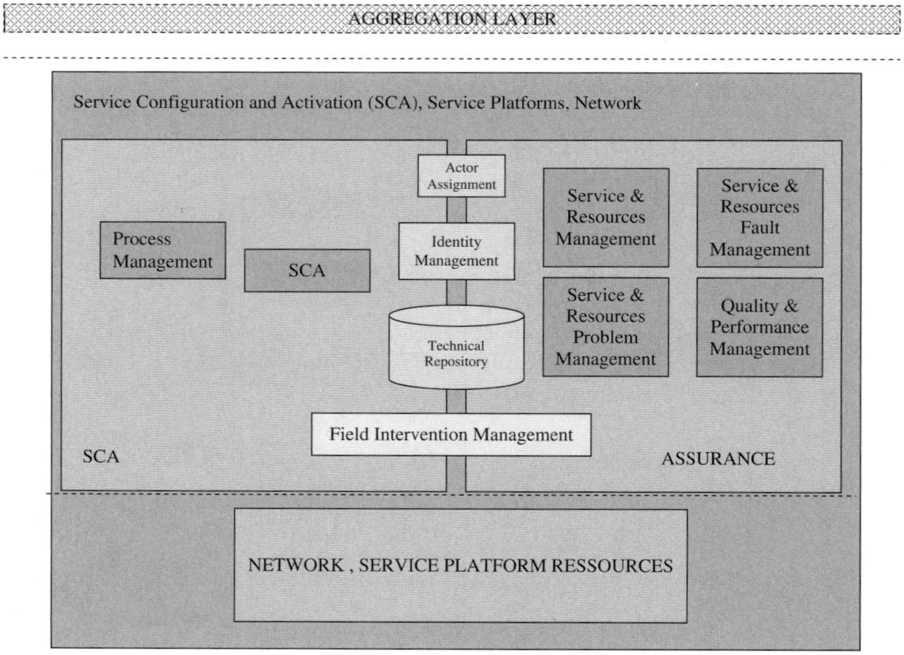

Figure 2-10. The SCA, service and network platform functional block

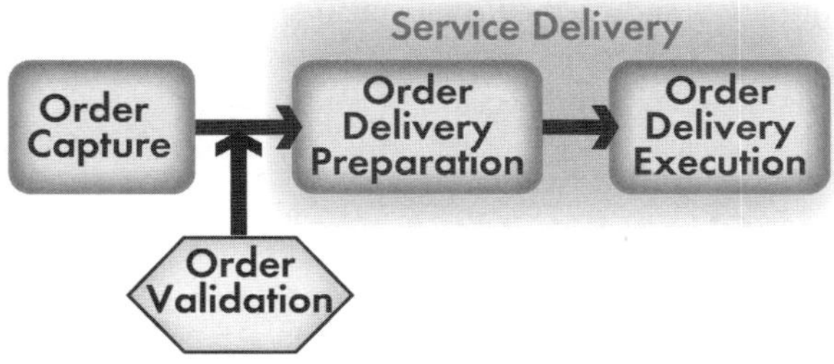

Figure 2-11. Order management

and standardized (standardized language). This interface supports only platform "administration" type of functions.

Every new platform is requested to use this common standardized language (eventually through an appropriate adaptor). During service execution, service platforms, for instance, communicate data to customer platforms through the "mediation bus" (see Figure 2-7). This covers passive mediation (e.g., post-event ticketing functions) and active mediation (e.g., credit information).

2.4.10.5 Order Management and Delivery As illustrated in Figure 2-11, handling an order requires three operations, order capture and validation, order delivery preparation, and order delivery execution (either direct execution for simple cases or execution through the delivery process manager for complex delivery).

2.4.10.6 A Crucial Cooperation between IS, Network, and Service Platform Today's triple play and quadruple play service offerings require smooth interaction between the three domains: Network, service platforms, and IS. This is more crucial than ever. Not only may the technological distinction between the three domains no longer be valid, but launching new offerings cannot be done without having designed and developed, in full service and Network cooperation, the associated IS function (delivery, supervision, etc.) and processes.

This requires adoption of a global approach based on common modeling and identification of customer data, manipulated by the different domains. This is particularly important for service behavior, scalability, and evolution. Additionally, identification of the network equipment data and the functions making use of them and clear specification of interfaces and mechanisms to allow updates of such data by IS are required. Planning of in-service life updates (e.g., when new service options are subscribed) needs to be defined from the beginning.

The data sharing scheme between different network elements and service platforms may be decided when traffic demand requires it. But such sharing requires that the point that will host the customer reference data be clearly identified.

It is also necessary, between the different domains, to agree on the required supervision and operation functions and on specification of the interface to Network equipment and service platform to perform alarm data collection, alarm correlation, network monitoring, etc. All these data will be used to provide, for instance, quality indicators related to transmission quality, packet loss, traffic volume, etc.

Production of billing information and correlation mechanisms, in case of different billing sources, induced by one customer service request, need also to be specified in common, early in the development process.

2.5 THE OPERATIONAL CHALLENGE

Operating triple/quadruple play services, when requests arrive from non-discriminated accesses, gives new challenges to Telecom operators. This requires revising current operating models in order to support business and infrastructure transformation.

An operating model is defined as the management scheme of Network, service platform, and IS infrastructure during the run phase in the service life cycle (three phases can be identified: think, build, and run). The run phase covers provisioning, delivery, monitoring, maintenance, performance analysis, billing, and management. It is a classical sensitive core operator activity when delivering services. In the context of a broadband network, supporting triple/quadruple play services, there are quite a few new aspects to take into account.

Multiple-play service characteristics imply more precision in the operation. Indeed, if the functional behavior of a set of interconnected equipment may be acceptable for one type of service, it might not necessarily be the case for another, even if requested by the same access. This is the reason why, besides traditional monitoring of interconnected pieces of equipment and functions, there is a need to look to individual services from an end-to-end perspective.

Therefore, in the context of a triple play service, it is required to support an end-to-end service customer view. This implies enhancement of the functional operating mode. This leads to introduction of a *service management center function (SMC)* for all technologies used by the services. This new function encompasses Network, IS, and service platforms from an end-to-end perspective, taking care of end-to-end QoS perceived by customers.

Because the technical architectures supporting triple/quadruple play services are becoming more and more complex, the SMC is where end-to-end vision of the technical chains used by these services should be maintained. *This is the right control tower to pilot service quality.*

In addition to the new service perspective function, other more traditional operational functions still exist in Broadband Triple Play even if the technology has changed.

This is the case for the *technical management center function (TMC)*, which is responsible for technical management of the operational Network, service platform, and IS infrastructures for a given "technology" (one TMC per "technology," i.e., transmission, IP routers, mobile radio access, fixed ADSL access, TDM-based switches, etc.). TMCs should have full knowledge and control of the technologies used in a given infrastructure. They perform a number of important actions such as resource control and problem resolution, corrective and preventive maintenance quality analysis, crisis management, technology integration management, and planned work coordination. TMCs have to perform operations in conformance with industry standards and security requirements. They drive the relationship with suppliers and are responsible for communication with the customer care center (CCC) and SMC about critical incidents impacting services.

This is also the case for *skill centers (SkC)*, which are responsible for the technical expertise related to a given technology (even restricted to a given supplier). They are in charge of preparing, validating, and accepting the different equipment releases to be implemented and operating such equipment. Depending on their internal organization and footprint (covering more than one country or region), operators may have to distinguish between an operational skill center (local operational skill) and a pure "technology" skill center (global operational skill).

The *provisioning function* is a traditional operation function responsible for the customer implementation using the installed resources ("technical management of customers"). It is usually located at the interface between sales and field intervention and can be supported by the field intervention function if needed.

The *field operation function*, also a traditional operation function, is responsible for the on-site interventions related to maintenance or provisioning and for customer interventions. The TMC remotely manages Network field operations actions, while the CCC manages customers field operation actions. The intervention dispatching function is part of the field operation scope.

The last traditional operation function is the *customer care center function (CCC)*, responsible for the customer relationship and the after-sales service. It is the single point of contact for customers for all matters dealing with services provided by the operator. The CCC performs actions such as customer call handling (hotline) and ticket management. For non-complex problems, in particular, when they concern provisioning issues, CCC handles trouble tracking and trouble shooting. For complex issues, CCC may escalate, when required, to SMC, TMC or, if relevant, to field operations. Communication to the customers on the basis of the information received from operational and technical functions is also its responsibility. Figure 2-12 illustrates the central role of the SMC between CCCs and TMCs.

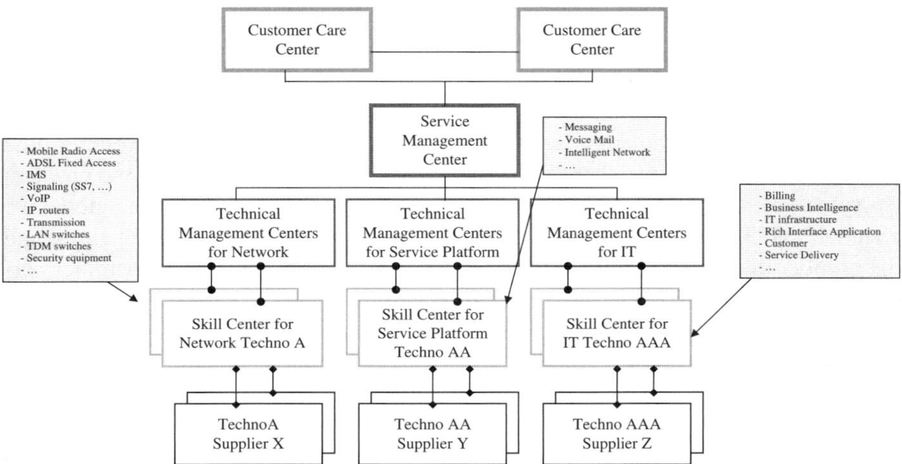

Figure 2-12. Relations between CCC, SMC, TMC, and SkCs

In a triple play/quadruple play service context, the customer relationship is enlarged and becomes more complex. This is due to the fact that what has been sold is one access but multiple services. As already mentioned, it could happen that one service does not work properly but the other does, and that situation is not always perceived by customers. Therefore, additional skills are needed in the CCC function.

2.5.1 Focus on the Service Management Center Function (SMC)

As previously mentioned, the SMC's responsibility is to control end-to-end QoS. SMC acts as an orientation and control tower able to translate, in technical terms for TMCs, a problem affecting customers. This end-to-end responsibility requires implementation of real-time service monitoring and perceived service quality analysis, service maintenance, service management, and operation functions and holds an essential role in internal and external communication.

The *real-time service monitoring function* includes proactive monitoring (e.g., via probes) of the behavior of the different services in order to have an understanding of end-to-end customer perception. Supervising and managing the service level agreement (SLA), e.g., monitoring thresholds, alert triggering, etc., monitoring, managing and optimizing the monitoring system are also performed. Service monitoring engineering consists of implementing and controlling all the tests needed to follow-up the QoS and adapt the monitoring tests and views to the customer perception and service evolutions. This task is essential in triple play services as technology and perception can evolve quite rapidly. The SMC is responsible for the technical QoS delivered to customers according to SLAs with the business owners. It is therefore the point of contact for the business owner for all matters concerning service management.

In case of service troubles, the *service quality analysis function* performs correlation between alarms delivered by the service monitoring system, qualifies incidents in technical terms but also in terms of impact on the customer perception, and gives a first diagnostic with involvement of the concerned TMCs. For voice services, traffic and voice quality need supervision and analysis. This covers the end-to-end vision of service accessibility and of traffic continuity (traffic efficiency) and, finally, checks the integrity of communications (voice quality). Supervision is a real-time exercise but some non-quality aspects, not seen in real time, may need some offline analysis. For TV services, there is a need, in case of trouble, to qualify the problem by identifying the part of the service concerned (premium channels or not, interactive channels, VoD, etc.), the area impacted (city, main frame, Giga Ethernet ring), and the number of customers impacted. It is also useful to get reports on customer perception (pixelization, black screen, abnormal zapping delay, abnormal log-in delay, etc.). Follow-up of problem resolution is also part of the SMC mission.

Service maintenance function covers service trouble resolution actions and crisis management. Service trouble resolution handles trouble tickets sent by CCCs and service alarms detected by the service monitoring system (trouble ticketing). SMC is also involved in the crisis management process as a service entity and is responsible for proposing palliative or corrective solutions to restore the service.

The SMC function is the single point of contact that provides information about the crisis status to the management line and business units.

Service management deals with service integration and with the build-to-run transition process. SMC is also involved in the build phase by ensuring that the run phase meets operational requirements, i.e., the monitoring function and interface specifications, statistics, test tools, etc. It also participates in go/no go decisions for integration of any new service in the operational environment with the business owner and all other concerned technical entities. SMC also coordinates, when relevant, all planned interventions or actions involving several TMCs.

Communication to the customer-facing entities (after sales, customer support centers, help-desks, etc.) and the management line is the responsibility of the SMC in terms of service impact level, expected repair time, engaged actions, etc. Technical information must be translated into messages understandable by the customer. In that respect, SMC is the operational link between customer care and the technical entities.

2.5.2 IS Tools for the SMCs

The SMC is responsible for maintaining QoS perceived by the end-customer. The information system for the SMC requires customer-orientation (end-to-end customer view) easy adaptability when the organization evolves, and becomes flexible enough to cater to business evolution.

Service impact analysis tools are at the heart of IS support to SMC. Based on the monitoring view provided by the TMCs, the SMC can build a global overview and be reactive and efficient to CCC requests. This can be done by the SMC function, feeding service impact analysis tools or by TMCs function feeding resources supervision tools. A set of common key performance indicators (KPIs), representing SLA performance, are defined for the whole SMC function. These KPIs aim to compare the different SMC activities on the same basis. The overall view of performance can be determined as the percentage of SLAs in each category. The SMC function supports the CCC function 24 hours a day and 7 days a week.

Figure 2-13 illustrates the process supervision role of the SMC after a customer has placed an order, in order to check the right resource allocation, delivery,

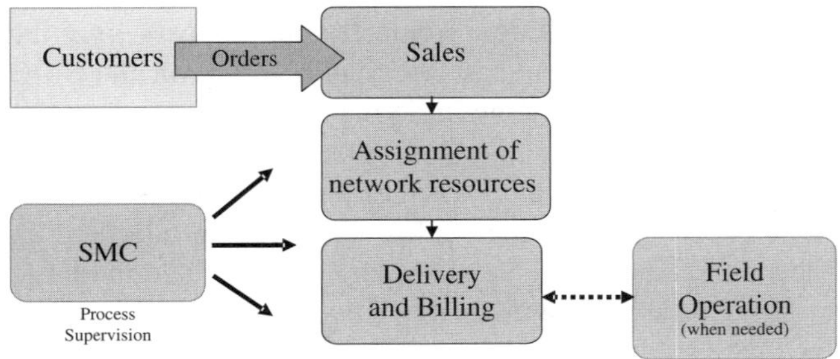

Figure 2-13. Role of SMC in delivery and provisioning process supervision

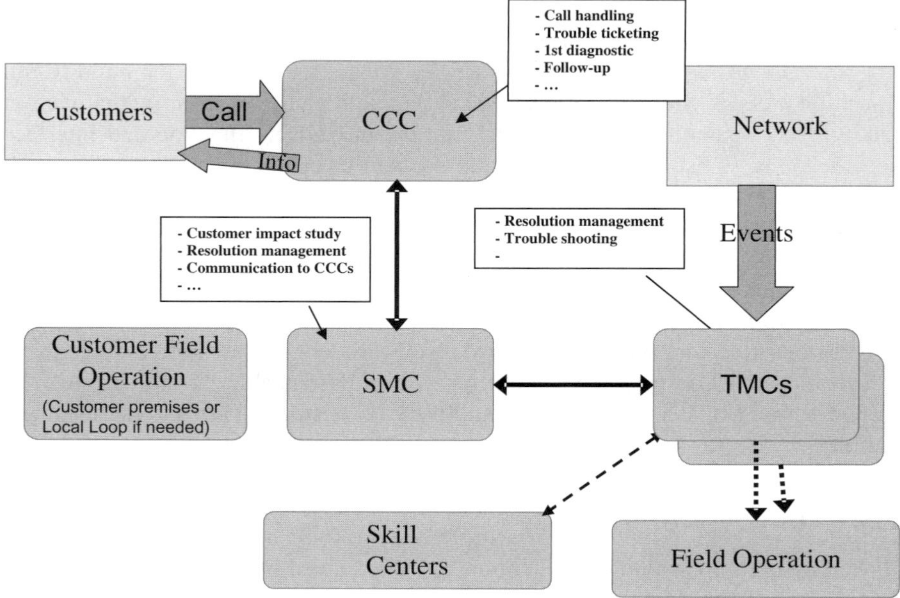

Figure 2-14. Role of SMC in maintenance and repair actions

and billing. Figure 2-14 illustrates the SMC role during service life. Upon customer complaints through the CCC, SMC is able to ask the appropriate TMCs to get identification and resolution of troubles and then to inform the CCC of the resolution progress. Through the TMCs, SMC is also able to alert the CCC of troubles that will affect a given service and/or a certain number of customers.

2.5.3 Operating IT and Service Platforms in Triple and Quadruple Play Contexts

It may be useful, at organizational and functional levels, to make a distinction between the IS that directly contributes to end-customer services (in other words, IS for external customer), such as online self-care, E-provisioning, service delivery, real-time billing for pre-paid, portal, service platform, etc., and IS not directly contributing to end-customer services and serving exclusively internal customers via specific call centers. If both can be managed by a single IS-related TMC, only the IS that is directly contributing to end-customer services is involved in the SMC processes and is strongly impacted by the new multiple play service offerings. The following focuses on this IS part.

Selling and operating triple/quadruple play services put a strong requirement on the new IS operations model, just as making triple/quadruple play working do for the Network model. Indeed the new IS operations model, as illustrated in Figure 2-15, needs to be built up around the same entities described above, i.e., the CCC, TMC, SkC, and SMC. One of the actions performed by the SMC is to forward CCC requests to the appropriate TMCs when multiple TMCs are concerned. Typical examples in IS are requests related to prepaid billing, data and content billing,

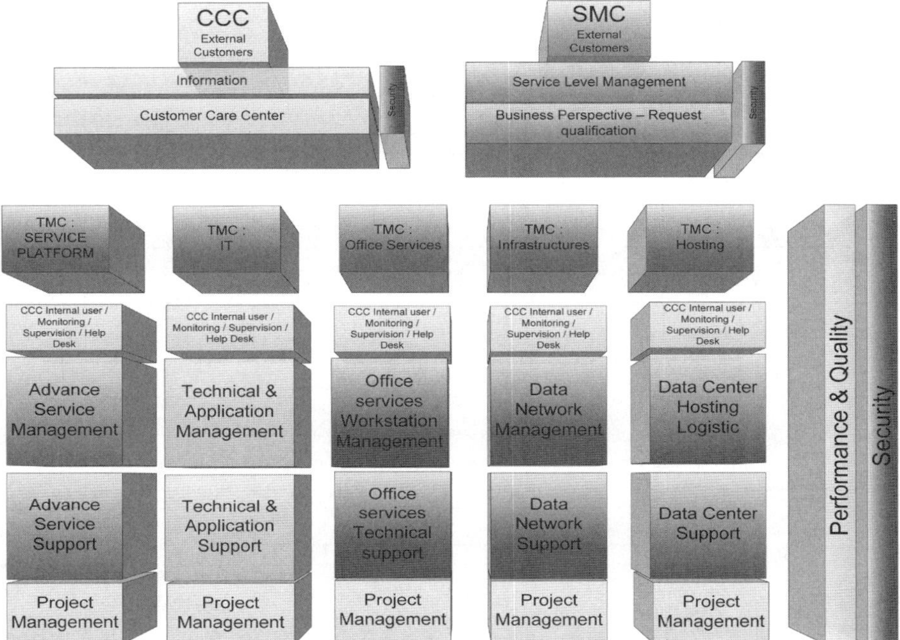

Figure 2-15. A new IS operating model for triple play/quadruple play

roaming, and any third-party applications. In less complex situations, the CCC directly addresses the TMC without passing through the SMC.

2.5.4 Roles and Responsibilities of the Different Functions

Based on the Information Technology Infrastructure Library processes (ITIL), a mapping of roles and responsibilities between CCC, SMC, and TMC needs to be done. Figure 2-16 shows a blank template for describing the various roles and responsibilities of the different IS operation functions. There is benefit to having a common and shared matrix for the service platform, the Network, and the IS.

The split of activities and processes between CCC, SMC, and TMC is globally the same for Network and IS. Differences may exist that come from the skill centers and from the level-three support missions defined.

SMC will suggest the right SLA level to the TMC. In a three-service level approach, for instance, a critical level, a non-critical level, and a best effort level could be defined.

The critical level (e.g., for IS customer service applications) implies high availability and disaster recovery planning (DRP). This service level should be applied to the customer front end (self-care, mail, messaging, SMS, MMS, etc.), real-time billing, or service delivery domains. The non-critical level (e.g., for traditional IT applications requiring high availability) should be based on local high availability. This service level should only be applied to domains not facing the customer service. The best effort level (for traditional IT applications not requiring high availability) may be applied to Intranet, human resources management tools, etc.

					Network Skill Center		IT Skill Center	
		CCC	SMC	TMC	OSkC	TSkC	ITinfra	ITappli
Application management	Operate							
	Optimize							
	Require							
	Design							
	Build							
	Deploy							
IT infrastructure	Design and plan							
	Operations							
	Technical support							
	Deployment							
Service delivery	Capacity management							
	Continuity management							
	Availability management							
	Financial management							
	Service level management							
	Customer equipment delivery							
Service support	Service desk							
	Incident management							
	IM in CRM							
	IM in service management and operation							
	IM in resource management and operation							
	Problem management							
	Configuration management							
	Release management							
	Change management							
Security management	Security management							
Customer equipment delivery	Customer equipment delivery							

Figure 2-16. Template to describe roles and responsibilities of the different IS operation functions

The following table below gives an example of service level objectives (SLO) that can be agreed upon the business entity.

Critical SLO	Non critical SLO	Best effort SLO
99.95%—1 shutdown max/ month 24/7	93%—2 shutdowns max/ month 18/6	12 × 5—no commitment
RTO incident: <20 minutes	RTO incident: <4 hours	RTO incident: <1 day
RTO disaster: <2 hours	RTO disaster: 8 to 24 hours	RTO disaster: <5 days
RPO: no data loss	RPO: 30 minutes to 4 hours	RPO: 1 day max
QoS: no degradation	QoS: degradation max 25%	QoS: degradation max 50%
RTO: Recovery Time Objective		
RPO: Recovery Point Objective		

2.5.5 New Skills in Operations

New skills and jobs are emerging in operations, specifically in supervision job lines, which require more autonomy and initiative to analyze technical information and bring right and pertinent recommendations for communications toward customers. There is a need to recruit or train new technical staff to take into account complex interactions, enhance help desk activity with ability to base priorities on SLA, make the right diagnosis, activate the right TMCs, elaborate proactive information to customers, etc. Another example is the service pilots in SMCs, who should be able to analyze offline QoS measurements and provide commented dashboards with explanations on service behavior evolution. Such service pilots should also pilot crisis management, including the communications, through the different internal entities of the company.

Upgrading or transforming OSS is critical for triple/quadruple play. This concerns the ordering platform, fault management, configuration management, service provisioning, network provisioning, performance management and surveillance, incident management, and traffic management.

Because the operations world is moving toward more concentration, security issues are becoming critical. Technology and service migration takes over and QoS improvement and stabilization become other stakes in the ground.

2.6 THE CUSTOMER EXPERIENCE IN BROADBAND TRIPLE PLAY

Even if it seems obvious to say it, operators must check that a triple play/quadruple play sale to a given customer is really possible in practice. That means, first of all, that clear identification and understanding, from commercial and technical view points, of the product to sell, exists. Feasibility of the requested service delivery must exist. This suggests that the true puzzle of delivery events can be synchronized (i.e., delivery of equipment, delivery of login information, agreed appointment with customer, installation, etc.). Then, as in any service sale, there is the assurance that the customer will be able to pay. Note that for the service delivery, this implies synchronization of the supply chain, local loop and DSL, service platforms, boxes (RGW and set-top box), etc.

In situations where technical products are still new, where service and infrastructure are complex, where IS and processes were not originally built for such complex offers, and where the learning phase has not yet happened from the operator and customer point-of-view, the main concern is to maintain control of OPEX induced by the volume of calls to call centers and by the number of interventions on the ground (including inevitable interventions that were not necessary).

There are four main phases in the customer experience: the sale, the delivery, the run and the aftersale.

During the *sale phase*, it is necessary to assure that customers understand what was sold to them (broadband access including the eligibility problem for copper access, VoIP with or without number portability, TV and VoD) in order to reduce the percentage of non-profitable calls from clients to the after sale desk.

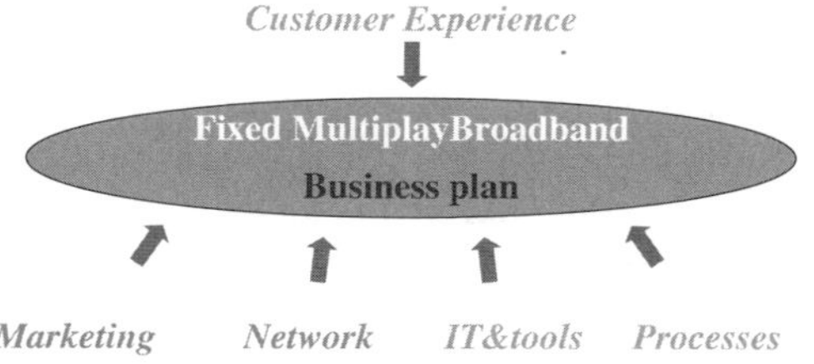

Figure 2-17. The customer experience should be anticipated

In the *delivery phase*, the main issue is to organize the delivery in line with what was said to the customer during the sale phase. This prevents an increase in calls to the after sale desk. While this is not trivial in terms of synchronization of tasks, even when copper is the operator's property, it can be very difficult to achieve when operators have no direct access to the copper line due to the unbundling regime, the process and discussion with the incumbent, the quality of the data, the legal aspects, etc. It should also be noted that TV play may double the cost in triple play service due to a more complex architecture (set-top Box, delivery platforms, installation procedures, etc.).

The *run phase* is also critical since it operates in real time "always on" access. The IP network is very different from past experiences on access and data networks. Today, in such infrastructure (not compartmented like previous infrastructures such as PSTN), every function located any place can interact with any other function, leading to a need for end-to-end interaction supervision.

The *after sale phase* deals with after sale processes, call center activities, field operation intervention, and SMC actions. Reducing repair time and preventing useless customer calls to call centers and the consequent useless interventions on the ground are key issues to be solved.

More specifically, in the ADSL local loop unbundling (LLU) context, support of the roll-out of multi-play and broadband services requires customer processes and customer experience to be analyzed in detail and continuously improved. This improvement should have as its goal to deliver a best-in-class customer experience, to roll out efficient customer facing and back office processes, and to keep operational costs under control.

As illustrated in Figure 2-17, customer experience should be anticipated when building a triple play business plan like marketing, network, IT and processes impacts. This needs to be done for each step of the customer journey. In order to minimize risk and prevent additional costs and planning delays, it is required, before launching multi-play offerings and broadband services, to anticipate the impact on the operational model and to define the pertinent performance indicators. For that purpose, it is necessary to rely upon appropriate and available operational tools. Analyzing the impact on a customer's journey and customer processes, however, requires a clear definition of the service offerings and a choice of the distribution channels.

2.6.1 Definition of the Offerings

Impact on processes depends on the type of multi-play services to be offered. Offering strategy has to be defined in advance to anticipate operational impacts.

Questions like Is triple play or quadruple play targeted? Is VoIP included in packages? If yes, is it as a first line or as a second line? Is TV proposed in offerings? If yes, is it as IPTV, satellite TV, VoD? need to be clearly answered.

2.6.2 Distribution Channels

Multiplay/broadband offers are usually distributed through different PoPs and channels like shops, partners, Web/e-shops, telesales, even indirectly through customer care, etc. For each of these distribution channels, the acquisition process is impacted. Line activation and configuration, equipment delivery, customer premises equipment (CPE) configuration, network/service activation, and different customer process options have to be considered (e.g., live box delivery via post office when e-shop channel is used versus live box delivery in shop for direct sale channels).

2.6.3 Relationship with the Local Operator

Operators proposing multi-play/broadband services may face different situations to get access to broadband networks whether or not they are incumbent. While incumbent operators access the broadband network with respect to national regulatory constraints, non-incumbent operators may have to decide, depending on the local situation, either to take over the access or to access the customer through the historic local operator (with local partnerships). In this last case, impact on the customer experience and on customer processes is more complex due to the relationship with the other operator or partner. Regulatory issues, quality-of-data provided by the partner (eligibility of the line to the service sold), commitment about installation and service activation procedures (portability duration), request response time for technical support (SLA), etc., need to be approved by contract before launching the services.

2.6.4 The Customer Journey

The customer journey, which can be analyzed along many lines (price, device, service, sale channels, customer contacts, etc.), can be divided into six stages (Fig. 2-18):

A. *Be aware.* This is the stage where clear information and understanding of services, conditions of delivery, usage and pricing are provided to the customer by the service provider.

B. *Join.* This is the stage where a simple and peaceful buying experience needs to take place.

C. *Set-up and first use.* This is the stage where a simple set-up and enjoyable first usage should take place confirming the previous buying stage.

D. *Use and get help.* This is the stage where products are used and where the customer's expectation is to get the agreed service performance and the understandable right invoice and to obtain a reliable support answering his/her requests.

E. *Evolve and renew.* This is the stage where customer's needs evolve and where his/her expectation is to be informed of the operator's propositions (proactively or not).

F. *Terminate.* This is the stage where the customer is willing to terminate his/her contract and needs to know how to proceed in a fair spirit.

Each step of the customer journey can be mapped into customer processes, operational performance management indicators (KPIs), mandatory tools to secure cus-

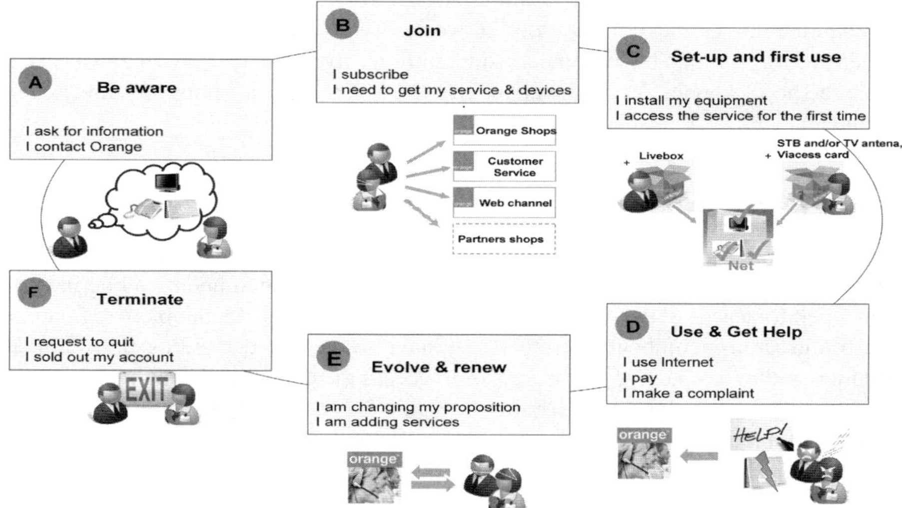

Figure 2-18. The six stages of the customer journey

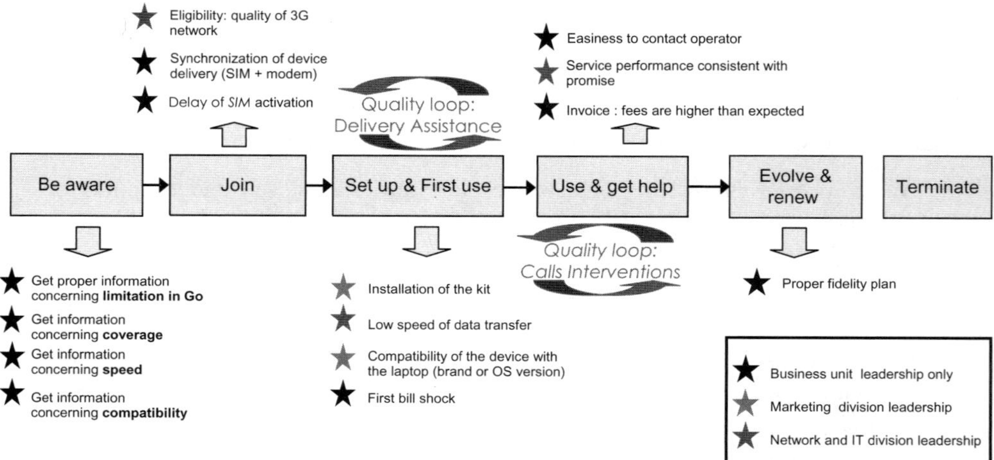

Figure 2-19. Implementing broadband and triple play main operational issues

tomer satisfaction and internal process performance, operation organization (i.e., dedicated teams, training, flows between entities, etc.) and financial aspects (i.e., to support the preparation of the business plan). Figure 2-19 illustrates such mapping and gives examples of entities within the company that have the responsibility to look after the corresponding processes, the indicators, the tools, etc.

2.7 THE ORGANIZATIONAL CHALLENGE

Telecommunications operators' historic organizations have produced specialized structures for fixed and mobile, network, and IS (each of them with their own efficiency optimization). Each could represent, from a global point of view, a lack of convergence, in terms of equipment and operations. Each of these specialized structures may have their own design and engineering, operation, field intervention, provisioning, customer care, etc. In the transverse approach required with triple play and quadruple play, this leads to processes that are too complicated and too costly.

With triple/quadruple play, it is therefore imperative to overcome the specialized silo approach and to simplify the complex organizational inheritance of structures and processes. In a new customer world of convergence and with multiple-play requesting more and more cross-innovations, robustness, and cost efficiency, the key words are more and more *simplification* and *convergence*.

2.8 CONCLUSIONS

During the IDATE's Digiworld event in November 2007, Didier Lombard Chairman and CEO France Telecom Orange declared:

> "We are no longer in the Triple Play we are already in the Multiple Play … more bandwidth on one side, more services on the other side but the two should move forward together […] All the actors are converging towards similar business models. Everybody is converging towards this (universe of) melting pot of networks and services"

Value chains will continue to blend themselves up, although the network layer will remain, for a while, the more important in terms of added value. But, certainly, the more dynamic ones will be the service layer and the content domain.

Telecommunications operators therefore need to walk on two legs. The network and the service accesses on one side, the relevant services using these networks on the other side. This is the necessary condition for operators to amortize, tomorrow, the huge network investments they are making today. This is especially so in access lines and access networks.

Operators are no longer pipes providers but are already in a position to be audience amplifiers. Telecommunication operators have essential and important characteristics—proximity and knowledge of the customer and capability to guaranty security regarding personal data usage while leveraging their potential business use.

An evolution has occurred from a utilitarian perception of the technologies to a more fundamental perspective, not just an always connected world but a world where technologies disappear behind usage and where no one can now imagine living without communications tools (especially mobile).

In that context, mutiple-play and all the surrounding technologies are giving everybody the potential to become an inherent active part of the network. This is going to happen if simplicity, quality of service, robustness, confidentiality, and security can be maintained for the long run at the adequate value for millions and millions of interconnected users.

There are indications that the key differentiators for tomorrow's communications world will be QoS and the ability to master tremendous technical and process complexity while providing more simplicity to the customer.

With that consideration, more and more standardization and interconnection between operators at national and international levels are required. This is an essential factor for societies to gain worldwide access to triple play and quadruple play. Triple and quadruple play are the very first steps for operators entering their second life!

2.9 ACKNOWLEDGMENTS

Warmest thanks to Vivek Badrinath for his essential support and advice; to Lucien Ducorney for the valuable working sessions around the customer experience improvement; and to Patrice Collet, whose great experience in overall IT&N architecture design contributed a lot to this chapter and its Q&A for graduate student instructors.

2.10 REFERENCES

1. Lombard D. 2008. *Le village numérique mondial: la deuxième vie des réseaux*. Paris: Odile Jacob.

2.11 SUGGESTED FURTHER READING

1. Boland M. 2007. Triple and quad play: who will win the bundled service battle? In *Beyond the Quadruple Play: Networking, Convergence, and Customer Delivery*. Chicago, IL: International Engineering Consortium.
2. Galyas P, Iacovoni D. 2007. Delivering IMS to the home: the role of CPE to provide full service convergence. In *Beyond the Quadruple Play: Networking, Convergence, and Customer Delivery*. Chicago, IL: International Engineering Consortium.
3. Pernet S. 2007. Bundles and range strategies: the case of telecom operators. In *Beyond the Quadruple Play: Networking, Convergence, and Customer Delivery*. Chicago, IL: International Engineering Consortium.
4. Roy M. 2007. Beyond Triple Play: In quadruple play, delivering exceptional customer service is more critical (and difficult) than ever. In *Beyond the Quadruple Play: Networking, Convergence, and Customer Delivery*. Chicago, IL: International Engineering Consortium.
5. Walker S. 2007. Service delivery in a quad-play environment. In *Beyond the Quadruple Play: Networking, Convergence, and Customer Delivery*. Chicago, IL: International Engineering Consortium.

CHAPTER 3

MANAGEMENT OF TRIPLE/QUAD PLAY SERVICES FROM A CABLE PERSPECTIVE

David Jacobs

3.1 INTRODUCTION

The roots of the cable industry can be traced back to 1949 when the first Community Antenna Television (CATV) system was installed to bring broadcast TV to rural areas in America [24]. Since that time, advances in technology and deregulation of telecommunications services has resulted in cable systems being deployed throughout the world that are not only capable of delivering high-definition TV but also advanced services such as video on demand, browsing information on the Internet, online gaming and unified communications. In countries that have a well run cable industry it is commonplace for subscribers to take the three major services, TV, Internet, and voice, also known as the triple play, from their local provider. In some markets it is possible to purchase wireless voice services as well.

Unlike deregulated telephone companies, cable systems throughout the world were licensed on the basis of non-overlapping geographic areas and as such did not compete with each other; however, consolidation has eliminated the majority of the smaller systems. It has only been in the last 10 to 15 years that some areas in America (usually cities with higher than average spending power) have seen competitive cable operators or over-builders appear.

Without doubt, the cable industry in North America is the largest in terms of homes served and revenue generated, primarily due to its rapid adoption of technology that enabled it to exploit usage of the Internet. As a result it is the most active in developing and standardizing technological advances through both CableLabs, which is funded by the cable operators themselves, and the Society of Cable Telecommunications Engineers (SCTE), an independent non-profit organization with strong ties to the American National Standards Institute (ANSI) for accreditation of its standards.

Next Generation Telecommunications Networks, Services, and Management, Edited by Thomas Plevyak and Veli Sahin

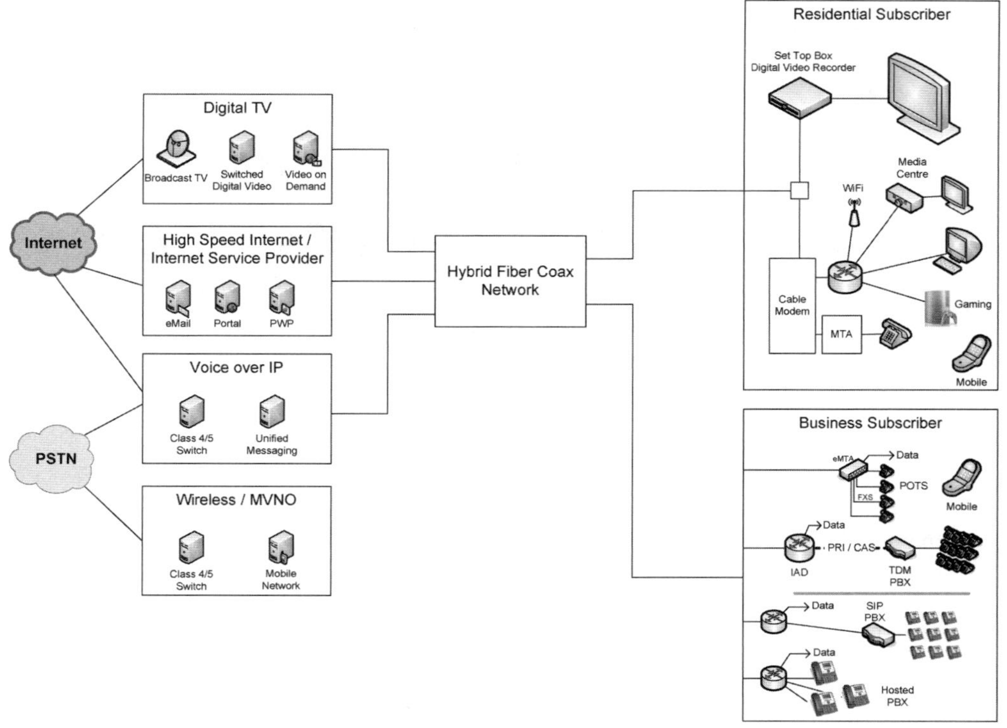

Figure 3-1. Functional elements of a cable system

Service expansion with offerings such as three-screen multimedia and broadband TV (or IPTV), along with penetration into new markets such as voice and data services for Small and Medium Enterprises (SME), is requiring the cable industry to accelerate the implementation of packet-based delivery and signaling infrastructure. Evidence of this can be seen with DTV infrastructures moving command and control functions to an IP transport with the introduction of the DOCSIS Set-top Gateway (DSG) as well as the adoption of IMS architectures in the form of PacketCable 2.0 to enable greater service convergence.

In contrast to their willingness to develop and adopt standards that bring new services to market, cable operators have been slow to adopt Operational Support System (OSS) and Business Support System (BSS) standards from industry bodies such as the TM Forum. However, with the increasing convergence of network control plane functionality and the ubiquity of services delivered over IP this trend is beginning to change.

This chapter provides an overview of how digital TV, Internet access, and voice services are delivered by an operator and the mechanisms used to manage them. Although intertwined at many levels, focus is given in subsections to the HFC network, digital TV, Internet access, and voice services as shown in Figure 3-1.

3.2 THE HFC NETWORK

Cable systems use a combination of fiber and coaxial cables for communication between a cable head-end and a subscriber's home. This type of network is called a hybrid fiber-coaxial or HFC network.

A fiber-optic transmission system is used between a cable head-end, or a regional hub, and a fiber-node that is situated near to a population of homes. A coaxial transmission system arranged as a tree and branch structure then provides connectivity between the fiber-node and the subscriber homes that are taking service. The amount of homes that can be served by a single fiber-path is a function of the bandwidth assigned to the DOCSIS channel, the number of connected fiber nodes, and the interactive usage per connected home. Homes are organized into serving groups, with a single fiber path able to support multiple fiber-nodes, which in turn can support multiple serving groups.

The bandwidth available between the regional head end and a subscriber's home varies by country and age of the cable infrastructure. In North America a typical system will have a downstream bandwidth of 816 MHz (54–870 MHz) and an upstream of 37 MHz (5–42 MHz). Using the NTSC analog channel bandwidth of 6 MHz, the system will support 136 channels. These channels are available to carry analog TV signals, standard definition digital TV (SDTV), high-definition digital TV (HDTV), video on demand (VoD), and data services for Internet and voice over IP services.

As with any large-scale distribution infrastructure that is required to provide consistent service to a mass population, an HFC network needs to be planned, maintained, and continuously upgraded.

3.2.1 HFC Planning and Inventory

Cable MSOs measure their addressable market (or maximum potential customer base) by counting the number of homes that their HFC network passes. In the same way that utility companies (electric, gas, water) ensure that their services are available to new housing developments, so too do cable companies ensure that they do not miss out on new potential subscribers. The typical considerations for any new network build will include:

- The type of subscribers who will be served—university campus, retirees, teleworkers, etc., and their projected bandwidth usage over 5–10 years.
- Town planning regulations—do cables need to be underground or is aerial suspension allowed (hung from telegraph poles), does the cable operator's franchise mandate that this particular location must be provided with service.
- The returns on investment—an estimation of the likely subscriber take up rate and what revenues can be expected over the life of the investment versus the cost of build.

Once the HFC network is built the cable MSO maintains an inventory of both the physical and logical elements that make up the network. As with any complex structure, various parties will want different views and have different needs of the inventory. For example:

- Assurance and repair departments will want to be able to identify the network components that serve a particular home in order to resolve a reported fault.
- Planning departments need to query the makeup of a serving group and view the relationship between CMTSs and fiber nodes to balance the network's traffic load.
- Fulfillment systems need to perform design and assignment functions to ensure the correct resources are utilized when provisioning a specific service; for example, assigning a new telephone number for a specific location.

In addition to these operational needs, the cable MSO also has regulatory responsibilities such as maintaining a database-of-record of the exact street address for all of the locations taking service. Law enforcement agencies (LEA), public service answering points (PSAP), and, in the U.S., the FBI have requirements for this information based on the service consumed. Primary-line telephone number to specific location is an example of the information used by PSAPs when alerting the emergency services of an urgent situation. In addition LEAs may need the information to trace an IP address to a specific location in order to deal with suspected criminal activity. The Master Street Address Guide (MSAG) is format used in North America for recording street addresses.

3.2.2 HFC Network Maintenance

Driven by local needs, most cable networks will have configurations of 25 to 2000 homes per fiber node, with typical values being around 500. However, the growth in the use of Internet peer-to-peer traffic and the increasing trend toward e-working is resulting in the homes served per fiber-node decreasing year-by-year.

The ratio of the number of fiber-nodes per downstream fiber-optic laser is known as the combining ratio; Figure 3-2 is an example of 1 : 2 combining. As interactive load increases and it is seen to be necessary to reduce the number of homes per fiber path, the cable operator will perform a node split. This is performed by either removing a set of homes from the fiber node or adding another fiber transmission path from the regional hub. In both circumstances the number of homes per fiber path is reduced.

North America in particular has a significant proportion of its HFC infrastructure implemented in an aerial configuration, i.e., suspended from telegraph poles. Although cheaper to deploy than underground solutions, aerial HFC can place a significant burden on the finances of an operator in locations, such as Florida, that are susceptible to violent weather conditions, as there is often a need to make repairs following a hurricane or violent storm.

3.2.3 HFC Network Upgrades

Network upgrades are driven by several factors such as new markets (e.g., business services), aging of the installed infrastructure, and demand for increased bandwidth.

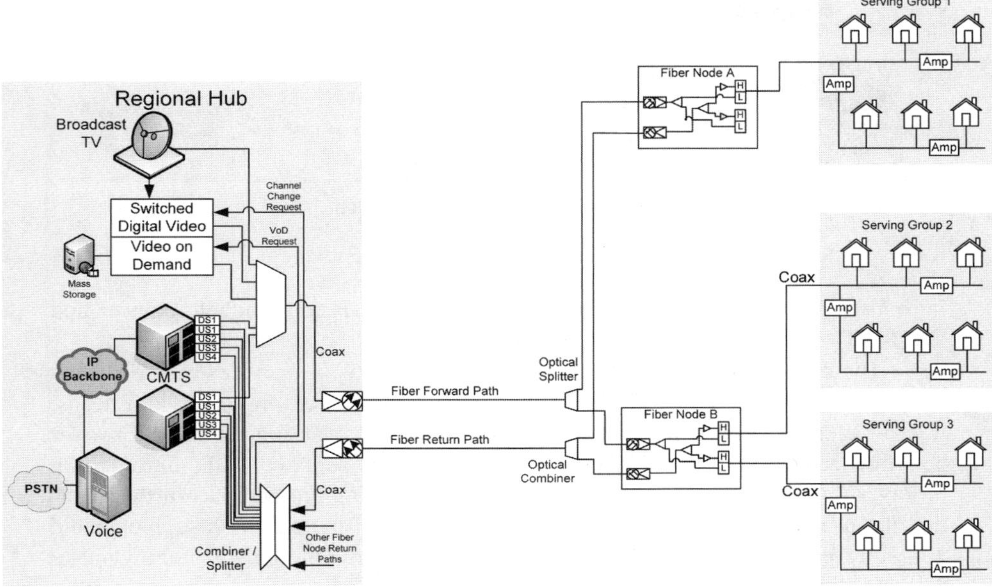

Figure 3-2. HFC network components

Increased bandwidth is the most potent driver as cable companies face the reality of a single home simultaneously consuming multiple HDTV channels and at least 20 Mbps for Internet services. This demand, as well as competition from operators who use digital subscriber line (DSL), fiber to the home (FTTH) and wireless technologies, will eventually result in a migration to an all-fiber network. Moreover as Next Generation Networks (NGN) are introduced as part of the PacketCable 2.0 initiative, see 3.5.1.4 below, service level requirements for new services that are either owned by the cable provider or sourced from third parties will also add pressure for network upgrades.

In the mean time, cable companies will attempt to get as much return on the investment as they can from their existing networks and we will see combining ratios reduce to 1 : 1, homes per node fall to single digits and techniques such as Switched Digital Video (SDV) and DOCSIS 3.0 being aggressively deployed.

3.3 DIGITAL TV

In September 1993, broadcasters, equipment manufacturers, and regulatory bodies established a European project known as Digital Video Broadcasting (DVB) to oversee the introduction of digital TV (see www.dvb.org). Three transmission standards were initially developed for delivery of consumer service: DVB-T for terrestrial networks, DVB-S for satellite networks, and DVB-C for cable networks. Technological evolution has seen many coding efficiency enhancements made to these standards in the form of DVB-S2 and DVB-T2 and has also prompted the

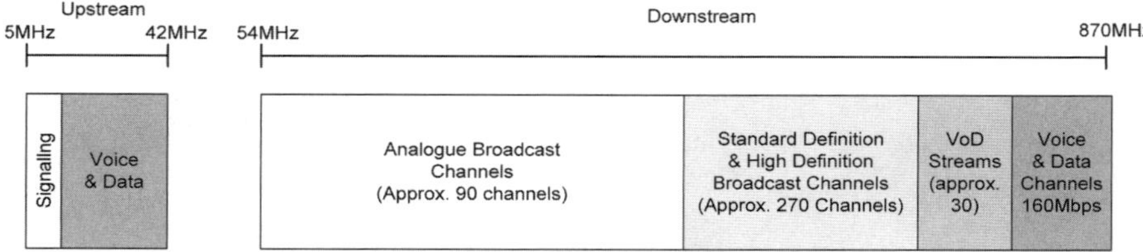

Figure 3-3. Typical North American spectrum allocation plan

introduction of several new standards, including DVB-H for handheld devices and DVB-IPTV for delivery of TV over IP networks.

The motivation for introducing digital broadcast of TV services was to be able to transmit a greater number of TV channels without increasing bandwidth needs, improving noise immunity of the transmitted signal, enabling transmission over heterogeneous networks and most importantly provide support for the introduction of Pay TV. The adoption of IP as the de facto networking protocol has further justified the use of digital technology as broadcast content is able to be supplemented with interactive services that improve the financial returns for the operator and make the user experience far more personalized.

ITU-T J.83 [21] details the framing structure, channel coding and modulation for transmission of audio-video and data services across cable networks. The four annexes of this specification define the implementations as used in various parts of the world. Annex A is used by European Cable MSOs and specifies the use of an 8 MHz program channel with 16, 32, 64, 128, or 256 QAM modulation. Annex B is used by North American MSOs and specifies 6 MHz program channels and 64 or 256 QAM modulation.

Figure 3-3 shows a typical spectrum allocation for services carried over a North American cable network. As can be seen, the use of digital coding and transmission increases the density of channels transmitted for a given bandwidth. Using 256 point quadrature amplitude modulation (QAM) 6 MHz of bandwidth can be used to deliver either, 1 analogue channel, 7–12 standard definition (SD) digital TV channels, or 2 high definition (HD) digital TV channels.

The increase in available content and the introduction of HD services is forcing the pace of conversion from analogue to digital transmission. In North America analogue terrestrial broadcast services are being turned off during 2009, with many NA Cable MSOs following suit. In the UK, cutover began in 2007 and is expected to be complete by 2012. Most other European countries have a similar timetable.

3.3.1 Digital TV: Coding and Transmission of Analogue Information

MPEG (Motion Picture Experts Group) is the name of a family of standards used for coding and compressing audio-video information for storage on accessible media such as a DVD, tape, and hard disk drives or for transfer via a communications medium such as a cable HFC or IP network.

The MPEG-2 (pronounced M-peg-2) system for encoding analogue audio-visual information is one of the key systems in any digital TV head-end [3]. Early systems used MPEG encoders to sample video at millions of times per second (720 samples/line for standard definition) and audio at thousands of times per second, quantize the samples into discrete sets of digital values, and then minimize the resultant digital information using a coding algorithm to produce a bitstream known as an elementary stream (ES). Modern-day systems now have the video and audio captured as an uncompressed digital stream with MPEG encoders used to compress the resultant data ready for transmission or storage.

The system part of MPEG-2 [19] specifies how the encoded or compressed information is packetized (to form a PES or packetized elementary stream) and multiplexed with other PES bit-streams, such as multiple audio languages for the video, as well as a program clock reference (PCR) and private data to form one of two output formats:

1. Program Stream—designed for use in low error rate (BER $< 10^{-10}$) environments such as DVDs and multimedia applications. Program stream packets are of variable size and can be of considerable length up to 64 Kbytes [22].
2. Transport Stream (see Figure 3-4)—designed for use in error prone environments (BER $> 10^{-4}$) such as transmission paths. Transport stream packets are 188 bytes in length or 204 bytes with Reed-Solomon forward error correction.

Unlike the MPEG-2 Program Stream, the MPEG-2 Transport Stream is a multiplex of several TV programs (audio, video, and data) that are effectively transmitted simultaneously via a transmission network (cable, satellite, terrestrial, etc.) to an

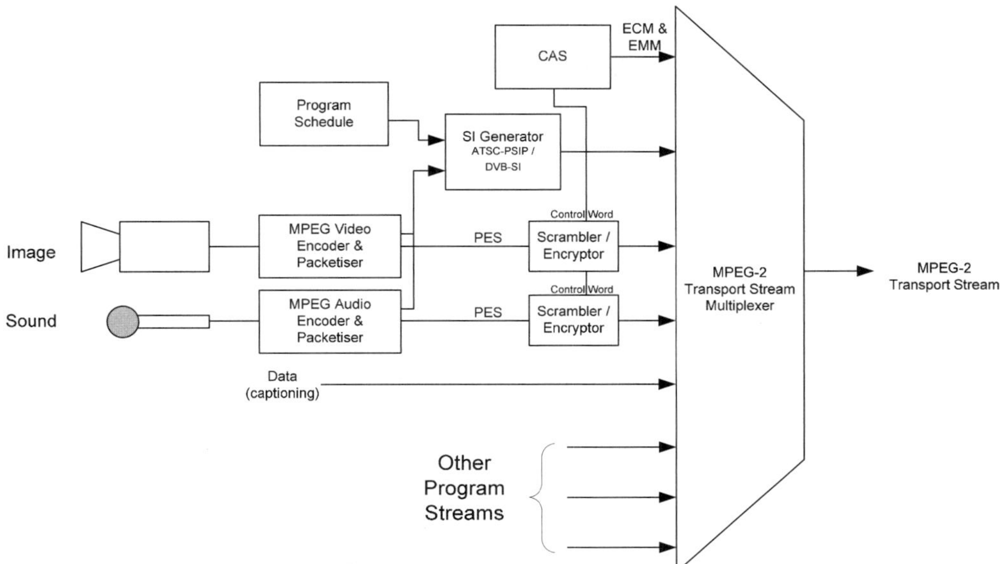

Figure 3-4. Components of an MPEG-2 transport stream (adapted from [3])

integrated receiver/decoder (IRD), also known as a set-top box (STB). Each transmitted packet contains a four-byte header to enable recovery of the transmitted data and identification of individual elementary stream. Thirteen bits of the header are used as the identifier which is known as a packet identifier or PID.

Following on from the success of MPEG-1 and MPEG-2, which won Emmy awards in 1996, MPEG-4 became an international standard in 1999 under ISO/IEC 14496. This is a multi-part specification that further improves compression ratios by around 30%, resulting in smaller file sizes when stored and reduced amounts of bandwidth required for transmission in an MPEG transport stream. Most HD transmissions make use of MPEG-4 as do many IPTV implementations and VOD services over the Internet.

We will see later on in this chapter that MPEG-2 Transport Streams are also used for transmission of data in a cable system and form the lower most layer of the DOCSIS (Data Over Cable Service Interface Specification) set of standards.

To enable the STB to locate and decode the required program, MPEG-2 defines four sets of information that form the MPEG-2 Program Specific Information (PSI):

- *Program Allocation Table (PAT).* Always located at PID = 0 its purpose is to indicate for each program carried in the transport stream the PID of the packets carrying its program map table.
- *Program Map Table (PMT).* Each program in the transport stream has a program map table that is transmitted on an arbitrary PID set by the operator. The PMT identifies the streams (video, audio, and private data) and their PID numbers that collectively make up the program. Among the private data is the optional entitlement control message (ECM), which is used for unscrambling programs that are only able to be conditionally accessed (most often access comes with payment).
- *Conditional Access Table (CAT).* This table is located at PID = 1 and indicates the PID of the packets carrying the Entitlement Management Messages (EMM) used to decode a scrambled conditional access program.
- *Private Data.* MPEG-2 provides facilities for the inclusion of a private data section within the transport stream. The private data is typically used by operators to deliver service information (SI) needed to construct a program navigator, in the form of an electronic program guide (EPG), and as a mechanism to provide interactive services such as online shopping and video on demand.

Digital TV SI information is standardized by two bodies: DVB adopted by most European networks and Advanced Television Systems Committee (ATSC) adopted in North America, Japan, and parts of Eastern Europe (Figure 3-5). The evolution of transmission and signaling technologies is resulting in the converging of standards from these two organizations.

DVB-SI

- The Event Information Table (EIT)—carries information regarding events that will take place in the transport stream such as now-and-next. It also contains

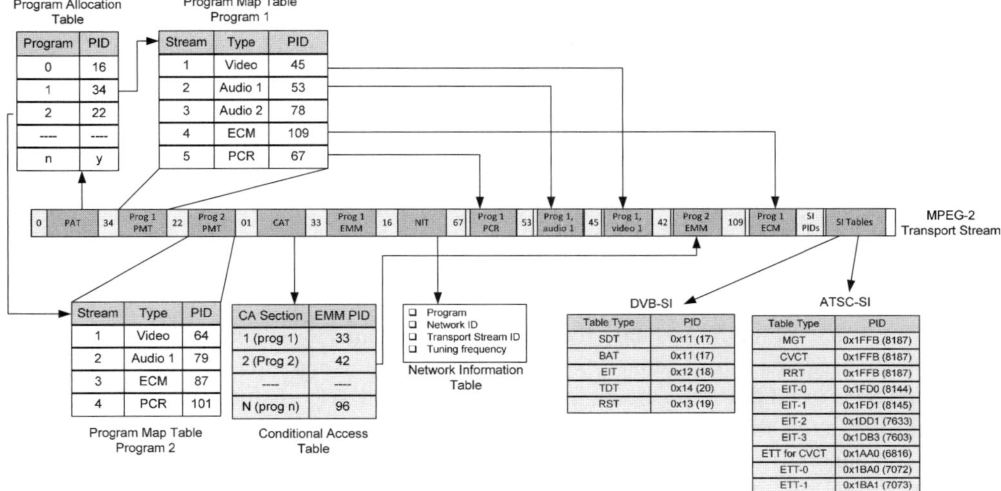

Figure 3-5. Fsormat of an MPEG-2 transport stream

a running status field that can be used by a recording function to start recording before the program actually starts [14].

- The Service Description Table (SDT)—lists the names and other parameters of the services carried in the transport stream.
- The Bouquet Association Table (BAT)—used to carry information to group programs and services. This table may be used by an EPG to group sports channels together.
- Network Information Table—used by systems that have more than one RF channel, such as a cable network.
- Time and Date Table (TDT)—used to set the real-time clock on the STB.
- Running Status Table (RST)—used to update the status of one or more events on an as needed but infrequent basis, e.g., update now-and-next status for a late running program.

In 2004 the US Federal Communications Commission (FCC) mandated the use of the ATSC A/65 PSIP standard [2] for the transmission of digital TV service information in North America. The standard states that the following PSIP data must be included in all ATSC compliant transport streams:

ATSC-PSIP

- The Master Guide Table (MGT)—defining the type, packet identifiers, and versions for all of the other PSIP tables included in this transport stream except for the System Time Table (STT).
- The Virtual Channel Table (VCT)—defining, at a minimum, the virtual channel structure for the collection of MPEG-2 programs embedded in the transport stream in which the VCT is carried.

- The Rating Region Table (RRT)—defining the TV parental guideline system (rating information) referenced by any content advisory descriptor carried within the transport stream.
- The System Time Table (STT)—defining the current date and time of day.
- Extended Information Tables 0-3 (EIT-0-3)—detailing program description and start and end times over a minimum of a 12-hour period.

3.3.2 Network Information Table (NIT)

The Network Information Table (NIT) is an optional table that lists the available networks and tuning information for the transport streams they contain. The NIT is typically used in a mixed delivery system such as satellite and terrestrial and is used by the STB to establish the tuning information for the transport stream that contains the program that the viewer is interested in. The PAT and PMT are then used to find the components to decode. Within a transport stream, the NIT is set to be Program 0 in the PAT and is located at PID = 16.

3.3.3 DVB-SI Program Decoding

1. The STB builds an EPG from the NIT, SDT, EIT, and BAT tables.

2. Once a program has been selected by the viewer the STB reads the PAT to locate and de-multiplex the program's PMT that contain the required audio, video, PCR, and, if necessary, ECM PIDs.

3a. When the viewer has selected an un-encrypted or in-the clear-program the STB de-multiplexes the video and audio packets, using the PIDs specified in the PMT, to recover and decode the various elementary streams.

3b. For an encrypted or conditional program the STB de-multiplexes the video, audio, and ECM packets using the PIDs specified in the PMT. It then uses the EMM specified in the CAT to decrypt the ECM and obtain the Control Word. The Control Word is then used to decrypt and recover and decode the video and audio elementary streams.

3.3.4 ATSC-PSIP Program Decoding

1. The STB builds an EPG from the NIT, VCT, and EIT tables.

2. Once a program has been selected by the viewer the STB reads the VCT and PMT to locate and de-multiplex the required audio, video, PCR, and, if necessary, ECM PIDs.

3a. When the viewer has selected an un-encrypted or in-the clear-program the STB de-multiplexes the video and audio packets, using the PIDs specified in the VCT, into the various elemental streams.

3b. For a conditional program the STB demultiplexes the video, audio, and ECM packets using the PIDs specified in the VCT and PMT. It then uses the EMM specified in the CAT to decrypt the ECM and obtain the Control Word. The

Control Word is then used to decrypt and recover the video and audio elementary streams.

It should be noted that ECMs and EMMs are not updated in real-time and are cycled relatively slowly, typically every 2 seconds and every 10 seconds, respectively, with the STB or smartcard responsible for caching a set of valid keys at all times. It should also be noted that the scheme described is illustrative as these mechanisms are proprietary to each vendor and are kept strictly confidential.

3.3.5 Conditional Access

More and more TV programs are only available on a conditional basis; the condition usually being chargeable subscription or pay-per-view. Enforcement of conditional viewing is accomplished by scrambling the conditional program's elementary streams with a 16-bit control word. The control word needed by the STB to unscramble program is encrypted with a service key and transmitted inside an entitlement control message (ECM) on a PID that is specified in the program's PMT. The service key (used to encrypt the control word) is encrypted with a user key that is also contained within the Smartcard of the STB and transmitted inside an entitlement management message (EMM) on a PID specified in the Conditional Access Table (CAT). Each user key is unique and is held within a Smartcard that is paired with an STB. A subscriber management system (SMS) maintains a record of STB and Smartcard pairs and their association to a subscriber and the programs that they pay for. The SMS configures the conditional access system, which generates ECM and EMM streams to create EMMs for Smartcards whose subscriber's have paid for access (see Figure 3-6 for the relationship between conditional access control word, service key and user key).

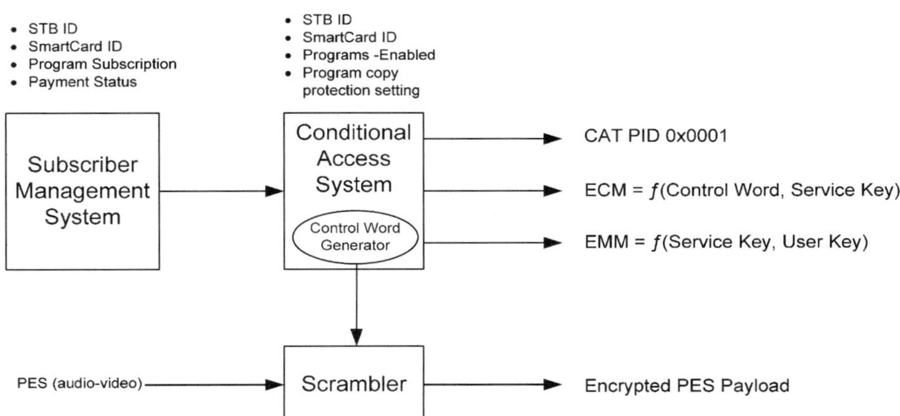

Figure 3-6. Conditional Access System (adapted from [3])

3.3.6 Out-of-Band Channels

Unlike DVB-T, cable systems are considered to have two transmission paths:

1. In-Band (IB)—this is a downstream path only from the head-end to the STB; used to carry PES packets containing audio-video bit-streams and PSI data describing the makeup of the transport stream multiplex.
2. Out-Of-Band (OOB)—This is a downstream Forward Data Channel (FDC) and optionally an upstream Return Data Channel (RDC) path used to carry signaling information between the head-end and STB.

The OOB FDC carries conditional access, SI, and other control information from the head-end to the STB. It has the advantage of simplifying the design, and lowering the cost of the STB, as it no longer needs to process control information from more than 100 transport streams. As the STB also has local memory, the CA and SI carousels can be cycled at a lower frequency, which further contributes to reducing overall cost.

The OOB RDC initially carried control information in the form of requests for events such as impulse pay-per-view (iPPV), near video on demand (nVOD), and information used to quantify their consumption. Use of this channel has now evolved to provide signaling for video on demand, switched digital video, and interactive services as well.

The Society of Cable Telecommunications Engineers (SCTE) published two standards for OOB FDC and RDC communications:

1. [25]—Introduced by General Instruments (now Motorola), FDC uses QPSK modulation and MPEG-2 frame structure to provide a 1.544 Mbps usable downstream bandwidth. RDC uses a differentially-encoded QPSK modulator and ATM frame structure with polling/ALOHA access to provide a 256 kbps usable upstream bandwidth.
2. [26]—Introduced by Scientific-Atlanta (now Cisco Systems), FDC uses differentially encoded QPSK modulation and ATM frame structure to provide a 1.544–3.088 Mbps usable downstream bandwidth. RDC uses a differentially encoded QPSK modulator and slotted TDMA access to provide a 256 kbps—3.088 Mbps usable upstream bandwidth.

Although virtually all cable systems have been upgraded to support two-way communications, there are still many early generation STBs in use today that have no return path functionality, which therefore reduces a cable operator's addressable market for video on demand and interactive services.

The success and widespread use of DOCSIS is now making it more cost effective as a mechanism for delivery of OOB information. In North America, the DOCSIS set-top gateway is the open standard developed by CableLabs dedicated for this purpose, see 3.3.10 below.

3.3.7 Digital Storage Media—Command and Control (DSM-CC)

DSM-CC, defined within [20], is used to enable the control and delivery of multimedia services over broadband networks. In cable systems DSM-CC runs over the OOB forward and return data channels and is used to control video on demand (VoD) and switched digital video (SDV); it is also used to deliver application data and command response information for the delivery of interactive services. DSM-CC is transport layer agnostic making it suitable for use within heterogeneous networks such as HFC, ATM, IP, etc. as well as combinations of these technologies.

In VoD applications, DSM-CC provides VCR-like control over a playing bitstream, providing the user with commands such as pause, fast-forward, goto etc. A key element of DSM-CC that enables these functions to be implemented is normal play time (NPT). NPT is a timecode that is carried in the private data section of an MPEG-2 transport stream that is used to tell the STB the location of the broadcast stream that is currently playing. Unlike the program clock reference (PCR), used to synchronize the audio to the video (i.e., lip-sync), NPT is used to refer to specific locations in the media stream. DSM-CC uses these references to implement the various VCR-like functions and also provide an event notification mechanism for triggering application functions at locations specified by the broadcaster in the program stream.

Switched Digital Video (SDV), described in 3.3.8 below, uses DSM-CC to implement the channel change protocol (CCP) and the mini-carousel protocol

Tru2way, described in 3.3.9 below, uses the DSM-CC "data and object" carousel to provide a mechanism for transferring applications software and application data to enable downloading and running of interactive TV services.

An alternative to DSM-CC for VOD control, which is targeted for use with IP only based media transmission, is the real time streaming protocol (RTSP). It is also used in Europe for DVB-based VOD control and is seen to be simpler to implement than DSM-CC.

3.3.8 Switched Digital Video

The bandwidth available to deliver compelling services to its subscriber base is one of the most precious resources a cable operator has. This became more evident with the introduction of high-definition (HD) broadcast services as a 6 MHz QAM can only support 2 HD broadcast channels compared with 7–12 for standard definition (SD) services. Furthermore, SD receivers cannot decode HD, resulting in the need to carry both broadcast streams simultaneously. In view of the fact that many cable MSOs had only recently completed a bandwidth upgrade of their cable plant and the threat was growing from satellite and Telco IPTV operators, switch digital video (SDV), also called switched digital broadcast (SDB), was introduced.

SDV enables cable operators to reclaim bandwidth by delivering TV services that are viewed on an infrequent basis only when they are required. Studies have shown that digital TV audiences view available channels broadly in line with the Pareto principle, whereby 80% of audiences only watch 20% of the channels available.

SDV is, in effect, a dynamic multicast mechanism, implemented over an MPEG-2 broadcast network. The DSM-CC channel change protocol (CCP) is used to receive commands from homes within a serving group. DSM-CC is then used

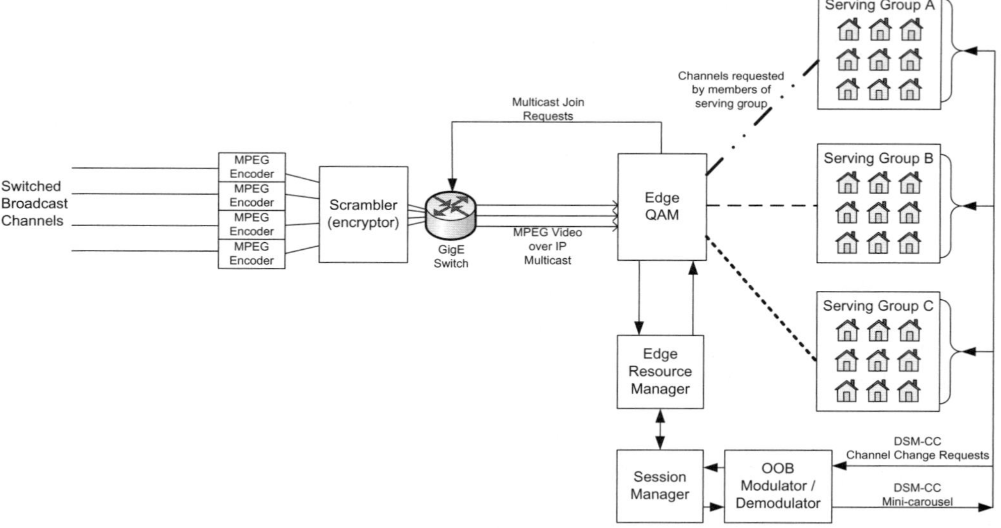

Figure 3-7. Functional elements of a switched digital video system

to transmit a mini-carousel to continuously update the serving group's STBs with the list of switched channels that are currently being transmitted to them; see Figure 3-7.

Although specific implementations may vary, SDV implementations typically work as follows:

1. The viewer selects a switched channel via the EPG (it should be noted that the viewer has no knowledge of whether channels are switched or not).
2. Assuming that the channel requested is not shown as available to the serving group in the mini-carousel, an SDV client on the STB uses the DSM-CC CCP to signal to the SDV session manager that a program stream is required.
3. The SDV session manager signals to the Edge QAM via the edge resource manager to stream a particular channel to the requesting STB's service group.
4. The Edge QAM sends an IGMP join to the GigE switch and joins a multicast stream that has the requested broadcast stream.
5. The session manager updates the STB virtual channel table with the tuning frequency for the requested channel.
6. The STB tunes to the channel and displays the requested content.
7. The session manager updates the mini-carousel to inform other STBs within the serving group that the channel is being streamed and its tuning frequency.

Many field tests have been conducted that confirm that the technology is largely transparent to the end user and that significant expansion of the number and types of channels is possible without the need for cable network upgrades.

3.3.9 Enhanced TV/Interactive TV

Enhanced TV (ETV) or interactive TV (iTV) refers to the ability for viewers to interact with content and services that are presented via their television set. ETV services can be related to the content being viewed, such as requesting a test drive during a car commercial, or playing along with a quiz show; they can also be unrelated, such as looking up local weather or ordering a take-out meal.

Several proprietary attempts at providing such services have been made in the past from companies such as Open TV and NDS with success limited to a specific operator or region. The limiting factors have been the lack of standards that would foster a widespread development community, the price-performance of set-top devices, and the unwillingness of operators to allow third-party access to their customer base.

ETV applications rely on an application platform known as TV-middleware that runs on top of the set-top device's real-time operating system. The TV middleware is there to provide an abstraction layer that shields applications from complexities such as hardware variations between different set-top device manufacturers, accessing a file system delivered via a broadcast stream, and rendering images on the TV screen that are synchronized with content being viewed.

Initial attempts at ETV standardization were fragmented as the standards bodies that drafted them also produced the local SI and broadcasting standards for their respective countries. In February 2000, ETSI released the Multimedia Home Platform (MHP) specification, which is the collective name for a compatible set of Java™ based open middleware specifications developed by the DVB project and is designed to work across all DVB transmission technologies. Globally Executable MHP (GEM) was subsequently developed to remove the DVB specific elements of MHP so that it can be used on non-DVB based delivery infrastructure such as IPTV.

To provide ETV open-standards for the North American cable market, CableLabs created the Open Cable Applications Platform also known as OCAP [35] with the DVB-MHP standard at its core. OCAP's primary benefit is that it is a common TV middleware platform that enables content developers to bring widespread television interactivity by virtue of its write once, run anywhere architecture. The consumer-facing brand name for OCAP and the OpenCable initiative itself is tru2way.

Although OCAP originally specified the use of Java, OCAP-J, extensions have since been made to support HTML, known as OCAP-H. These applications can be of two forms bound and unbound (Figure 3-8):

- Bound applications are connected to the channel that the viewer is watching; when the channel is changed the application is usually destroyed. Examples of bound applications are facilities to play along with a quiz show, voting as part of a talent show or purchasing items from a shopping channel.
- Unbound applications are independent of any channels being watched and can be initiated at anytime via a resident navigator program. Examples of unbound applications are facilities to look up local weather or view an online photo album.

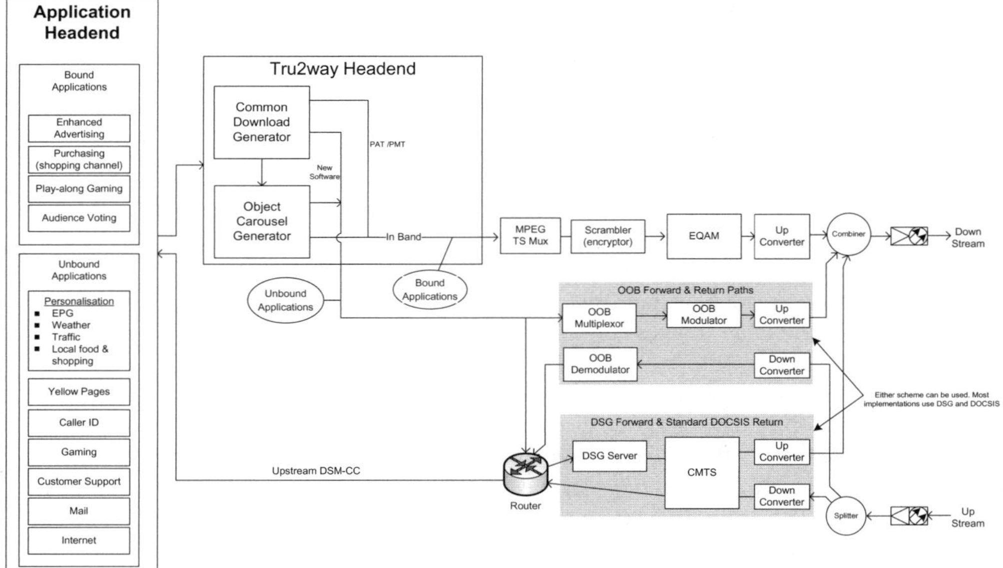

Figure 3-8. OCAP or tru2way functional elements

To enable OCAP to work with set-top devices or networks that support only one-way communication, application files and associated data structures are transferred to the TV middleware in a DSM-CC object carousel that itself is delivered via a broadcast MPEG-2 transport stream. This part of the DSM-CC protocol suite provides a broadcast file system enabling the TV middleware to access files as it needs them to run a particular application.

Similar to the reasons behind the existence of the PMT or VCT, the STB middleware needs to know which applications are available, how and when to start them, and other details for the viewer. It gets this information via two new SI tables, the application information table (AIT) for bound applications and the extended application information table (XAIT) for unbound applications. It is the content of these tables, together with normal play time (NPT) information, that enables the OCAP middleware to trigger actions within bound applications.

Clearly, in systems that have two-way networks and receivers, greater interactivity is possible when compared to one-way systems. In these environments the OCAP middleware is able to invoke applications that are hosted in the head-end using either HTML or Common Object Request Broker Architecture Remote Procedure Calls (CORBA-RPC). Examples of such applications include provisioning of new services, browsing the Internet, email, and sharing of photos.

As stated in 3.3.7 above, DSM-CC is transport layer agnostic and is therefore capable of working over the SCTE 55–1 and SCTE 55–2 OOB channels. However, as applications become more complex and interactivity volumes increase, higher upstream and downstream speeds will be required and we will begin to see greater use of the DOCSIS Set-top Gateway (DSG) as described in 3.3.10 below.

3.3.9.1 Enhanced TV Binary Interchange Format In providing a rich environment for creating interactive applications, OCAP requires a platform with sufficient hardware and memory to run a Java Virtual Machine (JVM) and the ability to perform processing of additional SI data. As of December 2009 the number of deployed devices that are capable of running an OCAP middleware stack is less than 10% out of a population of at least 50 million.

On the basis that it would be too costly and impractical to swap out the legacy population of set-top devices in any reasonable timeframe, CableLabs developed the Enhanced TV Binary Interchange Format (ETV-BIF) specification [15] to enable the support of interactive applications without the use of Java. ETV-BIF supports a limited set of functionality and unlike tru2way will not be supported by the consumer electronic manufacturers.

OCAP and EBIF are primarily focused toward the needs of the U.S. and Canadian cable markets. European cable operators who are deploying interactive services are either continuing with systems from proprietary vendors or are adopting systems based on MHP, which, as stated in 3.3.9 above, is the basis for OCAP anyway.

3.3.10 DOCSIS Set-Top Gateway

Competition, FCC regulatory pressure, and the success of the DOCSIS standard has led the cable industry to advance the mechanism used for OOB communication between the head-end and STBs.

The DOCSIS Set-top Gateway (DSG), developed under the CableLabs OpenCable™ initiative, is an open standard transport mechanism [29] for delivery of OOB FDC control information (i.e. conditional access, service information and emergency alerts) using IP and a DOCSIS downstream channel (Figure 3-9). To ensure compatibility with existing business models, the DOCSIS Set-top Gateway must provide a transparent transport of OOB traffic as defined in [25] and [26]. OOB RDC data (TV return path) is transported via a standard DOCSIS IP session.

OOB information is sent from the set-top controller to the CMTS inside UDP over IP datagrams where it is mapped to a DSG tunnel for delivery to the set-top device. CMTS to set-top device communication has the ability to use unicast, multicast or broadcast type datagrams.

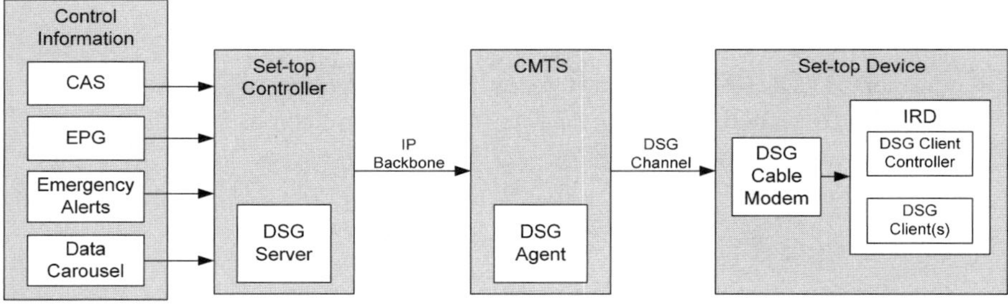

Figure 3-9. Out-of-band transport via a DOCSIS Set-top Gateway [29]

It should be evident that the DSG is another infrastructure enabler for cable providers to provide converged multimedia services to the TV and other consumer devices via a home gateway. In the near-term it will allow DTV command and control functions to be leveraged by adjacent applications running inside and outside of the STB. Longer term, and as Broadband TV and other multimedia services are more widely deployed, NGN signaling protocols will enable developers to create a richer and more device independent user experience.

3.3.11 Digital TV Head-End

Having looked at the major elements that comprise a digital TV head-end, Figure 3-10 shows how they are all put together and the relationships between them.

It is clear, even from this high-level diagram, that digital TV has moved beyond just being a more secure and better performing replacement for analogue TV distribution to now encompass richer content, high-definition picture quality, and interactivity. To complete the picture, we will now look at the composition of the receiving equipment that is used to recreate the original audio-video signal.

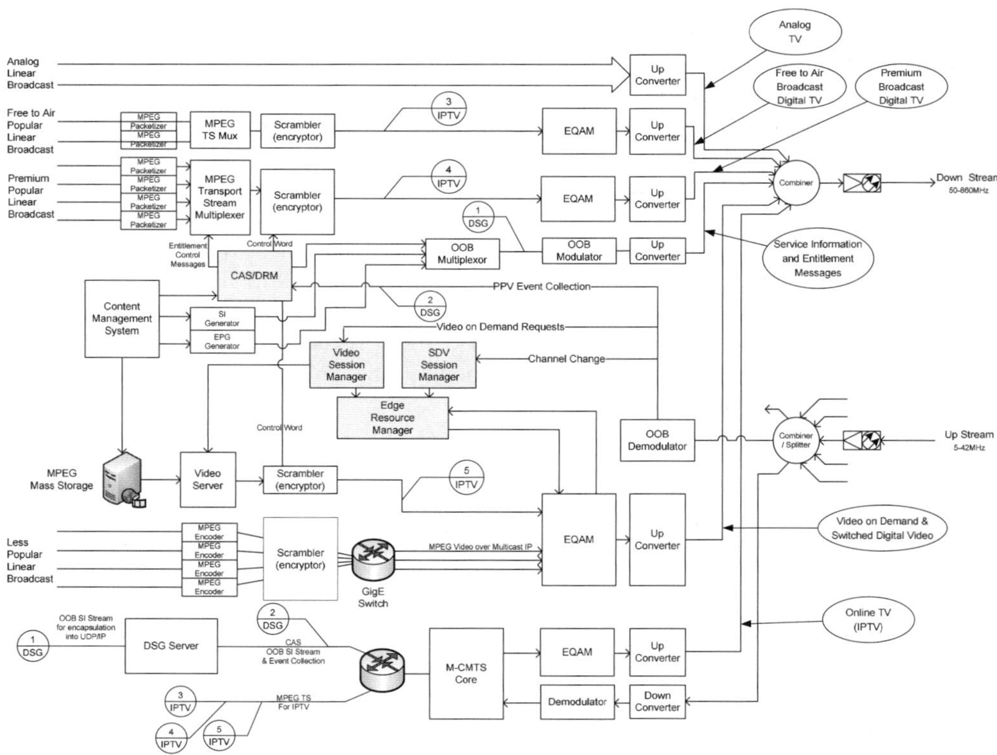

Figure 3-10. Functional elements of a digital TV head-end

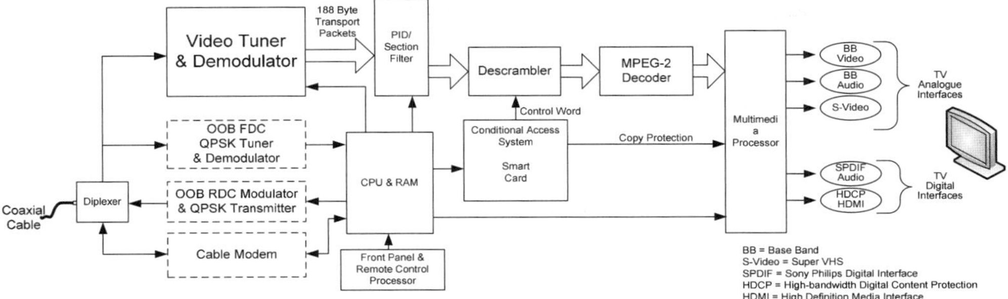

Figure 3-11. Set-top box functional elements, adapted from [13]

3.3.12 Integrated Receiver/Decoder or Set-Top Box

The Integrated receiver/decoder (IRD), more commonly known as a set-top box (STB), is the equipment located in a subscriber's home that is used to receive and decode digital TV signals.

Referring to Figure 3-11, TV programming is received by the STB and transferred to the television in the following manner:

1. The diplexer takes input from the HFC network and passes radio frequencies from 50 MHz to 870 MHz to the video and OOB FDC tuners.

2a. SI and return path via OOB tuner and transmitter: The OOB FDC tuner down converts radio frequencies (RF) in the range from 70 MHz to 130 MHz and recovers service information that is used by the CPU to construct the channel map, EPG, and other information to enable the viewer to select the required channel.

2b. SI and return path via cable modem: the cable modem receiver tunes to the frequency specified by its boot parameters. It recovers the individual data streams and directs the SI multicast stream to the CPU to construct the channel map, EPG, and other information to enable the viewer to select the required channel.

3. Once the viewer has selected the channel to be watched using a remote control or the STB front panel controls, the CPU instructs the video tuner to tune to the appropriate frequency (QAM channel). The video tuner down converts the RF to an intermediate frequency and recovers the QAM stream. It then decodes the QAM symbols, performs analogue to digital conversion and any required forward error correction to recover the 188 bytes per packet MPEG-2 transport stream.

4. The PID/section filter selects the required PES from the transport stream and passes it to the descrambler, which performs decryption using the control word that has been recovered using the appropriate ECM, EMM, and the user key contained in the smartcard.

5. The MPEG-2 decoder reconstructs the video and audio signals of the selected channel, and passes it to the multimedia processor for formatting into the various outputs that can be consumed by the viewer's TV set. If specified by the conditional access system copy protection is added to the output signal to conform to the distribution conditions of the content owner.

6. In networks that support two-way communication information such as usage events, VoD requests and SDV channel change commands are packetized by the CPU and sent either to the OOB RDC modulator for digital to analogue conversion and modulation or to the cable modem transmitter. The resultant RF signal from 5 MHz to 42 MHz is sent to the diplexer for transmission to the TV head-end via the HFC network.
7. Many current day STBs have hard disks incorporated within them to provide Digital Video Recorder (DVR) or Personal Video Recorder (PVR) functionality enabling the viewer to record one channel while viewing another and to pause and rewind broadcast TV. These STBs will have more advanced tuning and PID/section filtering capabilities and they need to be able to tune to more than one QAM channel simultaneously.

3.3.13 Point of Deployment Module/CableCard

In the United States the Telecom Act of 1996, section 629, led to the FCC mandating in 1998 that cable companies must provide separable security access devices to enable subscribers to use third-party equipment for accessing cable network services. After several delays and the FCC also mandating that cable companies use the same separable security themselves, the system came into effect on July 1, 2007 [37].

The solution that was implemented falls under the CableLabs OpenCable initiative and is comprised of two functional elements:

1. A host system that can perform service navigation such as an STB, PC, digital video recorder (DVR) or a TV with integrated STB functionality—this can be manufactured and distributed by non-cable companies such as consumer electronics manufacturers and retail distributors.
2. A point of deployment (POD) module that performs the security functions required by the Cable MSO—this is owned and distributed by the Cable MSO whose service is wanted by the receiving subscriber.

The POD module, also known as a CableCard, is built using a PCMCIA form factor and plugs into the STB or Digital Cable-Ready TV to provide the conditional access functions and the interface to the OOB channels via SCTE 55–1, SCTE55–2 or the DSG [30] (Figure 3-12).

The first generation of CableCard, known as an S-Card, is only capable of decoding a single stream or channel at a time. The current generation, referred to as an M-Card, is capable of decoding multiple streams simultaneously to facilitate the use of built-in DVR functionality whereby a subscriber can watch and record different programs at the same time.

As an alternative to the CableCard, CableLabs developed the Downloadable Conditional Access System (DCAS), which is a software solution that relies on a special chip being included with the host hardware. DCAS has yet to be deployed and it is still unclear whether it has been rejected or approved by the FCC.

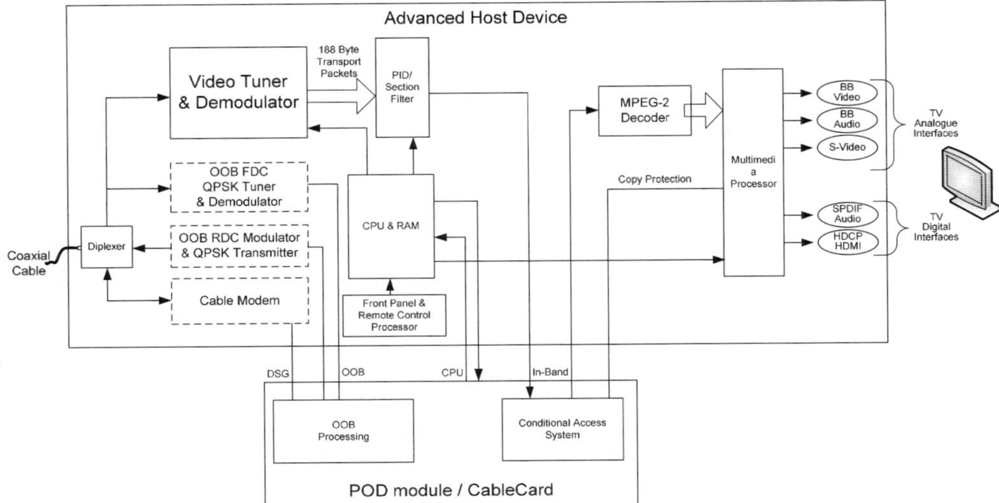

Figure 3-12. Advanced host and POD / CableCard supporting two-way networks, adapted from [13]

3.4 DATA OVER CABLE SERVICE INTERFACE SPECIFICATION (DOCSIS)

The rise in popularity of the Internet has prompted cable TV operators to use their networks to deliver broadband data services. The system that is implemented is comprised of a cable modem termination system (CMTS) in a regional hub and cable modems in each subscribing home, both of which are connected to the HFC network (Figure 3-13). In the downstream path the modulation frequency is set so as not to overlap with that of existing TV transmissions. In the upstream path, which has many transmitters talking to one receiver, a bandwidth sharing or multiplexing mechanism is used to avoid overlapping transmission bursts.

Manufacturers of the initial systems used proprietary hardware and proprietary transmission protocols with no interoperability between systems. However, as popularity grew, CableLabs, in collaboration with several network equipment vendors, developed the Data Over Cable Service Interface Specification (DOCSIS) with a view to lowering operating costs and enabling consumers to own the equipment that is connected to the cable network. DOCSIS specifications are now approved by SCTE and ANSI as the recognized standard for providing data services over a cable network for North American cable MSOs, and as part of the ITU-T J series in Europe, where it is commonly known as EuroDOCSIS.

Since standardization, DOCSIS has evolved through four versions of the specification: DOCSIS 1.0, DOCSIS 1.1, DOCSIS 2.0, and DOCSIS 3.0. At each increment there is a speed increase and from DOCSIS 1.1 onwards there are built-in dynamic quality of service mechanisms that are used to protect jitter sensitive and speed sensitive data services such as voice and streaming media.

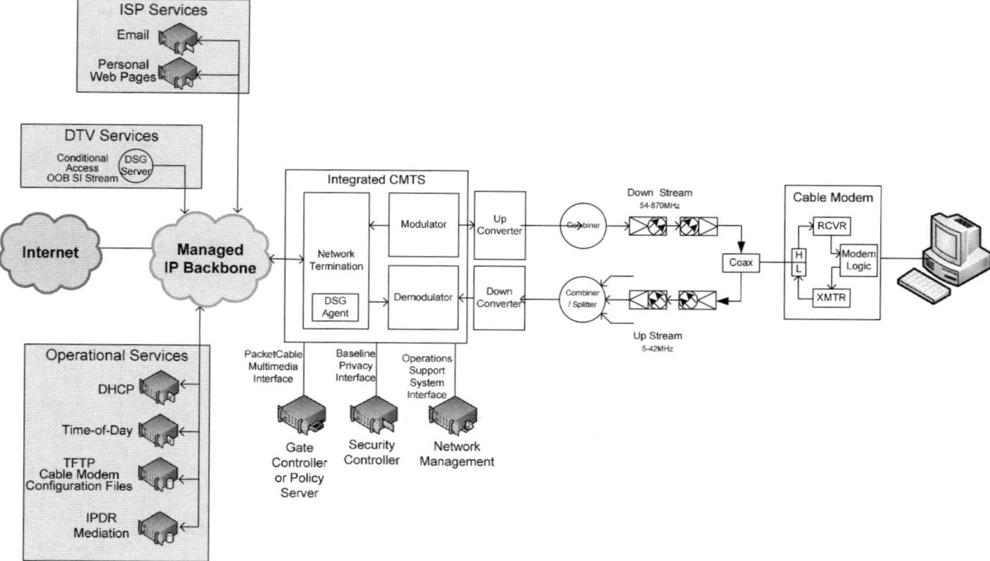

Figure 3-13. DOCSIS architecture

Table 3-1 highlights the advances that have been made since the introduction of DOCSIS 1.0 in 1997, with the most relevant increment being the increase in upstream and downstream useable data rates. In the early days of broadband cable a single CMTS could support multiple fiber nodes and multiple serving groups. However, as discussed in 3.2.2 above, the increasing use of streaming media and the rise of e-working is driving cable MSOs further toward 1 : 1 combining ratios.

With reference to the ISO seven-layer model, the DOCSIS specifications primarily define characteristics of the physical layer and the lower parts of the data link layer protocols with the network layer being the Internet Protocol. All DOCSIS versions support IP version 4 (IPv4) with DOCSIS 2.0+IPv6 and DOCSIS 3.0 supporting IP version 6 (IPv6).

3.4.1 Physical Layer

The DOCSIS physical (PHY) layer has two sublayers: the physical media-dependent (PMD) sublayer, which is applicable to both downstream and upstream data paths, and the downstream transmission convergence (DTC) sublayer, which is applicable to the downstream direction only. The PMD deals with the modulation, error correction, and electrical characteristics of the specification, and the DTC deals with downstream frame structure and service interleaving.

DOCSIS has a downstream channel spacing of 6 MHz, which is the same as the bandwidth allocated to an NTSC analogue TV channel and can use 64 or 256 quadrature amplitude modulation (QAM). As such the PMD must conform to Annex B of [21] for channel coding, channel modulation, and framing structure.

TABLE 3-1. Comparison of DOCSIS Versions, adapted from [38]

Parameter	DOCSIS 1.0	DOCSIS 1.1	DOCSIS 2.0	DOCSIS 3.0
Downstream				
- Frequency MHz	50–860	50–860	50–860	50–1002
- Modulation	64/256 QAM	64/256 QAM	64/256QAM	256QAM
- Bit rate Mb/s	30.342, 42.884	30.342, 42.884	30.342, 42.884	171.52 (4 Channel) 343.04 (8 Channel)
- Useable data rate Mb/s	38	38	38	152 (4 Channel) 304 (8 Channel)
Upstream				
- Frequency MHz	5–42	5–42	5–42	5–42
- TDMA				
○ Modulation	QPSK	QPSK, 16QAM	QPSK, 16/64 QAM	QPSK, 16/64 QAM
○ Symbol rate kbaud	160–2560	160–2560	160–5120	160–5120
- S-CDMA				
○ Modulation	N/A	N/A	64/128 QAM	64/128 QAM
○ Symbol rate kbaud	N/A	N/A	1280–5120	1280–5120
- Bit Rate Mb/s	5.12	5.12, 10.24	5.12, 10.24, 30.72	122.88 (4 & 8 Channel)
- Useable data rate Mb/s	9	9	9, 27	108
Standards				
- ANSI/ SCTE	22	23	79	135
- ITU-T	J 112	J 112	J 122	J222

EuroDOCSIS has 8 MHz channel spacing with the PMD conforming to EN300 429 [16]. Accounting for DOCSIS overheads, useable downstream data rates per RF channel are 38 Mb/s and 50 Mb/s, respectively.

The PMD sublayer for both DOCSIS and EuroDOCSIS requires the DTC sublayer to provide it with a continuous series of 188byte MPEG-2 packets. These packets have a 4 byte MPEG header whose format is only applicable on a DOCSIS data over cable well-known PID (0x1FFE) and indicates that the following 184bytes is a DOCSIS MAC frame. The DTC mostly exists to enable optional interleaving of video and data services over the PMD using a common framing structure. Using this mechanism it is possible to have common downstream modulators and demodulators for the delivery of both data and video services.

The upstream physical layer does not require a convergence layer as it is dealing with a different problem altogether; that of having multiple transmitters (cable modem modulators) sending information to a single receiver (CMTS demodulator) where the CMTS can instruct each CM transmitter to vary its transmission rate, power level, upstream frequency, modulation, etc. CM transmitters are informed of the time-slot in which they are able to transmit via Media Access Control (MAC) messages from the CMTS together with parameter adjustments on a transmission burst-by-burst basis.

DOCSIS 1.x supports two modulation formats, QPSK and 16QAM time division multiple access (TDMA), which for each upstream RF channel provides a maximum useable upstream data rate of 9 Mb/s. To support a more symmetrical traffic pattern, DOCSIS 2.0 supports a maximum useable upstream data rate of 27 Mb/s per upstream RF channel through the use of 64QAM TDMA and the introduction of 64QAM and 128QAM Synchronous Code Division Multiple Access (S-CDMA). S-CDMA also uses MAC messages to instruct CMs of the time-slot in which they are allowed to transmit, however its modulation format also enables multiple CMs to transmit simultaneously through the use of different orthogonal codes.

Finally DOCSIS 3.0 achieves higher downstream and upstream data rates (Table 3-1), through the use of channel bonding whereby multiple channels are combined to support a dynamic form of inverse time division multiplexing.

3.4.2 Data Link Layer

The data link layer sits above the physical layer and is comprised of three sublayers, which from bottom to top are media access control, link layer security, and logical link control.

3.4.2.1 Media Access Control (MAC) Sublayer A CMTS and its connected cable modems represent a MAC sublayer domain in which a single downstream transmitter in the CMTS sends MAC frames for reception by all CMs, and all CMs transmit data for reception by a single CMTS receiver. The MAC sublayer must support variable length Ethernet (802.3) Packet PDU format MAC frames; although not widely used, it must also support fixed-length ATM Cell MAC frames as well.

In the downstream direction CMs listen to all frames on the channel they are registered for and only pass them to upper layers where the destination address matches the CM itself or a device connected to its CMCI port such as a PC or MTA.

In the upstream direction multiple transmitters are sending information to a single receiver and as such must be controlled to prevent overlapping transmissions. The CMTS provides the controlling mechanism by transmitting a MAC control frame known as an upstream bandwidth allocation map (MAP for short) on a regular basis that informs each registered CM of the time periods when it is able to transmit data or request a time slot to transmit data. In order to reduce blocking of upstream traffic, a CMTS is typically capable of supporting multiple upstream RF channels. Each CM is instructed to transmit on a specific upstream frequency based on a setting in its configuration file or via an instruction in a MAP message.

To have a common reference for transmission times, DOCSIS characterizes the upstream transmission timeline as a series of mini-slots, where a mini-slot is a power-of-two multiple of 6.25 μs, i.e., 2, 4, 6, 8, etc., multiplied by 6.25 μs.[1] When a CM has data to transmit it refers to the MAP associated with the RF channel that it is registered on, to locate the time slot when it can contend for bandwidth. At the appropriate time, the CM, together with any other CM on the same RF channel that has data to send, transmits a request message to the CMTS requesting a number of mini-slots. The CM then looks at subsequent MAP messages to see its allocated time for transmission and the number of mini-slots it can use in a single burst. If the MAP contains no reference to a specific request it means one or more CMs also transmitted a request at the same time and hence the CMTS was not able to decode the requests. When this occurs all of the requesting CMs wait a random amount of time and then try their requests again when an appropriate mini-slot arrives. Clearly this mechanism provides a more efficient use of the upstream bandwidth when compared to a pure TDM mode of operation as CMs are only allocated upstream timeslots when they have data to send.

Service flows, which are characterized by parameters such as latency, jitter, and throughput assurance, were introduced as part of DOCSIS 1.1 as the mechanism for providing downstream and upstream quality of service (QoS). Service flow IDs are used to refer to a unidirectional flow between the CMTS and a CM. Upstream service flow IDs also have associated to them service IDs (SIDs), which the CMTS uses to allocate bandwidth to a CM for a specific stream of information via MAP messages.

One or more service flows are assigned to a CM during its registration process. Additional service flows can be established and managed by the CMTS or CM using dynamic service add (DSA), dynamic service change (DSC), and dynamic service delete (DSD) messages, collectively known as DSx messages. A service flow (configured or dynamic) must first be authorized before it can be established. Authorization can be provided within a CM's configuration file or is provided on an as-needed basis by an external gate controller that is attached to the CMTS (Figure 3-13). Prior to a QoS flow starting, DSA and DSC messages are generated by the originating endpoint under the direction of a higher layer application. Having checked that the service flow is authorized and that the requested resources are available, the CMTS makes an appropriate reservation—the service flow is now in an admitted state. Just as transmission is about to start, a DSC message is used to commit the resources—the service flow is now in an active state. When the service flow is no longer needed a DSD message is used to release the resources—unless de-authorized via the gate controller the service flow is now in an authorized state again.

To enable higher data rates in both directions, while continuing to support legacy CMs, DOCSIS 3.0 implemented downstream and upstream dynamic channel bonding, whereby a CMTS can manage the distribution of packets for a specific service flow over a set channels for a given CM. Packets that are transmitted across multiple channels are tagged with a sequence number to cater for latency differences between each channel. To ensure the best statistical multiplexing of traffic across bonded channels, the CMTS is able to dynamically modify upstream and

[1] In DOCSIS 1.x a mini-slot is typically sized to be able to carry 16bytes of information when using QPSK modulation. With four symbols per byte (QPSK) and a maximum symbol rate of 2560kbaud (see Table 3-1), 40,000 mini-slots are required per second (2,560,000 / (16 × 4)). Therefore, the number of microseconds required per mini-slot is 25 (1,000,000 / 40,000) resulting in four 6.25 μs ticks per mini-slot.

downstream bonding parameters for a given CM using the dynamic bonding change request (DBC-REQ) mechanism. Typical examples of its use include when a CM is required to tune to a different downstream channel to receive packets transmitted to a specific downstream bonding group or where a CM requests a high-bandwidth transmission slot and needs to transmit on multiple upstream channels simultaneously.

Classifiers are used to forward specific data packets to a QoS service Flow. The classifier is compared against the LLC header and the IP and TCP/UDP header of incoming packets; when a match is found the service flow ID associated with the classifier points to the service flow to be used. If no match is found the packet is forwarded to the primary service flow, which is a best-effort flow established during CM registration. The CableLabs' PacketCable Multi Media (PCMM) specification [7] refers to the functional component that performs classification of packets and enforcement of QoS as a gate and the component that authorizes and controls dynamic flows as a gate controller or policy server. When devices attached to CMs are incapable of sending DSx messages, gate controllers or policy servers are used as proxies to reserve and commit the service flow.

Once segregated into individual service flows, the CMTS or CM applies traffic shaping and queuing algorithms to ensure packets are transmitted at the appropriate time and within the constraints defined for that flow. CM requests and MAP messages are used to ensure upstream messages are allocated sufficient mini-slots to meet the required QoS characteristics.

3.4.2.2 Link Layer Security The link layer security sublayer, more commonly known as the base-line privacy interface plus (BPI+) [28], has two main purposes—ensuring data privacy for each connected subscriber and providing the cable operator with protection from theft of service.

As an HFC network is a shared medium, downstream traffic from a CMTS will appear at the data link layer of every cable modem that is connected to it. It is only after the cable modem has determined whether or not the destination address of the incoming packet is attached to its CMCI that it will discard or forward the received data. Today there are tools available such as protocol analyzers that can recover information transmitted over a DOCSIS network and see information that is destined for other users. Dishonest individuals are able to use these tools to not only look at the data but also to create a clone of an existing cable modem, with the exact same data link layer address, and go on to pretend to be a legitimate paying subscriber.

To prevent unauthorized viewing of subscriber data, BPI+ provides a mechanism to encrypt data flows between a CMTS and CM that uses a frequently changing shared key algorithm.

The key exchange mechanism itself uses public-key encryption with each CM having a unique X.509 digital certificate inserted at time of manufacture. During the registration process the CM presents its digital certificate containing its public key, MAC address, serial number, and other identifying information to the CMTS. Having verified the digital certificate the CMTS then uses the CM's public key to encrypt an authorization key and send it back to the CM. This authorization key is used to securely transmit the shared keys that are used to encrypt subscriber data.

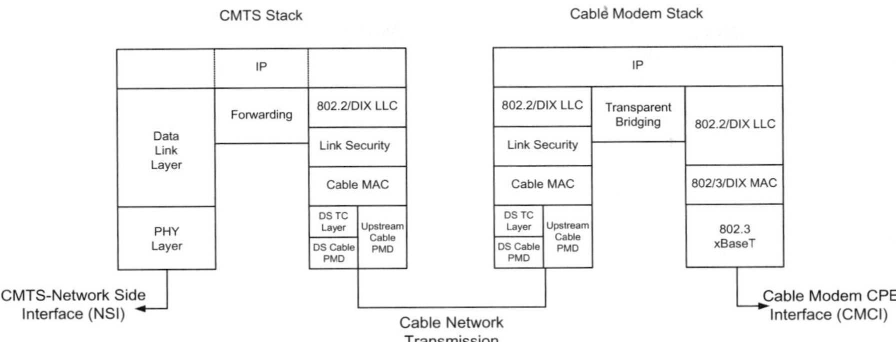

Figure 3-14. Data forwarding through the cable modem and CMTS [27]

This two-level key exchange is employed to remove the need for constant verification of a CM's digital certificate. DOCSIS 1.1 and 2.0 make use of 40-bit and 56-bit DES (data encryption standard) encryption for information that is exchanged between the CM and CMTS and triple DES for encrypting the traffic encryption keys themselves. As DES is now seen to be end-of-life, as a result of the availability of high-performance computing power at low cost, DOCSIS 3.0 has introduced the use of 128-bit AES (advances encryption standard) and EAE (early authentication encryption) mechanism to satisfy the needs of the various U.S. government agencies that use cable providers for their data access service.

Theft of service is prevented by associating the verified digital certificate with that of a paying subscriber. The digital certificate is also the mechanism that prevents cloning of a legitimate cable modem as only the cable modem's manufacturer has knowledge of its contents.

3.4.2.3 Logical Link Control (LLC) The LLC sublayer is responsible for address resolution functions and specifically for converting network layer addresses (Internet Protocol) to local addresses (48bit Ethernet addresses) in accordance with [23]. It is also responsible for providing bridging functions between the cable modem CPE interface (CMCI) and the network interface on the CMTS.

As the LLC is independent and above the MAC sublayer, it enables the Ethernet MAC sublayer at the CMCI to be connected to the cable MAC sublayer at the cable modem HFC interface (Figure 3-14). The same applies for connection between the CMTS HFC interface and the CMTS network side interface (CMTS-NSI). To help understand transmission behavior between CMCI and CMTS-NSI it is useful to picture the topology as a head-end LAN with many remote bridges to multiple customer premise LANs.

3.4.3 Network Layer

Although not mandatory for attached CPE, IP is the network layer protocol used for management and configuration of the CMTS and cable modems. Protocols used on top of, or in conjunction with, IP include:

- The Simple Network Management Protocol (SNMP) for network management, i.e., reading and setting values within the management information base (MIB) via the operations system support interface (OSSI).
- Trivial File Transfer Protocol (TFTP) for downloading software and configuration files.
- Dynamic Host Configuration Protocol (DHCP) for setting IP addresses and configuring remote devices.
- IP Detail Records (IPDR) for transmission of metrics and usage information via OSSI.

3.4.4 Multicast Operation

DOCSIS 3.0 improved upon the basic support for IP multicast that is implemented in DOCSIS 1.x and 2.0, with the main enhancements being support for source specific multicast (SSM), the use of channel bonding, and the provisioning of QoS for multicast traffic. Unlike any-source multicast (ASM), where the CM only subscribes to a multicast address, SSM allows the CM to subscribe to a multicast address and only be sent packets from a specific source. This has the advantage of reducing network load and improving overall security. The drivers for implementation of these features are to enable cable providers to efficiently deliver broadband TV over their DOCSIS infrastructure and lower the cost of the cable modems.

3.4.5 Cable Modem Start-up

On power-up, or after it has been instructed by its CMTS to reinitialize its MAC layer, a cable modem will need to perform a number of operations before it is able to forward or receive data packets. The following is a high-level list of the actions performed:

1. The CM acquires its downstream channel. It does so by either re-acquiring the last channel it tuned to or by scanning the downstream frequency band in 6 MHz steps (8 MHz for EuroDOCSIS).
2. The CM waits for an upstream channel descriptor (UCD) to retrieve upstream channel parameters for a possible upstream channel. The CM waits until it has obtained UCDs for all the available upstream channels. If it cannot find a suitable upstream channel to transmit on it goes back to step 1 to acquire a new downstream channel.
3. Once upstream parameters are acquired, the CM performs ranging, whereby it establishes its round-trip propagation delay to the CMTS and other transmission characteristics so that it can ensure it transmits at the correct burst time boundary, set its power level, and, if appropriate, set its S-CDMA parameters.
4. The CM transmits a DHCP request to obtain IP parameters. The parameters returned include an IP address for the CM itself, default gateway address, time-of-day server address, and configuration file name and location.

5. The CM retrieves the time of day and applies the time offset learned during ranging in step 3.
6. The CM uses TFTP to download the configuration file specified in the DHCP response. If the configuration file contains downstream and upstream frequencies that are different from the one that the CM is currently using, the CM must re-perform ranging as per step 3 using the new frequencies.
7. Once the configuration parameters have been validated and loaded, the CM then attempts to register with the CMTS by sending a registration request containing, among other things, service flow requirements. If able to do so, the CMTS responds with service flow IDs for all requested service flows.
8. After successful registration, baseline privacy is initialized and the modem is ready to receive and forward data.

3.4.6 IP Detail Records

To gather usage information from network and service elements for measurement of service consumption, quality of experience, etc., the cable industry adopted the use of Internet Protocol Detail Records (IPDR).

An IPDR is a record of information describing usage of an IP-based service and is similar in concept to that of a voice call detail record (CDR), which is generated by a voice switch after completion of a telephone call. The content of each IPDR is a function of the network or service element generating the information and the needs of the operator that is consuming the data. In broad terms the information that can be contained in each IPDR is categorized as follows:

1. Who—The element responsible for the usage of the service; e.g., host name, IP address, customer ID.
2. When—The time that usage took place; e.g., end time or event time.
3. What—The type of usage that took place; e.g., number of bytes sent or received, TV channel change, advertisement watched, etc.
4. Where—The context in which the event took place or the service was consumed; e.g., the device type or the program being consumed.
5. Why—The trigger for the IPDR being sent; e.g., an interactive advertisement or the start of a game.

The IPDR Streaming Protocol (IPDR/SP) was standardized [36] by the TM Forum (www.tmforum.org) to provide a reliable, scalable, and billing-grade protocol for the transfer of IPDR usage information between exporters and collectors. The key benefits of the protocol include:

High availability: There are features built directly into the protocol that enable automated fail-over in the event of connection or server failure. This allows operators to build redundant paths for valuable record streams.

Reliability: IPDR uses transmission control protocol (TCP) to provide connection-oriented transport reliability. In addition, record acknowledgment on the application layer provides further robustness.

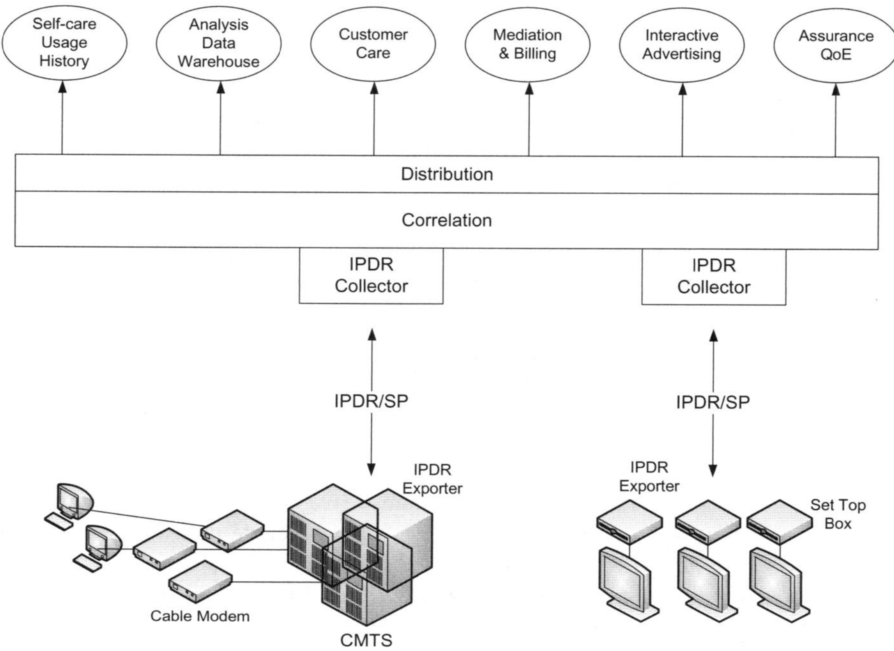

Figure 3-15. Usage Collection using IP Detail Records

Scalability: The stream-oriented behavior of the IPDR protocol provides a new way for data to be gathered from the network. The exporter "pushes" event-based records to the collection layer, removing the inefficiencies and cost of polling.

Efficiency: IPDR constructs messages using a binary representation. This results in very compact data records that occupy a minimal amount of network capacity while minimizing the expense of encoding and decoding.

As can be seen in Figure 3-15, network and service elements create IPDRs and export them to collectors, which in turn pass the information to a mediation function that correlates, transforms, and distributes the information to interested parties.

CableLabs and SCTE have mandated the use of IPDRs for usage collection within the OCAP 1.1 and DOCSIS 2.0 and 3.0 specifications. OCAP specifies that the middleware running on the set-top device support IPDR Exporter functions to report usage of enhanced TV services. DOCSIS OSSI specifications mandate the use of the Subscriber Accounting Management Interface Specification (SAMIS) for subscriber usage and billing.

3.4.7 DOCSIS Evolution

The increasing use of peer-to-peer multimedia applications, broadband TV, and greater levels of service convergence continue to drive the evolution of the DOCSIS specification. DOCSIS 3.0 has introduced higher speeds via channel bonding, improved security with AES, and better support for broadband multimedia services with support for source-specific multicast (SSM). Greater levels of service convergence, the growth of services for SMEs, and the increase in the number of third-party applications providers will put further pressure on standards bodies and vendors alike to better support packet-based transport and NGN signaling paths.

Competition from wire-line providers who are offering higher data speeds and IPTV via FTTx (fiber to the home, curb, etc.) is impacting DOCSIS infrastructure architecture. A prime example is the introduction of the modular-CMTS (M_CMTS), which separates the modulators and timing functions from the rest of the CMTS functions. The separation of the modulator has enabled the introduction of a universal edge QAM that is capable of modulating both DOCSIS and video traffic all under the control of a common resource manager, thereby significantly reducing capital and operational costs.

3.5 CABLE TELEPHONY

Subsequent to the privatization of national telephone services, and prior to the existence of voice over IP technology, cable MSOs wanted to provide competitive telephone services alongside their subscription TV offering. However, unlike the traditional telephone companies, who used individual pairs of twisted copper wires between the local telephone exchange and the telephone instrument in the home, cable MSOs had to find a way of using their shared heterogeneous network (fiber and coaxial cable) while ensuring call quality and individual subscriber privacy. The implemented solution consisted of using a network interface device (NID) in each telephone subscriber's home that would enable a standard analogue telephone to be connected to the HFC network (Figure 3-16). Inside the NID are a modem (*mo*dulator-*dem*odulator), ringing current and dial tone generator, and associated control logic for communication with a corresponding modem and control system in the

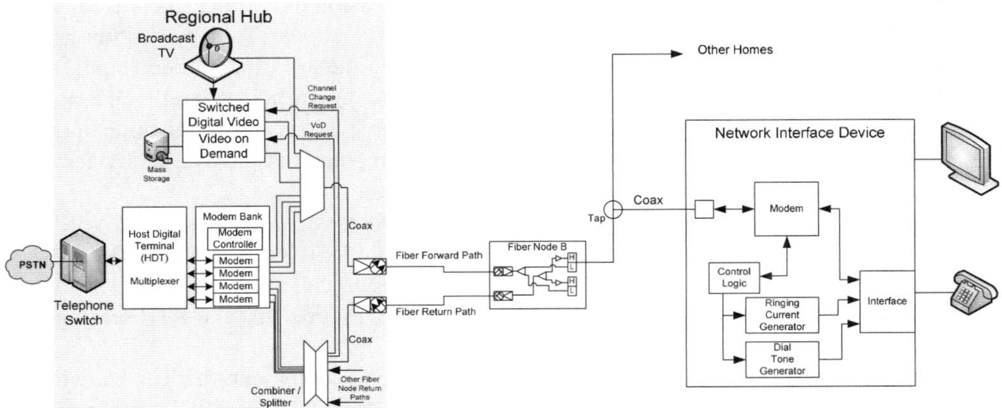

Figure 3-16. Cable telephony prior to voice over IP technology

regional hub. This in turn is connected via a time division multiplexor, known as a host digital terminal (HDT), to a Class 5 telephone switch connected to the public switched telephone network (PSTN).

A further problem that also needed to be overcome was powering of the NID itself [13]. Many countries stipulate that lifeline telephone service must be maintained when the utility power service has failed. Several solutions have been employed based on local regulations and layout of the network infrastructure; these include:

- Providing power using the center conductor of the drop cable
- Providing power using a twisted pair bonded to the outside of the drop cable
- Providing power from the home supply with a battery backup

With the implementation of DOCSIS for data communication and the maturing of VoIP technology, most of these systems have now given way to voice service as defined by CableLabs' PacketCable specifications.

3.5.1 Cable IP Telephony

IP telephony or voice over IP (VoIP) came into existence as a means of providing unregulated low-cost voice communications via the Internet. Following significant amounts of investment by network equipment vendors and operators alike, VoIP is now viewed by all types of network operators as the technology that will enable greater network efficiency, richer end-user features, and convergence with other online services.

The first generation of cable IP telephony call control, as defined by the PacketCable 1.0 and 1.5 specifications and the SCTE IPCablecom specifications [31], are based on the IETF's Media Gateway Control Protocol [1]. Other than lower subscription and call charges the subscriber should see no difference in the service when compared to the PSTN; hence, this generation of VoIP technology is referred to as PSTN emulation. The second generation is specified within PacketCable 2.0 and is based on more recent standards, including the Session Initiation Protocol (SIP) and 3GPP's IP Multimedia Subsystem (IMS) specifications. This generation is designed to enable convergence of IP services through the use of a common signaling mechanism (the SIP protocol) with endpoints having sufficient intelligence to interact with other services. To the subscriber, simple voice communication has similarities with the PSTN but is usually only one component of the overall service; hence this generation is referred to as PSTN simulation.

Encoding and transmission of the analogue voice signal between the VoIP endpoints and across the IP network makes use of the Real Time Protocol (RTP). The Session Description Protocol (SDP) is used by the call set-up mechanism to communicate to the endpoints the IP address and UDP ports that the RTP session will be using.

To enable the continued use of existing telephone equipment, devices known as media terminal adapters (MTA) are connected to the CM Ethernet interface to convert voice and signaling information between the analogue and VoIP domains.

To reduce cost and simplify operational procedures an MTA is typically housed in the same enclosure as the cable modem and is known as an embedded MTA or E-MTA. This combined device presents one Ethernet connection for data access and one or more analogue ports for connection to PSTN-type telephone equipment. As with HDTs, several mechanisms were considered for providing power to the MTA during times of local utility power failure. In North America, SCTE mandated [32] that E-MTAs must support options where power can be provided by cable operators via the HFC network or using local power with battery back-up. In the case of battery back-up, the MTA is required to send an alarm when there is a local power failure or when the battery is nearing the end of its life. A range of batteries are available that will provide power for between 4 and 24 hours.

3.5.1.1 Network Control Signaling PacketCable 1.0 and 1.5 The goal of PacketCable 1.x is to provide a highly reliable primary line voice service for residential subscribers that will scale using general purpose computer hardware. It differentiates itself from other IP telephony services by the fact that it is a phone-to-phone service that does not use the public Internet and that employs the quality of service characteristics of the DOCSIS access network. PacketCable 1.0 was the first specification to be produced by CableLabs that specified the mechanisms for voice service using embedded MTAs. PacketCable 1.5 made several enhancements to the original specification and added support for SIP signaling between core network elements such as CMS to CMS and CMS to unified messaging; it does not specify the use of SIP for endpoint signaling.

Referring to Figure 3-17, the call agent within the call management server (CMS) maintains control of all of its provisioned media terminal adapters (MTA) via an MGCP signaling path. This path is used by the MTA to signal to the call agent when the telephone is off-hook and when digits are selected on the telephone's keypad. It is also used by the call agent to instruct the MTA to generate various

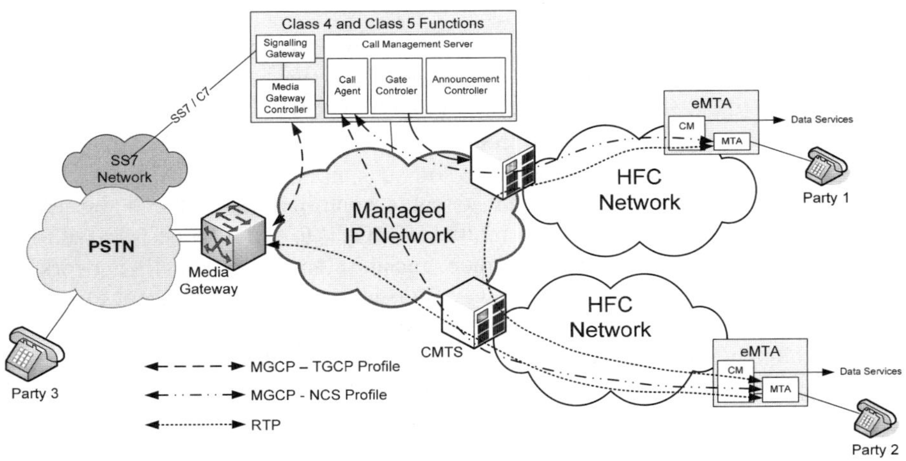

Figure 3-17. IPCablecom and PacketCable 1.0 architecture

tones such as dial-tone, ringback, and busy-tone, etc., and to generate ringing current to signal an incoming call. This type of centrally controlled signaling is referred to as Network-based call signaling or NCS.

The call agent refers to individual endpoints on an MTA via a unique reference that is a concatenation of the local name for the endpoint within the MTA, the "@" symbol, and the MTA's fully qualified domain name (FQDN). The PacketCable naming convention [34] for an FXS or analogue access line (aaln) endpoint on an MTA is `aaln/x` where "`x`" is a number from one to the number of endpoints on the MTA. The naming convention for the MTA itself is often the MTA's MAC address combined with the domain name of the operating Cable MSO. Using the convention described, an example of a reference to the second FXS port on an MTA with MAC address 00:11:22:33:44:55 that is operated by mycableco.com would be: aaln/2@0-6-00-11-22-33-44-55.mta.mycableco.net. To set up calls to the correct MTA interface or to know which telephone number is now off hook, the call agent maintains a table that maps directory number (i.e., telephone number) to FQDN and interface identifier.

The media gateway (MG) performs transmission adaptation between the TDM bearer network of the PSTN and Real Time Protocol (RTP) streams on the IP network. The PSTN uses 64 kbps of bandwidth that is time division multiplexed (TDM) within a digital hierarchy to carry digitized voice signals. In North America, each 64 kbps channel is known as a digital signal 0 or DS0 with 24 channels in a DS1 and 28 DS1s in a DS3. European networks refer to a 64 kbps channel as an E0 with 30 channels in an E1.

To provide signaling adaptation, the call agent is configured with a 1-to-1 relationship between the inter machine trunk (IMT) channel or circuit identifier used by the PSTN's SS7/C7 signaling layer and the corresponding FQDN used by the VoIP network. The IP naming convention [33] for each MG trunk channel or circuit is similar to that of an MTA with the local name for the endpoint defined as:

```
ds/<unit-type1>-<unit #>/<unit-type2>-<unit #>/.../<channel #>
```

Where the first "ds" indicates the naming scheme being used, the unit-type and unit number refers to the hierarchy of the connected trunk (such as ds1, ds3, E1, E3, etc.) and channel is the ds0 or E0 being referred to. Using this convention ds/ds1–4/14@ mg2.mycableco.net refers to the 14th channel on the 4th DS1 connected to media gateway 2.

The call agent uses the media gateway controller (MGC) and the Trunk Gateway Control Protocol (TGCP) profile of MGCP to provide the MG with the SDP parameters it needs for RTP voice transmission across the IP network to the relevant MTA. The signaling gateway (SG) is used by the call agent to inform the PSTN of the IMT bearer channel being used.

As already stated, consistently good call quality is one of the key design goals of PacketCable telephony. In a packet-based network voice quality is mostly affected by the variability of the time taken to transmit packets of information that contain encoded speech. This delay variability is more commonly known as jitter. As the path between the CMTS and the MTA is a shared DOCSIS medium and therefore subject to unpredictable traffic loads, a mechanism called dynamic quality-of-service, or D-QOS, is employed to protect the packets containing voice information. D-QOS makes use of dynamic service flow functions within DOCSIS and was first introduced in DOCSIS 1.1. After an MTA has registered with its call agent, typically following its initial boot sequence, the gate controller (in the CMS) uses a PacketCable Multi-Media (PCMM) interface and specifically the COPS protocol to signal to the CMTS to authorize and allocate service flow gates on behalf of the newly registered MTA. During the call establishment phase and just prior to sending ringing tone (or ringing current if being called), the call agent instructs the MTA to tell its attached CM to make dynamic service addition (DSA) requests to the CMTS to reserve some of the authorized resources by creating "service flows" for both the upstream and downstream paths; at this point the service flows are said to be in an "admitted" state. When the call is answered the MTA is instructed to request that the CMTS commit the reserved resources ensuring that the voice information is then protected; at this point the service flows are said to be in an "active" state. At the end of the call the MTA is instructed to tell its attached CM to inform the CMTS using a dynamic service deletion (DSD) request that the committed resources are no longer required.

The call set-up behavior of NCS cable IP telephony is best described using two simplified examples: an on-net call where Party 1 calls and communicates with Party 2 and an off-net call where Party 2 calls and communicates with Party 3. It should be noted that for clarity, responses to messages have been left out of the descriptions.

Referring to Figure 3-17, Party 1 is making a call to Party 2. Both parties are using embedded MTAs (E-MTA), where the CM and MTA are housed in the same enclosure and the MTA is able to signal QoS requests to the CM using MAC control messages. In this example both parties are connected to the same CMS. The call set-up and D-QOS behavior, adapted from [17], is described as follows:

1. At the end of the last call, the call agent sends a request notification (RQNT) to the MTA to instruct it to play dial-tone when the phone is off-hook, collect dialed digits as the user presses keys on the telephone key pad, and send the collected digits to the call agent via a notify (NTFY) message.
2. Party 1 takes the phone off-hook and dials Party 2. The MTA sends a NTFY to the call agent with the collected digits.
3. If not already done at MTA boot time, the gate controller in the CMS uses the COPS protocol to authorize the CMTS that is supporting Party 1's MTA to allocate resources when requested by the MTA. The CMTS returns a Gate-ID.
4. The call agent sends a create connection (CRCX) to Party 1's MTA telling it to create a connection and reserve resources on its attached CMTS. The connection is created in an inactive state as, at this time, the Session Description Protocol (SDP) parameters for the party being called are unknown.
5. Using the Gate-ID in the CRCX message Party 1's MTA uses the MAC control service, which is only available to embedded MTAs (E-MTA), to

instruct the CM to send a dynamic service add message to the CMTS. The CM sends the DSA to the CMTS, which confirms authorization and reserves resources.

6. If not already done at MTA boot time, the gate controller in the CMS uses the COPS protocol to authorize the CMTS that is supporting Party 2's MTA to allocate resources when requested by the MTA. The CMTS returns a Gate-ID.
7. The call agent sends a CRCX to Party 2's MTA with Party 1's SDP parameters and the Gate-ID to reserve resources. The connection is created in an inactive state where Party 2's telephone does not ring.
8. Using the Gate-ID in the CRCX message the MTA uses the MAC control service to instruct the CM to send a dynamic service add message to the CMTS. The CM sends the DSA to the CMTS, which confirms authorization and reserves resources.
9. The call agent sends a modify connection (MDCX) to Party 1's MTA with Party 2's SDP parameters.
10. The call agent then sends RQNT messages to both parties' MTAs instructing Party 1's MTA to play ringback and Party 2's MTA to send ringing current to sound the ringer or bell on the called party's phone.
11. When the called party answers the telephone, Party 2's MTA sends a NTFY to the call agent to indicate the called telephone is now off-hook.
12. The call agent sends MDCX to both MTAs to change the connection from inactive to "sendrecv" mode and to instruct their respective CMs to commit the reserved resources.
13. Both MTAs instruct their CMs to send dynamic service change (DSC) messages to their respective CMTSs to commit the reserved DQOS resources, thereby protecting the voice traffic.
14. Real Time Protocol (RTP) messages flow between the MTAs across the IP backbone.
15. RTP traffic may have begun to flow in an unprotected manner prior to the resources being committed.
16. At the end of the call the calling party goes on-hook causing Party 1's MTA to send a NTFY to the call agent.
17. The call agent sends a delete connection (DLCX) to both MTAs.
18. Both MTAs instruct their respective CMs to send dynamic service delete (DSD) messages to their respective CMTSs to release the DQOS resources.
19. The call Agent then sends RQNT to both MTAs to put them in a state that is ready for the next call.

An off-net call is where a telephone on the VoIP network calls a telephone on the PSTN or where the PSTN is used to connect VoIP networks that are operated by different operators. Call set-up behavior is similar to an on-net call, but with the media gateway controller (MGC) and the media gateway acting as a proxy MTA to both the call agent and the on-net MTA. The call set-up behavior, adapted from [17], is described as follows:

1. At the end of the last call, the call agent sends a request notification (RQNT) to the MTA to instruct it to play dial-tone when the phone is off-hook, collect dialed digits as the user presses keys on the telephone key pad, and send the collected digits to the call agent via a notify (NTFY) message.
2. Party 2 takes the phone off-hook and dials Party 3. The MTA sends a NTFY to the call agent with the collected digits.
3. If not already done at MTA boot time, the gate controller in the CMS uses the COPS protocol to authorize the CMTS that is supporting Party 2's MTA to allocate resources when requested by the MTA. The CMTS returns a Gate-ID.
4. The call agent sends a create connection (CRCX) to Party 2's MTA telling it to create a connection and reserve resources on its attached CMTS. The connection is created in an inactive state as, at this time, the Session Description Protocol (SDP) parameters for the party being called are unknown.
5. Using the Gate-ID in the CRCX message Party 2's MTA uses the MAC control service to instruct the CM to send a dynamic service add message to the CMTS. The CM sends the DSA to the CMTS, which confirms authorization and reserves resources.
6. Having looked in its routing database the call agent establishes that the called party is off-net and instructs the MGC to establish the call via the PSTN.
7. The MGC determines the correct MG for the call (typically an MGC will control several media gateways) and sends a CRCX for a specific channel/circuit.
8. The MGC also instructs the SG to send an SS7 message that informs the PSTN switch at the far end of the PSTN of the circuit that is being used.
9. The MGC and CA send MDCX messages to the MG and Party 2's MTA, respectively, to enable one-way communication so that Party 2 can hear any tones or announcements sent back from the PSTN.
10. The MGC is informed via the SG that Party 3's phone is ringing. In addition, ringback tone is sent by Party 3's switch along the trunk to Party 2's MTA and phone.
11. When Party 3 answers the phone, the SG informs the MGC and the MGC and call agent send an MDCX to the MG and MTA to put the connection into "sendrecv" mode.
12. Party 2's MTA instructs its CMs to send a dynamic service change (DSC) message to its CMTS to commit the reserved DQOS resources, thereby protecting the voice traffic.
13. Real Time Protocol (RTP) messages flow between the MTA and MG across the IP backbone.

14. RTP traffic may have begun to flow in an unprotected manner prior to the resources being committed.
15. At the end of the call the calling party goes on-hook, causing Party 2's MTA to send a NTFY to the call agent.
16. The call agent sends a delete connection (DLCX) to Party 2's MTA and instructs the MGC to tear down the call.
17. Party 2's MTA instructs its CM to send dynamic service delete (DSD) messages to its CMTS to release the DQOS resources.
18. The call agent then sends RQNT to Party 2's MTA to it in a state that is ready for the next call.
19. The MGC uses the SG to signal Party 3's switch that the call has ended and to release the trunk. It also sends a DLCX to the MG to release the trunk.

3.5.1.2 Distributed Call Signaling As discussed earlier, network controlled signaling is effectively an emulation of the behavior of the PSTN with all functions and resources controlled centrally. While this mechanism was effective in proving that voice communication could be provided using Internet protocols for a mass market, it is directly at odds with the goals of the Internet, which is largely based on a distribution of intelligent endpoints that are able to autonomously initiate communications with multiple services on an as-needed basis.

In 1999 the IETF published RFC2543 [18], which describes the operational behavior of the Session Initiation Protocol (SIP) that is used by endpoints on the Internet to signal each other to establish a session for multimedia communication. Since first publication, SIP has grown in popularity and is now the de facto signaling protocol for all packet-based multimedia communication services. As a precursor to PacketCable 2.0, CableLabs produced a draft specification for distributed call signaling [6]. Although this specification is now deprecated by CableLabs and was never standardized by SCTE, several implementations were put into operation, mostly outside of North America, and are still in use today.

DCS uses many of the same components as an NCS network with the primary difference being that the call agent in a CMS is replaced by a DCS-proxy (DP) that does not maintain state and is only involved in the call set-up phase of a multimedia communication session. In a typical call set-up, the MTA or SIP endpoint will signal to the DP (using SIP) that it wishes to contact another SIP endpoint. The DP performs the role of a trusted decision point and routes the request based on internal routing rules to either the PSTN via an MGC and MG or another SIP endpoint that is either directly connected or contactable via another DP (Figure 3-18).

3.5.1.3 Embedded MTA Start-up As an embedded MTA or eMTA is an IP device, it must first wait for its associated cable modem to complete its boot process, see 3.4.5 above, before it can commence its own start-up sequence. For embedded MTAs the PacketCable 1.5 MTA device provisioning specification [8] extends the

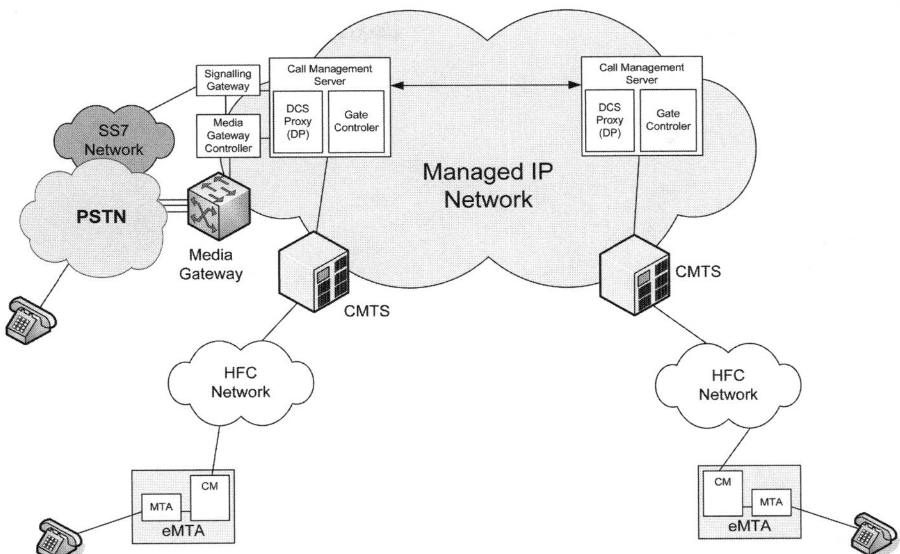

Figure 3-18. Simplified distributed call signaling architecture

DOCSIS CM boot process to include the use of DHCP Option 122 to provide the MTA element with its required network server addresses (DHCP, Provisioning and Kerberos server) and to instruct it to perform either a secure, basic, or hybrid start-up flow.

In all three start-up flows the eMTA is ultimately instructed to download a telephony configuration file that contains the parameters it needs for operation on the cable provider's network and, once configured, inform the provisioning server that telephony provisioning is now complete. The differences are the mechanisms used to ensure that the provisioning process is secure and to prevent malicious masquerading of another user's telephone device. The primary characteristics for each flow are:

1. Secure flow—implements Kerberos mutual authentication between the MTA and the provisioning system and makes use of SNMPv3 for secure messaging.
2. Basic flow—similar to DOCSIS CM provisioning and makes no use of Kerberos or SNMPv3 security features.
3. Hybrid flow—similar to the secure flow but without the use of Kerberos and with SNMPv2c instead of SNMPv3.

In view of the complexities associated with operating Kerberos servers and diagnosing fault conditions, many cable MSOs have opted to use basic and hybrid start-up flows only.

3.5.1.4 PacketCable 2.0 The desire to combine voice, video, data and mobile services in a manner that will generate new service revenues and protect existing

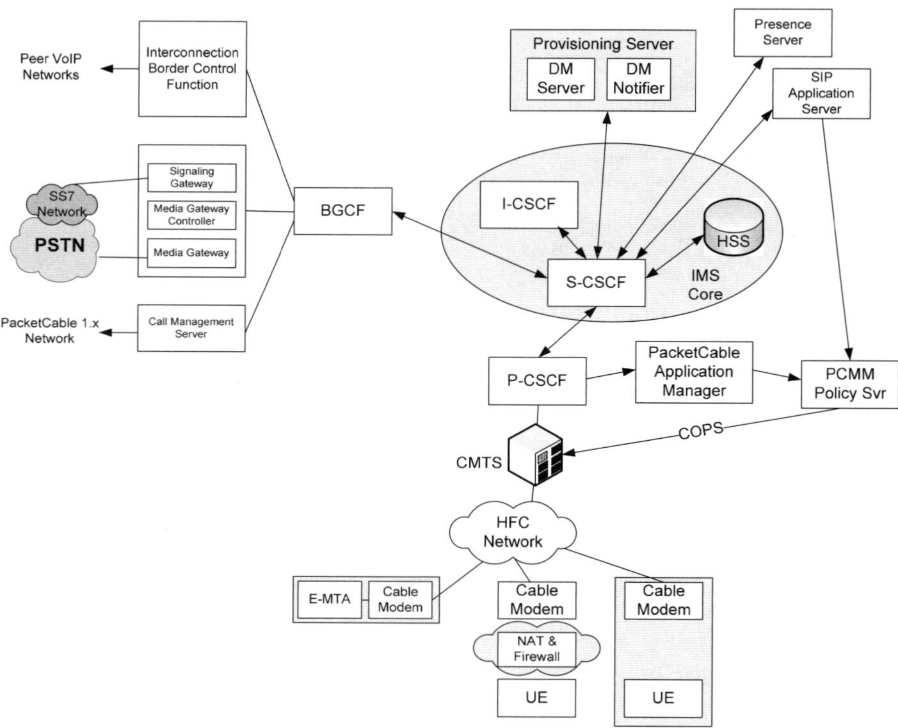

Figure 3-19. Simplified PacketCable 2.0 reference architecture

ones is the primary motivation for the cable industry adopting the 3rd Generation Partnership Project's (3GPP) IP Multimedia Subsystem (IMS) specification under the umbrella title of PacketCable 2.0 (Figure 3-19). PacketCable 2.0 is based on Release 7 of IMS and will track major changes as the specification matures [4].

To adapt IMS to a cable environment several extensions were made to the originating specification such as support for QoS using PacketCable Multimedia, provisioning, and activation of user equipment and the requirements of the regulatory authorities.

PacketCable 2.0 formally introduced SIP as the endpoint signaling protocol for cable network user equipment (UE) and replaced the intra-network functions of a CMS with a:

- Serving-call session control function (S-CSCF) and a SIP application server (AS) in place of a call agent to perform the function of a trusted decision point
- PCMM policy server and PacketCable Application Manager in place of a gate controller to provide QoS for DOCSIS attached user equipment.

It also introduces the concept of public and private user identities to enable subscribers to roam across all network access types and to use other operators' networks while still accessing their home provider's services.

Rather than describe the behavior of an IMS network, which is already available in many excellent publications, this chapter will focus on the functions that are specific to a cable network; namely configuration of User Equipment and setting of QoS across the DOCSIS network.

PacketCable 2.0 caters to two classes of user equipment, embedded user equipment (E-UE) and stand-alone user equipment (UE).

1. Embedded devices are provisioned using similar mechanisms to that of PacketCable 1.5 eMTAs, see 3.5.1.3 above and PacketCable 2.0 E-UE provisioning framework [9].
2. Stand-alone and soft UE provisioning is based on the Open Mobile Alliance Device Management (OMA-DM) initiative and is specified in the PacketCable 2.0 UE Provisioning Framework [10].

OMA was established in 2002 by mobile operators, interested software vendors, and mobile device manufacturers to advance the development of mobile service interoperability enabling specifications. 3GPP IMS specifies the use of OMA-DM as the mechanism for managing device firmware updates, parameter configuration, diagnostics, and security.

As PacketCable 2.0 is based on SIP and IMS, it uses relevant aspects of the OMA-DM specifications to address the additional challenges of stand-alone UEs such as NAT (Network Address Translation) traversal, managing devices that are not directly connected to the managing cable provider's network and data synchronization for devices that are designed to work in both online and offline modes.

The UE Provisioning specification refers to two provisioning models, "pre-configured," where the UE is pre-configured with basic service provider information prior to making contact with the network, and "dynamic configuration," where the UE has no pre-configuration. To satisfy the needs of initial PacketCable 2.0 deployments, the UE Provisioning specification currently caters for the pre-configured model of provisioning only.

Referring to Figure 3-20, management sessions are started by the DM client using HTTP requests to the DM server. Server-initiated sessions are established by the DM server sending a notification to the client via an OMA Push that contains either content in a short message or an instruction to start a management dialogue. PacketCable 2.0 delivers notifications via a SIP Push mechanism that uses the SIP MESSAGE method as described in RFC3428 [12].

The DM server implements the protocols required for communicating with UEs and with other servers used for the management of the UE population. It also implements the SyncML (Sync Markup Language) representation and synchronization protocols, and associated XML DTDs to perform synchronization of information between itself and the UE.

SyncML[2] is a vendor initiative that provides a symmetric mechanism for synchronizing mobile device data to networked data and vice versa. It is designed to cater to devices that can be frequently offline and where either networked data,

[2]The SyncML initiative is now merged into the Open Mobile Alliance. Specifications are available at http://www.openmobilealliance.org/tech/affiliates/syncml/syncmlindex.html.

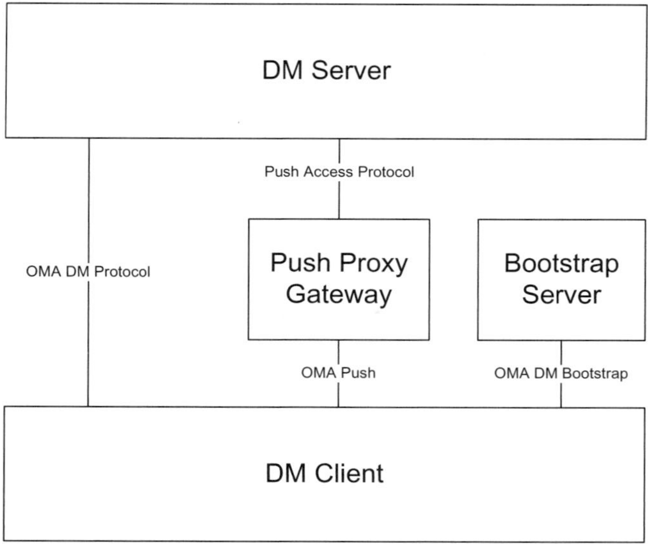

Figure 3-20. OMA DM logical architecture [10]

such as configuration parameters, or device data, such as contact lists, have been updated and need to be reconciled between endpoints. Although SyncML is transport independent it uses HTTP as the transport protocol for IP-based devices and therefore is the mechanism used for PacketCable2.0 UEs.

The Push Proxy Gateway is responsible for delivering push content or notifications to the DM client. In doing so it may need to translate the client address as provided by the DM server to a meaningful network address such as a SIP URI. If the UE is offline it is also required to perform a store and forward operation.

The bootstrap server is a logical function that is responsible for providing the DM client with credentials to enable it to establish connectivity with the DM server. DM clients can either be pre-configured with connectivity details for the bootstrap server or DM server credentials such that it does not need to contact the bootstrap server at all.

The combined OMA-DM entities of the DM server, Push Proxy Gateway and the bootstrap server collectively form the PacketCable 2.0 Provisioning Server (Figure 3-21). As previously stated, DM sessions are established by the UE making an HTTP request to the DM Server. If there is no session established and the provisioning server needs to update the UE, the provisioning server sends a notification known as Package #0 to the UE. In this model the entity that sends notifications to the UE using SIP push via the SIP MESSAGE method is known as the DM notifier.

The following sequence outlines the start-up dialogue between a PacketCable 2.0 UE and its associated provisioning server:

1. At power-up the UE reads its internal pre-configured bootstrap and performs DHCP functions to obtain an IP address and any associated DHCP options.

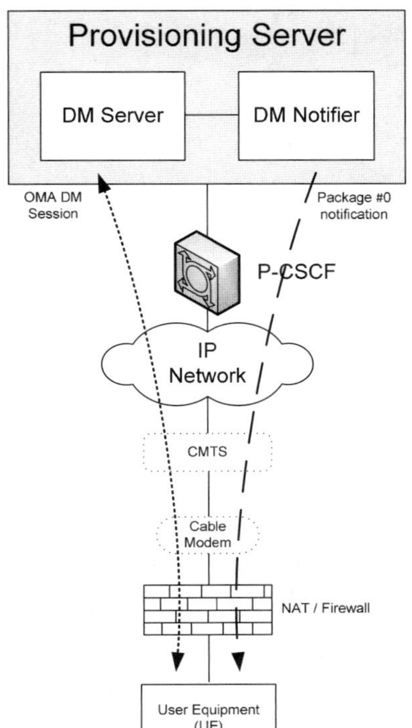

Figure 3-21. Packet Cable 2.0 Provisioning Server and UE Interaction

2. Having established IP connectivity the UE registers with the P-CSCF having discovered its address either via bootstrapping or DHCP options.
3. The IMS core notifies the provisioning server that the UE has SIP registered. It does so either via a third-Party REGISTER (as the provisioning server is effectively an IMS application server) or via a "reg" event package if the UE has performed a subscription to the registration-state event package, see Session Initiation Protocol (SIP) and Session Description Protocol (SDP); Stage 3 Specification, [11].
4. Having received a notification from the IMS core that the UE is registered, the provisioning server's DM notifier sends a Package #0 message, using a SIP MESSAGE request, to the UE instructing it to start an OMA DM session. The DM notifier is able to send an unsolicited message to the UE and perform any required NAT traversal as a NAT pinhole was created at the time the UE Registered with the P-CSCF and optionally subscribed to the registration-state package.
5. The UE sends an HTTP request to the DM Server to initiate an OMA DM session whereby any required configuration and or updating of information occurs.

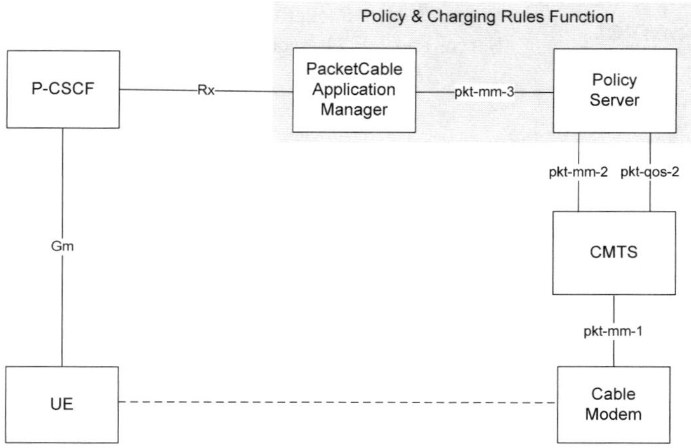

Figure 3-22. PacketCable 2.0 QoS architecture reference model

During session initiation, such as making a voice call or starting a video conference, a SIP INVITE is sent by the UE to the P-CSCF. This in turn sends a DIAMETER request to the PacketCable Application Manager (PAM), which, in conjunction with the Policy Server (PS), determines whether there is a matching policy for the session being established. If there is, the PS pushes commands to the CMTS using PCMM COPS messages, to authorize reservation and commitment of resources based on the configured policy.

The combined capability of the PacketCable Application Manager and the policy server fulfills similar functions to that of the Policy and Charging Rules Function (PCRF) defined within the 3GPP specifications. Within IMS, the PCRF authorizes resources and instructs the network as to how it should treat specific application streams.

The QoS reference model in Figure 3-22 and the descriptions in Table 3-2 from [5] show the relationships between the various elements that are responsible for implementing QoS functions.

Comparing the architecture to that of PacketCable 1.x, the PAM and PS perform a similar function to the gate controller, with the CM still issuing DSx commands as a result of MAC control messages from the UE. As with the NCS call agent, the P-CSCF is the initiator of the QoS authorization and reservation process.

3.6 WIRELESS

Having succeeded in bundling data and voice services with TV services to create a triple-play bundle, many cable MSOs looked to bring to market a quad-play offering by including cell phone service as well. However, other than notable exceptions such as Rogers Communications Inc. in Canada and Cox Communications in the United States, the majority of cable MSOs function as mobile virtual network operators (MVNO) as they do not own licensed frequency spectrum. As

TABLE 3-2. PacketCable 2.0 QoS interface descriptions

Reference Point	Description
pkt-mm-1	Used by the CMTS to send DSx messages to the CM and vice versa to add, change, and delete DOCSIS service flows
pkt-mm-2	Used by the policy server to push policy decisions onto the CMTS. It uses the COPS protocol
pkt-mm-3	Used by the PacketCable Application Manager to request that the policy server install a policy decision on the CMTS on behalf of the UE
Gm	SIP-based interface used by the UE for registration and session control
Rx	Used by the P-CSCF for session-based policy set-up. This is the same interface as used by the P-CSCF when communicating with the PCRF in a GPRS environment
pkt-qos-2	Used by the policy server to discover the serving CMTS for a given UE

such, for most operators their quad-play offering is primarily a billing phenomenon and purely exists to prevent subscribers from moving to competing services from wire-line operators.

With the deployments of IMS increasing and the adoption of PacketCable 2.0, cable operators now have the potential to form network peering relationships. As such it is only a matter of time before more cable MSOs either purchase spectrum of their own, or merge with existing mobile network operators, to deliver their service portfolio across any form of access network technology.

3.7 CABLE FUTURES

The increasing speed of Internet access and the convergence of network signaling will inevitably minimize the significance of the access mechanism that consumers use to obtain their online services. As a result, greater focus is being placed on the content that makes up a service and the devices that will be used to consume it. In many cases content originators and telecom companies refer to their offering as being available via three screens—the TV, PC and mobile handset—with several operators embracing the mantra that services should be able to be consumed everywhere, and all for a single subscription fee.

Historically, cable MSOs have shown strength in packaging third-party content, and originating content of their own, for delivery to a widely distributed customer base. More recently they have demonstrated their ability to adopt new technology and grab significant market share in the areas of broadband Internet access and the provision of consumer voice services. Now with the availability of additional spectrum for broadband wireless, greater interactivity to generate new advertising revenues, and the transformation of services aimed at the business community the opportunity exists for them to grab an even bigger market share.

As they rise to this challenge we can expect a more rapid adoption of packet-based transmission and Next Generation signaling under the umbrella of PacketCable 2.0, and we are likely to see that:

- More content will be put online that will be accessible at no extra charge for existing subscribers and that will be pay-per-view for non-subscribers. This will give rise to increased demand for higher access speeds resulting in greater penetration of DOCSIS 3.0 as well as reducing combining ratios still further.
- Business services will open another battlefront between cable and wire-line and will potentially see cable operators competing with each other and mobile network operators as well. To enable them to better serve medium and large-scale enterprises, cable operators will adopt greater use of fiber to the premise and will also bring to market services including hosted unified communications, carrier Ethernet and IT functions such as security and file backup and restore.
- Changing lifestyle and working behaviors will see greater emphasis on place shifting where leisure and business services are accessible from any location. This will see cable operators place a greater emphasis on identity systems and PacketCable 2.0 to enable users to access content via any network provider.
- The need to obtain a greater share of advertising revenues will increase focus on enhancing TV interactivity as well as collecting usage data to better target programming and advertising. This may prove to be the catalyst that will see greater use of IPDR as a usage collection mechanism and more development in the area of IPDR mediation systems.

Finally, it should be evident from the various services and technologies described in this chapter that cable MSOs can no longer be considered as minor players in the telecommunications industry. Moreover, in several parts of the world they are well positioned to become leading providers themselves.

3.8 REFERENCES

1. Andreasen F, Foster B. 2003. Media Gateway Control Protocol (MGCP) Version 1.0. RFC 3435. Fremont, CA: Internet Engineering Task Force (IETF).
2. Advanced Television Systems Committee. 2006. ATSC Standard: Program and System Information Protocol for Terrestrial Broadcast and Cable. ATSC A/65. Washington, DC: Advanced Television Systems Committee.
3. Benoit H. 2002. *Digital Television: MPEG-1, MPEG-2 and Principles of the DVB System*, Second Edition. Woburn, MA: Focal Press.
4. CableLabs. 2007. PacketCable™ 2.0 Architecture Framework Technical Report. PKT-TR-ARCH-FRM-V03–070925. Louisville, CO: Cable Television Laboratories, Inc.
5. CableLabs. 2008. PacketCable™ 2.0 Quality of Service Specification. PKT-TR- PKT-SP-QOS-I02–080425. Louisville, CO: Cable Television Laboratories, Inc.
6. CableLabs. 2000. PacketCable™ Distributed Call Signaling Specification. PKT-SP- DCS-D03-QOS-000428. Louisville, CO: Cable Television Laboratories, Inc.

7. CableLabs. 2008. PacketCable™ Specification Multimedia Specification. PKT-SP-MM-I04–080522. Louisville, CO: Cable Television Laboratories, Inc.
8. CableLabs. 2009. PacketCable™ 1.5 Specification MTA Device Provisioning. PKT-SP-PROV1.5-I04–090624. Louisville, CO: Cable Television Laboratories, Inc.
9. CableLabs. 2009. PacketCable™ 2.0 E-UE Provisioning Framework Specification. PKT-SP-EUE-PROV-I03–090528. Louisville, CO: Cable Television Laboratories, Inc.
10. CableLabs. 2008. PacketCable™ 2.0 UE Provisioning Framework Specification. PKT-SP-UE-PROV-I01–080905. Louisville, CO: Cable Television Laboratories, Inc.
11. CableLabs. 2008. PacketCable™ IMS Delta Specifications. Session Initiation Protocol (SIP) and Session Description Protocol (SDP); Stage 3 Specification 3GPP TS 24.229. PKT-SP-24.229-I05–090528. Louisville, CO: Cable Television Laboratories, Inc.
12. Campbell E, Rosenberg J, Schulzrinne H, et al. 2002. Session Initiation Protocol (SIP) Extension for Instant Messaging. RFC 3428. Fremont, CA: Internet Engineering Task Force (IETF).
13. Ciciora W, Farmer J, Large D, et al. 2004. *Modern Cable Television Technology*, Second Edition. San Francisco: Morgan Kaufmann.
14. Digital TV Group. 2009. http://www.dtg.org.uk/reference_si.html. http://www.dtg.org.uk. The Digital TV Group Limited (accessed March, 1, 2009).
15. CableLabs. 2007. OpenCable™ Specifications ETV. Enhanced TV Binary Interchange Format 1.0. OC-SP-ETV-BIF1.0-I04–070921. Louisville, CO: Cable Television Laboratories, Inc.
16. ETSI. 1998. Digital Video Broadcasting (DVB); Framing structure, channel coding and modulation-for cable systems. EN 300 429 V1.2.1 (1998–04). Sophia Antipolis, France: European Telecommunications Standards Institute (ETSI).
17. Evans DR. 2001. *Digital Telephony Over Cable: The PacketCable™ Network*. Indianapolis, IN: Addison-Wesley Professional.
18. Handley M, Schulzrinne H, Schooler E, et al. 1999. SIP: Session Initiation Protocol. RFC 2543. Fremont, CA: Internet Engineering Task Force (IETF).
19. ISO/IEC. 2000. Information technology—Generic coding of moving pictures and associated audio (MPEG-2 system, video, audio). ISO/IEC-13818–1, -2 & -3. Geneva, Switzerland: International Organization for Standardization (ISO).
20. ISO/IEC. 1998. Information technology—Generic coding of moving pictures and associated audio information—Part 6: Extensions for DSM-CC. ISO/IEC-13818–6. Geneva, Switzerland: International Organization for Standardization (ISO).
21. ITU-T Recommendation J.83: Cable networks and transmission of television, sound programme and other multimedia signals, 2007. Geneva, Switzerland: International Telecommunication Union Telecommunication Standardization Bureau.
22. Kucera D. 2002. *Introduction to MPEG-2 Compression and Transport Streams*. Beaverton, OR: Tektronix, Inc.
23. Plumer DC. 1982. An Ethernet Address Resolution Protocol. RFC 286. Fremont, CA: Internet Engineering Task Force (IETF).
24. Robichaux M. 2002. *Cable Cowboy: John Malone and the Rise of the Modern Cable Business*. Hoboken, NJ: John Wiley & Sons.
25. Society of Cable Telecommunications Engineers. 2009. Digital Broadband Delivery System: Out of Band Transport Part 1: Mode A. ANSI/SCTE 55–1. Exton, PA: Society of Cable Telecommunications Engineers.
26. Society of Cable Telecommunications Engineers. 2008. Digital Broadband Delivery System: Out of Band Transport Part 2: Mode B. ANSI/SCTE 55–2. Exton, PA: Society of Cable Telecommunications Engineers.
27. Society of Cable Telecommunications Engineers. 2005. DOCSIS 1.1 Part 1: Radio Frequency Interface. ANSI/SCTE 23–1. Exton, PA: Society of Cable Telecommunications Engineers.
28. Society of Cable Telecommunications Engineers. 2007. DOCSIS 1.1 Part 2: Baseline Privacy Plus Interface. ANSI/SCTE 23–2. Exton, PA: Society of Cable Telecommunications Engineers.
29. Society of Cable Telecommunications Engineers. 2007. DOCSIS Set-Top Gateway (DSG) Specification. ANSI/SCTE 106. Exton, PA: Society of Cable Telecommunications Engineers.
30. Society of Cable Telecommunications Engineers. 2007. Host-POD Interface Standard. ANSI/SCTE 28. Exton, PA: Society of Cable Telecommunications Engineers.

31. Society of Cable Telecommunications Engineers. 2006. IPCablecom 1.0 Part 1: Architecture Framework for the Delivery of Time-Critical Services Over Cable Television Networks Using Cable Modems. ANSI/SCTE 24–1. Exton, PA: Society of Cable Telecommunications Engineers.
32. Society of Cable Telecommunications Engineers. 2007. IPCablecom 1.5 Embedded MTA Analog Interface and Powering. ANSI/SCTE 24–14. Exton, PA: Society of Cable Telecommunications Engineers.
33. Society of Cable Telecommunications Engineers. 2006. IPCablecom 1.0 Part 12: Trunking Gateway Control Protocol (TGCP). ANSI/SCTE 24–12. Exton, PA: Society of Cable Telecommunications Engineers.
34. Society of Cable Telecommunications Engineers. 2006. IPCablecom 1.0 Part 3: Network Call Signaling Protocol for the Delivery of Time-Critical Services over Cable Television Using Data Modems. ANSI/SCTE 24–3. Exton, PA: Society of Cable Telecommunications Engineers.
35. Society of Cable Telecommunications Engineers. 2005. SCTE Application Platform Standard OCAP 1.0 Profile. ANSI/SCTE 90–1. Exton, PA: Society of Cable Telecommunications Engineers.
36. TM Forum. 2009. IPDR Streaming Protocol (IPDR/SP) Specification Release 1.0. TMF8000-IPDR-IIS-PS. Morristown, NJ: TM Forum.
37. Wikipedia contributors, "CableCARD," Wikipedia, The Free Encyclopedia, http://en.wikipedia.org/wiki/CableCARD (accessed April 11, 2009).
38. Wikipedia contributors, "DOCSIS," Wikipedia, The Free Encyclopedia, http://en.wikipedia.org/wiki/Docsis#Speed_Table (accessed August, 26, 2009).

CHAPTER 4

NEXT GENERATION TECHNOLOGIES, NETWORKS, AND SERVICES[1]

Bhumip Khasnabish

4.1 INTRODUCTION

This chapter presents an overview of Next Generation (NG) technologies, networks, and services with particular reference to their architecture and Management (FCAPS[2]) requirements. It is argued that *convergence* is the driving force behind the emerging generation of technologies, network elements, architectures, services, and their management. Therefore, both networks and services management must take an end-to-end view in design, implementation, and operations. For example, the IP Multimedia Subsystem (IMS) is being incorporated at the core of all NG services architectures, be it wireline (cable and telco) network evolution or for supporting fixed-mobile convergence in wireless networks. Similarly, NG network elements are becoming more aware of the functions across their usual layers of operations. Finally, devices are becoming fully capable of supporting any media, i.e., voice, IM/SMS (instant messaging and short message service), real- and non-real-time video, etc. These applications and services must be supported and offered in an access-agnostic fashion.

Consequently, the overall architecture and service operations/management are becoming increasingly complex. These call for the utilization of appropriate normalization and interworking modules for effective operations and management of services without directly affecting the ongoing workforce development and deployment in corporations. Consequently, industry is witnessing a newer mode/paradigm of networks and services management.

A brief overview of each of the above-mentioned trends is provided in this chapter.

[1]The ideas and viewpoint presented in this Chapter belong solely to Bhumip Khasnabish, Lexington, Massachusetts, USA.

[2]FCAPS stands for 'Fault, Configuration, Accounting, Performance, and Security.' It is ISO's (www.iso.org) model and framework for network management.

Next Generation Telecommunications Networks, Services, and Management, Edited by Thomas Plevyak and Veli Sahin

4.2 NEXT GENERATION (NG) TECHNOLOGIES

As suggested before, convergence at all levels will be the driving force behind NG telecommunications technologies. At the device level, customers want to use the same device (cell phone, TV, computer, wireline phone) for voice, data, video (both real-time and streaming), and gaming services. At the network element level, the edge and core routers are becoming more aware of applications and services (like security, fire-walling, and service adaptation). In addition, optical networking devices will be more aware of IP-based services and will need to support alignment with NG networks and services management, e.g., bandwidth-on-demand. Figure 4-1 shows high-level interaction among NG technologies, services, and workforce with push and pulls. The push usually comes from the marketing and technology groups within a company to introduce new technologies and services with an objective to generate new streams of revenues. The pull comes from the workforce (personnel) that maintains the legacy networks, elements, and the operations support systems. They like to stay within their comfort zones of network management and maintenance and are usually reluctant to take any risk.

4.2.1 Wireline NG Technologies

Wireline-based NG technologies include both NG wireline access technologies and NG devices that seamlessly support emerging converged services, irrespective of whether these services are hosted in wireline domains or in wireless domains. Telco's NG wireline access and transport technologies will include the following options. However, other technologies for packet-based NG wireline services can include advanced cable modem, NG multimedia terminal adapter, cable modem termination system, and high-end set top box as well (see, for example, Ref. [10]).

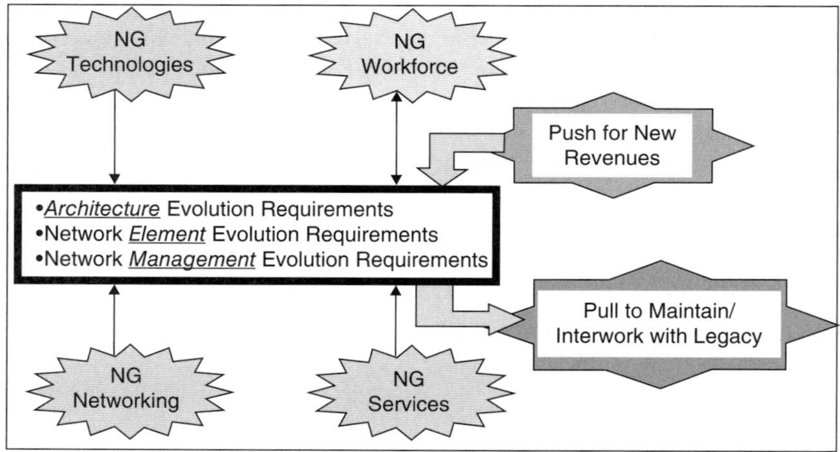

Figure 4-1. Interaction between legacy and emerging forces of network and service evolution

4.2.1.1 Fiber to the Premises (FTTP) FTTP service includes *mostly* passive optical network (PON) fiber based all-/any-purpose communications services to single-family homes, multi-tenant apartment buildings, business and education campuses, LTE[3] towers (for backhauling), etc. Broadband PON (BPON) has been specified in ITU-T's G.984 standard with the following bands and speeds: 1310 nanometer for upstream data at 155 Mbps (1.2 Gbps with Gigabit PON or GPON), 1490 nanometer for downstream data at 622 Mbps (2.4 Gbps with GPON), and 1550 nanometer for radio frequency (RF)-based (excluding IPTV) video with 870 MHz of bandwidth. In addition to supporting a higher rate and a choice of layer-2 protocol, GPON also supports enhanced security and reliability. IEEE's 10 Gigabit Ethernet PON (10G-EPON, IEEE P802.3av) will be backward compatible with Ethernet PON (EPON or GEPON, IEEE 802.3ah) and will use separate wavelengths for 10 Gbps and 1.0 Gbps downstream, but will continue to use a single wavelength for both 10 Gbps and 1.0 Gbps upstream with TDMA (time division multiple access) separation. If a hybrid TDM/WDM (time/wavelength division multiplying) system with each wavelength to feed a GPON or EPON is used, then utilizing 32 wavelengths and 1:64 split-ratio GPONs, more than 2000 customers can be served.

GPON and GEPON based FTTP are being deployed today in the United States, Europe, and Asia. The NG PON (NGPON) including 10 Gigabit Ethernet PON (10G-EPON) and wavelength-division multiplexed PON (WDM-PON) will be available within next three to five years.

As of year-end of 2009, Japan had more than 15 million FTTP/H (home) sub scribers (with more than 50 million homes passed). Europe and the United States have installed wiring and infrastructure to support more than 10 million homes with subscriber numbers around the 3 million level.

4.2.1.2 Long-Haul Managed Ethernet (over Optical Gears) This is being driven by at least two key factors. First is the emerging economic requirements to reduce the complexity (and hence capital and operational expenses) of long-haul transmission by incorporating lightweight provisioning and packet-routing capabilities in the optical transport layer. This cross-layer mechanism enables cost effective and efficient management of switching and transport resources. Second is the advancement of key photonics technologies including commercial support for higher-speed modulation, forward error correction, and wideband filtering as well as availability of standardized photonic integrate circuit modules. Once deployed, these 100 Gigabits/sec (or more) capabilities will eliminate a large number of intermediate network elements and interfaces, essentially reducing overall power/space requirements and streamlining the provisioning, operations, and management of end-to-end service quality.

NG wireline devices will support convergence, not only in terms of interfaces but also in terms of supporting features and functions to maintain seamless services. For example, routers and switches will be increasingly aware of application (including media) and service layer features and functions. Physical layer elements like optical transport devices will be increasingly aware of Ethernet and IP layer features

[3]LTE: Long Term Evolution, as defined later in this Chapter.

and functions. Application and consumer layer devices will be increasingly aware of any or all types of human machine interfaces (HMIs). For example, televisions will support bidirectional voice communications and video conference calls in addition to supporting instant messaging and email services. Wireline phones will support Web services and low-bit-rate television services in addition to continuing to support plain old telephone service (POTS). Devices that are used commonly to perform adaptation and integration functions will be more integrated with the terminal devices or will be supporting a multitude of features and functions inherently.

4.2.2 Wireless NG Technologies

Wireless NG technologies (near and beyond third generation) find their home mostly in the new generation of hand-held or wearable devices supporting real-time voice/data/video, gaming, and high-bandwidth access to location-based services and health/pollution/traffic monitoring services. Technologies that support high-bandwidth wireless access and interconnections are discussed in this section. In addition, since these devices are increasingly becoming aware of the Internet protocol, along with context, proximity, and data-driven services in both home and foreign domains, they create tremendous challenges for both handset and network equipment manufactures.

4.2.2.1 Broadband Bluetooth and ZigBee The emerging Broadband Bluetooth (IEEE 802.15.1 standard, www.ieee802.org/15/pub/TG1.html) will utilize multi-band orthogonal frequency division multiplexing (OFDM) over unlicensed 2.4 GHz frequency band to support secure, high-speed (up to 480 Mbps) and low-power communications over short distances[4] for personal/body area networking applications. These services will range from high-definition phone conversations to file transfer to gaming to highly secure transactions, as in Near Field Communications (NFC). Major challenges are related to automated device pairing, privacy and security, topology management, and preventing hijacking of the service.

Zigbee (www.zigbee.org), the IEEE 802.15.4/4b Standard, as can be found at www.ieee802.org/15/pub/TG4.html, and grouper.ieee.org/groups/802/15/pub/TG4b.html, is another low-power and low-complexity wireless communications. Zigbee operates over unlicensed 2.4 GHz frequency band for interaction between smart gadgets/toys, home automation, sensors, personal home/hospital care, etc., supporting a data rate of up to 250 Kilobits/sec. Current challenges include reducing ambiguities and complexity without compromising (a) performance over the newly available frequency bands and (b) flexibility in security key usage.

4.2.2.2 Personalized and Extended Wi-Fi Wi-Fi (see, for example, www.ieee802.org/11/ and www.wi-fi.org) can use both single-carrier (the direct sequence spread spectrum one) and multi-carrier (the orthogonal frequency division multi-

[4]Class-1 Bluetooth devices can support communications over a distance of 300 feet using power of no more than 100 milli-watts. Class-3 devices use up to one milli-watt for communication over three feet distance.

plexing one) wireless radio technologies. Wi-Fi supports both secure (using, for example, Wi-Fi protected access version 2 or WPA2) and unsecure wireless communications typically over a personalized local area (or the last mile with WiMax as the wide area network or WAN link), wirelessly over unlicensed spectrum for a variety of services. The services include automated (or with minimal human intervention) inter-device communications within home, at office, or in designated hotspots (public spaces) like airport waiting areas, parks, shopping malls, trains, planes, etc. Devices include any Wi-Fi enabled gadget like cameras, laptops, printers, digital storage, photo display devices, TVs, cell phones, alarms, utilities, and monitoring systems. Devices within the home may include hygrometers, refrigerators, etc. The objective is to support communications, e.g., upload/download files, check the status, or update of a system, tetherlessly and without the hassle of setting up of connections at designated spots (home, office, and waiting areas) in order to save time and improve lifestyle.

Although Wi-Fi has been originally designed and performance-optimized for communications over the local area using high-gain, multi-input–multi-output antennas and repeaters, researchers and early adopters have demonstrated Wi-Fi communications over tens of miles. Traditional cellular service providers may use Wi-Fi for voice/data/low-resolution-video communications for fixed-mobile convergence or FMC (see for example, www.thefmca.com) and femtocell-based[5] service at home and at service hotspots. However, the problem of fitting the antennas for a variety of wireless communications services over the surface of a mobile device (e.g., a handset) becomes very challenging. Therefore, the problem of fitting antennas for Wi-Fi, GPS, Bluetooth, WiMax, 3G/4G, etc. communications must be resolved before it can be reality. In addition, although the performance and availability of devices and services matter, ultimately the price of the device and its seamless interoperability determine the overall acceptance of the technology. It is expected that the price will drop as the technology and specification mature, and the Wi-Fi enabled devices are mass-produced. The Wi-Fi alliance (www.wi-fi.org) is focusing on testing and certifying Wi-Fi capable communications and monitoring devices in order to ease interoperability related problems.

4.2.2.3 Mobile Worldwide Inter-operability for Microwave Access (M-WiMax) Mobile WiMax is based on the IEEE 802.16m standard (www.wimaxforum.org, www.ieee802.org/16).It has evolved from the point-to-point microwave data communications standard, IEEE 802.16. It can use both licensed and unlicensed spectrums (2 to 6 GHz), and has widespread support from both chipset manufacturers and end point/system suppliers. M-WiMax use of OFDMA based access, and with 256 OFDM (scalable from 128 to 2048 with OFDMA), can support up to 15 Mbps with 5 MHz channels. NG M-WiMax will use MIMO to improve throughput. Many of the M-WiMax features (security, frame format, QoS, etc.) are based on cable TV service operators' requirements and specifications.

[5]See for example, 3GPP's Release 8 and Broadband Forum's TR-069 based Femto Forum's spec. at www.femtoforum.org.

4.2.2.4 Long Term Evolution (LTE) This telecom wireless technology is expected to provide further simplification of the radio access and core packet network with an objective to support greater spectrum efficiency and reduced latency in over-the-air interfaces. LTE uses a multiple-input multiple-output (MIMO) system in which up to four antennas can be used in both terminals and evolved edge-nodes to support a downlink (to the user terminal) data rate of 100 Mbps using orthogonal FDMA[6] (O-FDMA) and an uplink (to the network) data rate of 50 Mbps using single carrier FDMA (SC-FDMA) over 20 MHz of bandwidth. The orthogonal frequency division multiplexing or OFDM utilizes digital modulation based frequency division multiplexing or FDM. In addition to supporting space and frequency, multiplexing MIMO also supports beam forming and beam steering to improve throughput and reduce latency. In terms of architectural support and evolution, LTE is expected to reduce IP network complexity significantly so that end-to-end latency can be reduced and throughput can be simultaneously improved, thus reducing network costs.

4.2.2.5 Enhanced HSPA This represents an evolution of wideband CDMA for enhancing HSPA (High Speed Packet Access). HSPA utilizes time-domain sharing of a 5 MHz wideband CDMA channel and 16 QAM modulation. E-HSPA uses MIMO antennas and higher order (64 QAM) modulation in order to achieve higher bitrates in both up and downstream. It is expected that the target downlink speed of 42 Mbps will be supported by E-HSPA (and uplink speed would be approximately 11.5 Mbps).

4.2.2.6 Evolution Data Optimized (EVDO) and Ultra Mobile Broadband (UMB) EVDO is a part of the CDMA2000 family using a 1.25 MHz channel for Revision A (actual downlink speed is less than 1 Mbps), 5 MHz channel for Revision B, and up to a 20 MHz channel for Revision C (or UMB). UMB is based on the widely used CDMA, TDMA, OFDMA, and spatial division multiple access (or SDMA) based MIMO techniques, and is expected to improve the data-rate supported by CDMA2000's 1xEVDO Revision C. The target downlink speed of up to 280 Mbps (over up to 20 MHz) is expected to be supported. However, since CDMA carriers are aggressively moving toward LTE deployment, there may not be any widespread use of the EVDO-Rev. C technology and network.

4.2.2.7 Mobile Ad Hoc Networking (MANET) and Wireless Mesh Networking (WMN) These are purpose-built networks that are made to fit in the existing terrain irrespective of whether the application is sensor networking, personal-/body-area communications, vehicular communications, or battlefield communications. The challenges include determining appropriate placement of the nodes with an objective to guarantee signal strength, facilitate topology sharing and maintaining quality of service, security, and robustness. Many experimental testbeds and early prototypes are being developed to further investigate these issues (e.g., see the

[6]FDMA: Frequency division multiple access, a multi-user access scheme where each user is assigned a channel within a frequency band with adjacent channels separated by a guard band to avoid interference.

IETF website on Mobile Ad Hoc NETworking or MANET at http://www.ietf.org/html.charters/manet-charter.html for details).

Wireless mesh networking (WMN) is a special case of MANET. For example, the wireless nodes (access points, routers, gateways to the Internet) can be bridged using a regular or irregular mesh topology as defined in IEEE 802.11's specifications (grouper.ieee.org/groups/802/11/Reports/tgs_update.htm). The objective here is to install a self-organized, self-healing, secure and high-bandwidth metro-area wireless backbone network for private or public use without the hassle and cost of deploying fiber over the wide area. WMN can be implemented using various wireless technologies including 802.11, 802.16, cellular technologies or combinations of more than one type of radio communications technologies. WMN uses directional smart antennas, flexible spectrum management, and cross-layer design to achieve power- and cost-efficient routing.

4.2.2.8 Cognitive (and Software Defined) Radios and Their Interworking

In this scheme, wireless devices are provided with intelligence to utilize the available spectrum, i.e., allocated but not being used, spectrum in the surrounding areas with an objective to improve QoS. Consequently, a convergence of heterogeneous radio communication services is achieved by continuously monitoring the dormant and idle spectrums and adjusting the transmission and reception channels, including parameters. Although the cognitive radio system uses mostly the unlicensed wireless spectrum, it may use licensed spectrum through special arrangement with the title-holder of the spectrum. Technical challenges of implementing a cognitive radio system include spectrum sensing with use of the pilot channel for joint and dynamic spectrum management, wireless-enabled context awareness, neighborhood and local resource discovery, embedded and instantaneous sharing of spectrum information, self-organization, and wireless/mobile peer-to-peer sensing/networking. Although the cognitive scheme sounds very effective and useful from technology viewpoints, there are many regulatory, policy, and socio-economic issues that need careful resolution for deploying the cognitive radio-based interworking capable appliances.

For cognitive radio-based handset manufacturers, the challenges are: (a) support of multi-band transmission and reception, (b) efficient management of power consumption, (c) cost-effective service (including error and noise) management without sacrificing antenna size, and (d) testability and certification for seamless operation.

For networking equipment manufacturers, the challenge is to find innovative ways to utilize the same hardware and even the radios irrespective of the supported Standards and networking methods—(be it WiMax or LTE) by simply updating the modules' configurations and software. In addition, the fact that different carriers will prefer different sizes, varying level of smartness, location of antenna, base stations, and amplifiers, etc. will remain valid in the near future.

4.2.3 Software and Server NG Technologies (Virtualization)

With an objective to accomplish efficient utilization of resources, virtualization will be heavily utilized in emerging systems, and NGNs are no exceptions. Although it is commonly used in semiconductor and Information Technology (IT) and Web-based services design, virtualization is rapidly gaining popularity among networking and communications system design engineers as well. Virtualization not only offers flexibility, convenience, and improved performance; it also reduces power consumption and achieves the same results or goals. For example, support of virtualization can help run multiple instances of application-/service-specific utilities, implementing border-wares and/or middle-wares, in the same router blades (line cards).

Similarly, a hand-held device supporting virtualization of storage, processing, and interfaces can be reconfigured for communications to entertainment to remote medical diagnosis purposes. Consequently, any software-defined service can reap the benefits of virtualization techniques and the network elements can be made aware of both the services being offered and the media that these services are carrying. The major challenges in implementing a cost-effective virtualized storefront include network-, service- and secrecy-/privacy- aware management of computing, communications/networking and storage resources.

Many organizations are enthusiastically working to resolve these issues. They include ATIS Service Oriented Network or SON Forum (http://www.atis.org/SON/index.asp), Cloud Computing Interoperability Forum (CCIF, http://www.cloudforum.org/), and the OpenCloud Forum (http://opencloudmanifesto.org/).

4.3 NEXT GENERATION NETWORKS (NGNs)

There is no doubt that Internet Protocol (IP), Ethernet, and optical technologies will dominate the NGN scene for some time in the future. For wireline broadband access, digital subscriber loop (DSL) technology may still have a few years of lifetime left. However, optical fiber is going to be the dominant technology. In addition, for mobile broadband access, both LTE and M-WiMax look promising. Existing investments and knowledge of technology may very well determine future direction and evolution. As always, operators will make significant attempts to reduce overheads resulting from deployment of competing and complimentary networking technologies. As a result, a new look at the FCAPS requirements is required.

Figure 4-2 shows NGN architecture where functions are divided into applications, service, and transport strata [1–4]. In this architecture, the end user functions are connected to the network by the user-to-network interface (UNI). UNI includes both transport and control functions. Other networks are interconnected through the network-to-network interface (NNI). NNI also includes both transport and control functions. Interoperability between different domains can use NNI functionality. The application-to-network interface (ANI) facilitates third-party application developments. With suitable programmability and network policies, new applications can be developed at a faster pace for consumption by users hosted in any network/domain [2, 3].

4.3.1 Transport Stratum

The transport stratum provides the physical termination, adaptation, bearer functions, and port functions for signal and bearer traffic connections. Within the transport

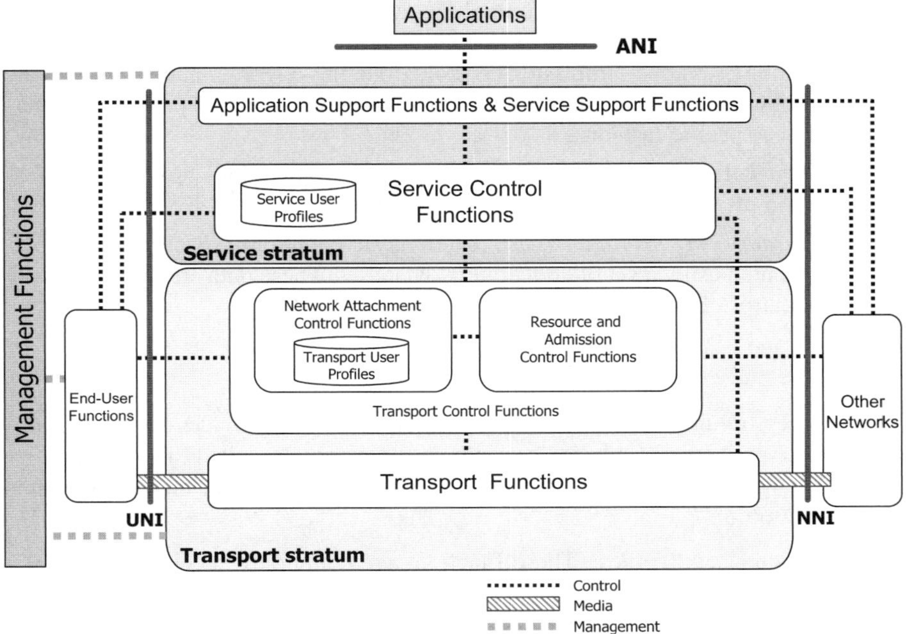

Figure 4-2. NGN architecture overview (Source: ITU-T Y.2012, 2006)

stratum, adaptation elements are responsible for providing interconnection to the large variety of access and trunk interfaces that the Switching/Routing plane may support. This layer provides IP-based connectivity and can support the QoS. It is divided into access and core networks. Other components of the transport function include the following:

Access Functions: These functions manage access to the network. A variety of wireline and wirleless accesses must be supported by the NGN.

Access Transport Functions: These functions manage transport of information across the access network. Based on service requirements, a variety of QoS control mechanisms may be invoked by these functions.

Edge Functions: These functions process signaling, media and management traffic for further aggregation (if needed) and delivery to the core network.

Core Transport Functions: These functions are responsible for carrying traffic over the core networks. QoS control mechanisms invoked by these functions must be light weight in nature. This is equivalent to saying that the core transport is equivalent to a multi-lane highway, and that no traffic control signals are required there.

Gateway Functions: These functions provide and support capabilities to interwork with other networks.

Transport Control Functions: These include the functional entities as described below:

Resource and Admission Control Functions: These functions are involved in admission control and gate control activities for voice, video, data, and mobile sessions. Admission control commonly involves authentication and authorization. Gate control may involve enforcement of policies in a service-specific fashion.

Network Attachment Control Functions: These functions provide initialization of end-user functions for accessing NGN services.

Transport User profiles: These functions are commonly specified and implemented using a set of cooperating databases. These data are utilized during control of transport of information from the user.

4.3.2 Service Stratum

The service stratum stores service profiles for users and provides service control functions. Applications and service support functions also reside in this stratum. These functions transparently support services from applications and other domains to the end user.

Service User Profiles: These functions are commonly specified and implemented using a set of cooperating databases. These data are utilized for allowing and managing access to NGN services by the user.

Service Control Functions: These functions include service level registration, authentication, and authorization functions and may include NGN session control data as well.

4.3.3 Management

Management functions include the network and service management functions, management information base (MIB) and interfaces within the network. The objective is to guarantee expected level of security, reliability, availability, and QoS for the billable NGN services. This provides the following services across the network:

4.3.3.1 Fault Management Fault management refers to the management of services and sessions at the agreed-upon levels even when there are faults, including overloads and disasters, in the service, application, and transport strata. Functions may include monitoring and control of utilization of resources during setup, maintenance, and release of sessions for NGN services. Since NGN supports a multitude of services, it is recommended that appropriate filtering and correlation be used to manage service- specific faults. When services span multiple technology and administrative domains, as would be the case for networks and services interoperability, one or more fault management mechanisms are required per bilateral agreement. This will help maintain service transparency across the NNI.

4.3.3.2 Configuration Management This refers to developing, monitoring, and managing hardware and software configurations of devices, elements, and systems with an objective to maintain network operations without negatively affecting services and revenues. For example, it is often desired to store the tested and approved configurations of the end-user terminals in a networked server so users can download them as they sign in for new features and services. Although this practice is very common in the cellular phone industry, Internet and IP-based television (IPTV) service providers are also finding it increasing useful. Similarly, for network elements' configuration management, specialized on-line and off-line servers are commonly utilized.

4.3.3.3 Accounting Management Accounting management in the context of NGN commonly refers to recording the utilization of services and network resources with an objective to create a billing record. The recording can be done in various formats including the raw comma-separated-values that can be fed to format data in other acceptable standard formats in order to create customer readable bills. Measurement of the use of services and resources can be done in multiple ways. For example, per-service per-user paradigm is routinely utilized for cell phone users for voice data/text-messaging, video download, gaming, etc, unless flat-rate billing is assumed. For enterprise customers, recording of network and service utilization and events like service outage and repair time for managing the service level agreement (SLA) are more important than documenting the service usage.

4.3.3.4 Performance Management Performance management is concerned with monitoring the performance of networks elements, both transmission links and nodal devices, with an objective to maintain the desired level of service quality or SLA. Both active and passive monitoring devices and techniques are commonly used in NGNs. The challenge, however, is to locate and harden performance monitoring and management systems uniformly in the network without overburdening service creation, management, and delivery modules.

Passive monitoring requires the use of splitters in the transmission links, and special-purpose hardware for off-line filtering/storing/analyzing the captured data. Active measurements can be conducted without significant overhead, at any desired time, and for any desired time period. For transmission links, the parameters of interest are throughput and utilization, uptime and downtime, time to recover gracefully from overloads and disasters, etc. For nodal device like switches and routers, parameters like delay, response time, local or remote switchover times for service quality maintenance during failures and overload, are of paramount importance.

4.3.3.5 Security Management NGN security management includes managing the user's identification, authentication, authorization, certificate, etc. in an access-neutral fashion. Otherwise, it will be very difficult to maintain service continuity when the user (or session) moves from one access network to another or roams from one service provider to the other. Since NGN uses IP-based transport, additional mechanisms are required to protect both service and network from worms, viruses, intrusion, denial of service, etc. Simple monitoring-based mechanisms may not be sufficient. Proactive measures must be invoked. ITU-T Study Groups 13 and 17 addresses these issues. Once again, when services span multiple technology and administrative domains, as would be the case for networks and services interoperability, one or more security management mechanisms are required per bilateral agreement. This will help maintain service transparency securely. In terms of the Telecommunications Management Network (TMN) functions, this encompasses both the Element and Network Management layers.

4.3.4 Applications Functions

This functional block supports service application programming interfaces (API), session control, service logic, translation, and routing logic, directory and policy management functions across the network. Some of the specific applications functions provide:

1. Messaging services, such as those used in e-mail and voice mail
2. Processing services, such as automatic speech recognition and credit card processing
3. Value-added IP telephony services, such as virtual second line, Web-based toll-free calling
4. Directory enabled services, such as Freephone/8xx number translation, local number portability, and single number follow-me services for voice telephony
5. IP naming and addressing services including DNS, DHCP, and RADIUS (DNS stands for Domain Name System, as defined in IETF RFC 1035, Dynamic Host Configuration Protocol or DHCP has been defined in IETF RFC 2131, Remote Authentication Dial In User Service or RADIUS as defined in IETF RFC 2865)
6. CLASS-5/CLASS[7] services, such as call waiting, call forwarding, conference calling for voice communications service (telephony applications)
7. Virtual Private Networking (VPN) for voice and data
8. Bandwidth Services, Optical VPN (IETF RFC 2547), etc.

4.3.5 Other Networks: Third-Party Domains

Domains are defined by network administration to give structure to the network, and to make management of the network as simple and easy as possible. Domain structure may:

1. represent the reach of a domain-specific address allocation
2. define the reach of an Interior Gateway Protocol (IGP)
3. define the boundaries of an Exterior Gateway Protocol (EGP)
4. define a level of traffic aggregation and management
5. correspond to a region of guaranteed QoS
6. correspond to a region defined by a level of guaranteed security ("trust" domain).

[7]CLASS: Custom Local Area Signaling Services [5].

Domains should be defined hierarchically. If they are "multi-domain," management shall act as the highest-level domain, managing resources and supporting traffic that must traverse multiple domains. Multi-domain management also should be capable of acting as a back-up resource for any individual domain under its management.

Third-party networks include other service providers and carriers as well as medium to large enterprise customers who maintain their own networks and IT departments. The borders or boundaries define the transition points for signaling, media, and OAMP[8] messages. It is very important to manage signaling, security, policy, QoS, and SLA in order to support end-to-end transparency of the services across borders or boundaries.

4.3.6 End-User Functions: Customer Premises Devices and Home Networks

Traditionally this domain contains equipment, wiring, and functions that are required in the residential customer premises or in small/medium business' premises. Since traditional Telcos are entering the entertainment business, they have upgraded their twisted copper wire pairs to high-speed DSL or fiber- or coaxial cable-based links so that they can offer broadband services (video and Internet) to their customers as well. These call for installation of additional modems, router, security devices, video converters, and set-top boxes to homes. Similarly, traditional cable TV service providers are offering digital telephone and Internet-based services using the emerging version of the DOCSIS (Data Over Cable Service Interface Specification) and others standards (for details, see www.cablelabs.com). These types of evolution to support multimedia services create a multitude of networks and devices inside home, making the problem of debugging, diagnosis, and service installation difficult.

4.3.7 Internet Protocol (IP): The NGN Glue

Irrespective of whether it is wireline or wireless at the physical layer, the networking[9] layer is going to ubiquitously use the Internet Protocol or IP (IETF RFC 791 for IPv4, and RFC 2460 for IPv6). The following variants of IP and the add-on features are currently available in the industry to keep it as the most useful glue at the Networking layer.

4.3.7.1 Internet Protocol version 4 (IPv4) IPv4 (IETF RFC 791) uses a 32-bit dotted decimal based addressing scheme (e.g., 132.197.34.181), and can support up to 4 billion (2^{32}) addresses including those reserved for private networks and for multicast communications. However, by using network address translation (NAT, which converts private IP addresses to one or more public IP addresses, IETF RFC 2766 and 2767), the possibility of address space exhaust can be minimized. The

[8]OAMP: Operations, Administrations, Maintenance, and Provisioning.

[9]The third layer from the bottom of International Standards Organization's (ISO's) seven-layer Open System Interconnection (OSI) model.

problems related to manual configuration and to supporting privacy, security, quality of service, and mobility remains open for IPv4. In most cases, these issues are resolved by using additional adjunct devices or network element, but these add to complexity and overhead of both network and performance management mechanisms. Consequently, the emerging NG networks would most likely avoid using IPv4-only elements and devices.

4.3.7.2 Internet Protocol version 6 (IPv6) IPv6 (IETF RFC 2460, and 4294) uses a 128-bit addressing scheme expressed using (colon-separated) hexadecimal strings (e.g., 4ffe:4700:2100:3:510:a4ef:fda0:ba97), and can support up to 2^{128} addresses for devices using IPv6-based communications. IPv6 has built-in support for stateless auto configuration, mobility, efficient routing, and traffic engineering, security, quality of service, privacy, and multicast services. However, since IPv6 is not backward compatible with IPv4, the network elements (including routers) and devices must be upgraded to support the additional processing and memory requirements in order to support IPV6 address-based communications.

4.3.7.3 Mobile Internet Protocol version 6 (MIPv6) In order to better support the mobility of devices, a set of extensions to the original IPV6 has been proposed in IETF (RFC 3775 and 3776). The extensions include caching the binding of a mobile node's home address with its care-of address, and then sending packets directly to the care-of address. This helps an IPv6 device to transparently (via tunneling) maintain transport layer connection when it moves from one subnet to another. Additional mechanisms to support local handoff and global mobility using a network-based mobility management entity called proxy mobile IPv6 (as defined in IETF RFC 5213) are also being discussed in IETF.

There are some discussions in the industry forums about the complexity, and hence the delay and costs it adds, of the IP layer. Attempts are being made to explore cross-layer optimization. This may make the upper (applications and transport) layer devices smarter and lower (link and optical) layer elements more capable and cost-effective.

4.4 NEXT GENERATION SERVICES

This section presents a high-level description of Next Generation (NG) services. The architectures that are required to support these services are briefly discussed along with essential transport and application plane requirements. Only a sample of emerging NG services is presented below. It is prudent to remember that given the openness of emerging technologies and the flexible regulatory environment we live in, it is only inventors' imaginations that can and will determine future innovative services [6].

4.4.1 Software-Based Business Communications Service

A new trend in the Communications industry is enabling complete separation of the capability to support Business communications services, i.e., telephony, auto-attendant, video/Web conferencing, instant messaging, white boarding, etc., from the hardware platform or device that is hosting the service. This allows traditional software companies to focus on developing business/office communications software suites that can be installed in general-purpose Internet connected servers for service-specific execution and service implementation. As a result, knowledge workers can enjoy the benefit of leading-edge communications services, cost-effectively, irrespective of where they are and when they want to use the services.

4.4.2 High-Definition (HD) Voice

A combination of availability of wide-band voice codecs (like G.722 and adaptive multi rate, wide-band, or AMR-WB) and ubiquitous broadband access in both wireline and wireless network will be making widespread availability of HD and stereo voice communications service. These codecs support a sampling rate ranging from 14 KHz to 22 KHz to reproduce very clear and highly intelligible voice sounds. To support HD voice service, it is required to maintain codec-transparency and bandwidth availability (by enforcing transmission policy) across all segments of network access and transmission. Current industry activities are directed toward making availability of the codecs and their settings (configurations) uniformly in handsets, soft-clients, and IP phones in a standards manner over all segments of the network so that the consumers can enjoy the benefits of using this service seamlessly.

4.4.3 Mobile and Managed Peer-to-Peer (M2P2P) Service

Traditional P2P service is used for applications ranging from file sharing to real-time streaming and video communications (file sharing). Since these contents are distributed throughout the Internet nodes, unmanaged P2P can generate tremendously large volume of redundant traffic especially in the transport links. Managed P2P service attempts to alleviate this problem by allowing *trusted* (through a broker) sharing of information related to the nearest logical location of the content and the best path (topologically and from routing viewpoint) to deliver the content to the requester. It is expected that service provider, content provider, and network provider will all benefit from using the mobile and managed P2P.

4.4.4 Wireless Charging of Hand-Held Device

Viable commercial technologies are being developed to recharge hand-held communication and entertainment device wirelessly over a few miles via for example electromagnetic induction and other beams, as has been demonstrated by Nikola Tesla in 1893. Once these radiation-receiving and recharging interfaces are incorporated in the hand-held and other devices, consumers will be able to get rid of the variety of power cables that they need to carry with their hand-held devices. These hand-held devices must be equipped with wideband (500 MHz to 10 GHz) receiver to capture the electromagnetic radiations that are emitted from mobile and TV antennas. MIT Labs recently conducted experiments to demonstrate wireless charging of hand-held devices. University of Washington in Seattle and Intel jointly demonstrated the use of TV signal to power a small sensor over a distance of few miles. A Wireless Power Consortium is being formed to develop viable business and regulatory models to commercialize this technology.

4.4.5 Three-Dimensional Television (3D-TV)

The NG television (TV) systems will utilize viewer-controlled clusters of cameras to capture scenes and views including depth and shadows in real-time and will transmit these images over very-high-bandwidth access and transport lines to the users. Researchers have recently developed a system to capture live scenes by using a cluster of 64 video cameras connected via a local area network to a PC. The PC converts input from all of these HTTP-enabled video cameras into JPEG sequence of images for display per user-controlled viewing requests. Challenges remain in the areas of capturing 3D images, processing, and storing these images without losing information and fidelity, and visualization without impairments (parallax, segmentation, etc.) for applications like virtual reality-based gaming and entertainment, remote collaboration and tele-diagnosis, etc.

4.4.6 Wearable, Body-Embedded Communications/ Computing Including Personal and Body-Area Networks

These are being enabled by miniaturization of computing and communications devices due to the emergence of nanotechnology. These devices can be worn and/ or implanted in the human body, effectively to support unidirectional (monitoring) and interactive sensing or communications. Potential applications and business opportunities include (a) monitoring and diagnosis of patients' conditions, (b) graceful rehabilitation, (c) automatic dispatching of first responders in case of emergency, and (d) tactile sensing and interactions for advanced entertainment (gaming) and education services. However, many biomedical and bioethical challenges remain to be resolved before making these happen in reality. These include development and commercialization of (a) intuitive interfaces, (b) extremely low-power implantable circuits, and (c) physiological and ambient intelligence gathering sensors. Some of the short-reach, ultra-low-power communications technologies like ZigBee and Bluetooth (as discussed earlier in this chapter) may be useful for this purpose.

4.4.7 Converged/Personalized/Interactive Multimedia Services

Convergence of services refers to the ability to deliver real- and non-real-time voice/ data/video/tactile information to a single device in a context-sensitive and interactive manner. The device could be wired or wireless or could be a hand-held phone or portable computer or a television. The context can be a combination of network, content, location/social environment, etc. The interactivity can be dictated by user-preference, social setup, adaptive semantics, etc. One such example is the mobile TV and infocast service using the digital video broadcast-handheld (DVB-H, www.dvb-h.org) standards. DVB-H device uses low power (less than 100 milliwatts) transmission of coded OFDM, OFDMA based access, and transmits digitized IP packets in 100 milliseconds time slots. A DVB-H terminal can receive a 15 Mbps bit stream over an 8 MHz channel in the 700 MHz band. Note that the DVB-H system can be easily adapted to operate with 5 MHz bandwidth in L-band (1670–1675 MHz) as well.

4.4.8 Grand-Separation for Pay-per-Use Service

This refers to supporting separation among access, transport, application, services, networked-resources (CPU, storage, etc.), networked-contents (generated and managed by anyone), security services, content subscription and exchange, transaction capability, etc. with well-defined open interfaces. This will drive users and developers alike to continuously build and market innovative services to improve the lifestyle of human beings for both work and play.

4.4.9 Mobile Internet for Automotive and Transportation

This refers to the use of high-bandwidth wired communications among different sensors and devices within the vehicle, and wireless communications between vehicles, between vehicle and curb-side fixed or mobile units, and between neighboring vehicle and infrastructure or fleet maintenance units (for vehicle maintenance, locating vehicles), etc. Note that high-speed mobility of the vehicle adds to the existing challenges of supporting delay- and fault-tolerant communications in different weather and road conditions.

4.4.10 Consumer- and Business-Oriented Apps Storefront

This refers to the opening of the network and application programming interfaces (NPIs and APIS) to the developer community so that anyone with access to the Internet can develop and upload an application for a useful service, to consumers and businesses. The applications can then be downloaded to wired/wireless handheld devices including cell phones for games and entertainment, learning, health-care, or business. This enables an applications developer, in, for example, a remote village in India or Brazil or China, with Internet access to develop and upload an application to the storefront of a service or network provider for downloading by using wireline or wireless access. A host of computer and software companies, handheld device manufacturers, and service providers, have announced their enthusiasm and participation in this effort (see, for example, www.appstoreapps.com, forge.betavine.net, plaza.qualcomm.com/retail, www.BlackBerry.com/AppWorld, www.jil.org, etc.). Some service providers are developing alliances to support all of the services one can think of; gaming, communications, mobile-TV/entertainment, wellness, mobile wallet/payment, home security, hotel check in including door key (yes, that can be uploaded when check-in to the hotel is completed!) and remote control services.

It is interesting to note that popular APIs like Amazon's eCom and payment, eBay's buying/auction/checkout, Facebook's markup/query, Google's Maps, Calendar, OpenSocial, and Android, etc. are very much proprietary in nature.

However, standards activities, along the line of network and service API development, are making progress as well, and these activities include ATIS SON (www.atis.org/SON/index.asp) group, Parlay group's Web services and presence API, OMA's (www.openmobilealliance.org) Web and mobile service enabler (with APIS yet to be developed), etc. These standards will provide guidance on creating and delivering services rapidly and seamlessly using enablers in multi-domain (Web, IMS, IT, Telco, etc.) environment with reduced costs.

4.4.11 Evolved Social Networking Service (E-SNS)

Today's social networking services use audio, image, video, and data files along with Skype-based (www.skype.com) communications services over the Web for sharing of unverified information of common (among those in the group) interests. This essentially helps build an online community of geographically dispersed Internet users where there is no strict boundary between consumer and producer of information and content. Facebook, Twitter, MySpace, and LinkedIn are a few popular SNS Websites. If used properly, SNS can be great way to instantly disseminate context-sensitive information in an orderly fashion to organize teams and activities. These can be any one of the following: development of a business model, accelerating experiment-based research works that are conducted non-stop by volunteers spread all over the world, providing medical and/or education services to the underdeveloped communities anywhere in the world and connecting a summer intern to a prospective employer. However, many issues related to privacy, authenticity of information, safety of the users of SNS and abuse and misuse of SNS services for subversive activities must be resolved.

4.4.12 NG Services Architectures

It is rapidly becoming apparent that all versions of NGN architectures are using, or will be based on Third Generation Partnership Project's (3GPP's) IP Multimedia Subsystem (IMS) architecture. In addition, since it is based on IP and supports multimedia services (voice, video, gaming, conferencing, etc.), IMS can seamlessly integrate with other systems via appropriate adapters/mediators or interworking elements. These other systems could be based on Web or any legacy system.

The concept of IMS originated from 3GPP to support seamless mobility and user identity across wireless carrier domains. IMS is currently being touted as the *de facto* standard[10] for the signaling and control plane of the emerging NGN architecture for converged mobile, fixed, and fixed-to-wireless services. ITU-T defined NGN as an IP-based network. The target is version 6 of the Internet Protocol, or IPv6. IPv6 will seamlessly support QoS, mobility and provisioning for Telecom and other broadband revenue-generating services. The functional entities (FEs) in the IMS layer interact with the applications and feature/services layer FEs to provide advanced or enhanced services to the endpoints. Interconnection of the FEs in the

[10]Almost all of the national and international standards organizations (ATIS, ETSI, Cable Labs, WiMax Forum MSF, ITU-T, etc.) are using IMS as the basis of their NGN architectures.

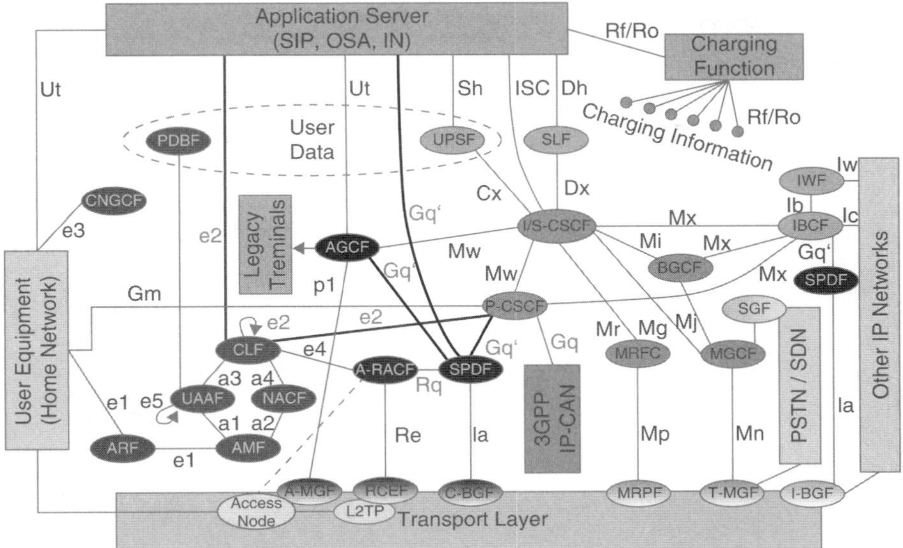

Figure 4-3. A common IMS for Next Generation Networking (Source: 3GPP and ETSI. Please see http://www.3gpp.org/ftp/Specs/html-info/21905.htm for a detailed vocabulary of 3GPP specifications)

IMS layer with NGN transport, legacy PSTN, and IPv4 transport systems is supported by using appropriate border, gateway, and middle-ware functions. In addition to mobility and seamless provisioning, there is a need to support reliability/availability, security/privacy, and regulatory requirements as well as to achieve realistic deployment of new/NGN services using IMS.

Figure 4-3 shows common IMS where IMS core is being augmented by a select set of interfaces and functional elements in order to satisfy wireline, wireless and mobile users requirements. The IMS core consists of user database (user profile server function or UPSF, subscriber location function or SLF, etc.) and proxy/edge, serving, and interrogating call sessions control functions (or P-, S-, and I-CSCF).

Further details on IMS can be found in the 3GPP Website (ftp://ftp.3gpp.org/specs/latest). In addition, note that the following technical specifications (TSs), noted in parentheses, contain information on Architecture (23.002, 23.221), Interworking (29.162, and 29.163), Charging (23.815), Security (33.102, 33.103, 33.203, and 33.210), Evolution of Policy and Charging (23.803), WLAN access (23.234, 24.234, and 29.161), and Fixed Broadband access (24.819). These documents can be readily obtained from the FTP site, ftp://ftp.3gpp.org/specs/latest. ETSI also updates the IMS-based NGN specifications as the requirements change and technologies to implement the requirements are updated (e.g., see portal.etsi.org/docbox/TISPAN/open).

With a common IMS core, it is expected that issues related to Interoperability, i.e., when services cross wireline to wireless domains and vice versa, will be minor. The results of interoperability-of-services when IMS is used in both wireline and wireless domains can be found in MSF's (MultiService forum's) Website (www.msforum.

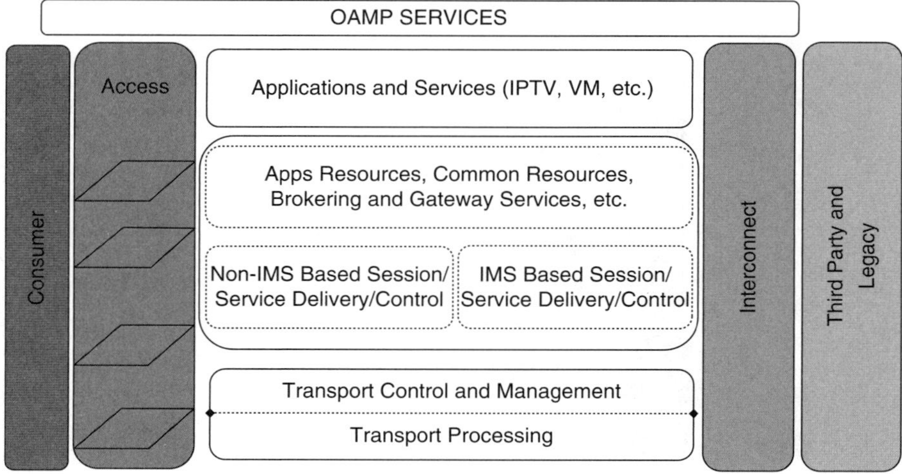

Figure 4-4. MSF Release 4 (R4) architecture template (Source: Multi-Service Forum)

org/interoperability/GMI.shtml). Figure 4-4 shows the planes (strata) and domains of MSF's Release 4 architecture template for converged services implementation across multiple-domains with the option to support consumer access over a variety of wireline and wireless technologies (as shown by the tiles in the access domain). Elements in the interconnect domain provide peering-of-services over untrusted network boundaries and sometimes even support disparate technologies. For voice service interconnection, these disparate elements include voice over IP (VoIP) to time division multiplexing (TDM) voice interconnection, VoIP to VoIP interconnection for different IP addressing (version 4 and version 6) mechanism, and so on.

4.4.13 Application Plane's Requirements to Support NG Services

This section discusses application plane requirements to support NG services, primarily in one administrative and/or technology domain. The network elements in this domain include, for example, location and presence servers, voice mail server, calendar and instant messaging server, call redirect server, IPTV applications server including enhanced program guide server, etc. Users registered in the IMS and non-IMS domains interact through standard interfaces with the servers in this domain via brokers and/or resources/service gateways in order to receive the services they subscribe to or are authorized to use (as shown in Figure 4-4).

4.4.14 Transport Plane's Requirements to Support NG Services

This section discusses transport plane requirements to support NG services primarily in one administrative and/or technology domain. Elements in this domain administer access control and resource management with an objective to satisfy end-to-end service quality management, effectively guaranteeing a higher quality-of-experience (QoE). As shown in Figure 4-4, transport processing is managed from transport control and management, either by using router-/switch-embedded access/resource controller or by using adjunct devices. The adjunct or embedded devices could be policy server, and session border gateway/controller, as discussed in MSF technical reports and implementation guidelines (available at http://www.msforum.org/techinfo/reports.shtml and http://www.msforum.org/techinfo/implementation.shtml).

4.5 MANAGEMENT OF NG SERVICES

As discussed before, development and deployment of Next Generation services, in a cost-effective manner, is becoming increasingly complex. This is because users are demanding a multitude of services over various traditional (wireless and wireline) access and device Interfaces irrespective of the capability or domain of the service providers. NG service providers are expected to deliver voice and video calls over regular TV screens in addition to continuing these services seamlessly to screen-based POTS phone and hand-held devices (PDA, cell phones, etc.) per user's convenience. The task of managing security, quality-and-continuity-of-service, mobility, and billing therefore become enormously convoluted [7].

4.5.1 IP- and Ethernet-Based NG Services

This section discusses next-generation services over IP and Ethernet. Services that are of interest include VoIP, IPTV, and mobile multi-media services and their evolution. Real-time audio, video, and data services demand predictable delay, jitter, and loss (and recovery from loss) in addition to reliability and scalability so that the expected user experience can be satisfied.

IPv4 and IPv6 have many capabilities to support QoS and transmission resiliency for real-time services like voice and live video transmission. Multi-Protocol Label Switching (MPLS), a sub-IP-layer-protocol, has a set of built-in capabilities to support transmission of real-time services. However, the EoS or Ethernet over Synchronous Digital Hierarchy (SDH) or Synchronous Optical NETwork (SONET) is getting more attention recently.

EoS uses encapsulation based on Generic Framing Procedure (GFP) and a combination of virtual concatenation (VCAT) and Link Capacity Adjustment Scheme (LCAS) protocols to efficiently transport real-time traffic over wide area networks. This may help service providers migrate to purely packet-based layer-2 networks. In addition to using the bonding technologies with existing SDH/SONET based circuits, both pseudo-wire-based circuit emulation and synchronized-Ethernet and timing standards such as IEEE 1588 v2/v3 can be used for seamlessly transporting VoIP and IPTV traffic over carrier Ethernet (both wireless and wireline) very reliably. Since EoS is based on layer-1 and layer-2 protocols, it does not need any intervention of higher-layer protocols or equipment to support hitless restoration for maintaining higher service availability.

Since EoS uses GFP with fixed overhead, it is more efficient for bandwidth utilization. GFP supports the ability to send management frames that are used for OAM. In addition, the use of VCAT and GFP can provide byte-level granularity of data to lower latency and jitter by reducing the buffering needs.

The end points in EOS typically do not require expensive and extensive buffering and memory within the equipment because they operate as pure layer-1 and layer-2 transport termination devices to maintain the connections and perform re-convergence using traditional SONET or SDH protocols. EoS can also multiplex bandwidth to deliver higher speed carrier Ethernet based WAN (Wide Area Network spread over geographically dispersed regions) services over multiple service providers networks when used in conjunction with the virtual private LAN service or VPLS.

EoS can offer quality and reliability based on the types of services being supported. For multimedia applications such as streaming media, interactive gaming, or broadcast TV, features like hitless bandwidth adjustment, low latency, and efficient use of bandwidth are of critical importance and EoS can readily support these features.

4.5.2 Performance Management of NG Services

This refers to monitoring and maintaining an acceptable level of performance, not only for the network, but also for the services that are being supported by the network.

Since emerging services use a combination of real- and non-real-time sessions, it is becoming a norm to utilize the deep packet inspection (DPI) feature to securely monitor and manage end-to-end performance of both enterprise and service providers' networks. DPI offers visibility of the session and service, i.e., peer-to-peer, music/game/video download, voice over IP or VoIP, IPTV, etc., that are generating the packets, so that these can be throttled or prioritized. Throttling prevents unauthorized and illegal use of network resources. Prioritization helps maintain appropriate quality of experience for a variety of services to the premium customers. DPI features can be embedded in the existing network elements or separate appliance-based system can be used.

Traditionally, network-level performance parameters that need to be monitored include throughput, utilization, information loss, information transmission delay or response times, variation of information transmission delay, mean-time-to-failure (MTTF), mean-time-to-repair (MTTR), etc., of network links under both nominal and overload conditions. However, since emerging IP-based networks carry both signaling and media (payload) traffic for both real-time and non-real-time sessions and services, the performance management requirements get a bit more complicated.

Commonly used reactive mechanisms are not sufficient. A combination of predictive and proactive performance management mechanisms is required. This is because the emerging converged services networks are expected to support mobility across a variety of wireless and wireline access networks, service quality as the session and/or device roams from one administrative or technology domain to the other, and authentication and security of services seamlessly, without affecting the performance requirements of the session/service.

Finally, interworking with the legacy performance motoring system is also mandatory because it may be frequently required to use data from, or to feed data to, the legacy performance management system for one or more segments of the service.

4.5.3 Security Management of NG Services

Managing security in emerging NG networks is a very complex task. This is due to the drive to support IP-based convergence in both networks and services areas at the same time that hackers are becoming increasingly smart due to openness and ubiquity of the Internet.

Information security solutions must address user, end-point, service, and administration level security without compromising the flexibility and simplicity of use of the network and service. Certain popular networking and service developments or offerings, e.g., peer-to-peer services over Internet, create more vulnerability in networks. Legislative measure alone cannot protect consumers, networks, and services because attacks on networks and services are often triggered by personal frustration and other factors. The challenge is how to operate the networks and services efficiently and cost-effectively without compromising privacy, security, and vulnerability of the services.

What is required here is an open and flexible framework to define service-specific network security requirements, incorporate these requirements into network nodes' and transmission links' design and performance specifications, and test and certify the network and nodal security solutions before deploying these in an operational network. Then continuously upgrade the deployed protective mechanisms to *outsmart* the hackers and network attackers as the technologies evolve!

4.5.4 Device Configuration and Management of NG Services

Managing capability and configuration of customer premises NG devices remotely, including those in enterprise, is an overwhelming task. The situation is more manageable in medium and large enterprises because of an existing process that is routinely followed for upgrading and adding new devices to the system or network. However, in small business and residential locations, the users add/move/modify devices sometime knowingly and on other occasions download plug-ins for the target services even without any direct knowledge of those plug-ins. The latter situation often causes malfunction and system-level crashing of the devices.

To overcome these problems, various standards organizations are creating forums and focus groups, and a few of these are described below. ATIS recently established the Home Networking or HNet Forum (www.atis.org/HNET). The objective is to develop specifications and guidelines for interconnecting IP-based NG home appliances/devices/system by using the emerging technologies so that the services can be delivered seamlessly.

The Broadband Forum (www.broadband-forum.org/technical/trlist.php) has released a number of technical reports (see, e.g., TR-064, TR-069 and its addendums,

TR-098, TR-104, TR-106, TR-111, TR-135, TR-140, and TR-196) covering networked server-based management of configuration of customer premises devices including those at homes. Home-based devices include voice over IP gateway, set-top box (STB), etc.

The HomeGrid Forum (www.homegridforum.org) is working to promote ITU-T's G.hn Standards for using unified wireline MAC/PHY specifications for enabling service providers and consumer electronic device manufacturers to deliver services to connected homes cost-effectively. These will allow the network access and service providers (Telcos or cablecos) to actively monitor home networks and devices at home to identify and resolve problems proactively.

Note that for wireless devices, service providers commonly utilize proprietary mechanisms to activate and upgrade devices through over-the air interface to the device and by using special codes (keys) and ports, and this trend probably needs to be reversed to provide more flexibility to the customers.

4.5.5 Billing, Charging, and Settlement of NG Services

Various paradigms of billing and charging are being discussed in Standards organizations, i.e., online (real-time) and offline (batch processing) methods of charging are the most common and useful ones. In this era of globalization, no service provider is an island, and hence it is highly desirable that one unified settlement scheme be used among service providers to support seamless mobility and consistency of services. Both policy-based service management and service-type based policy can be used to openly settle payment among the service providers. However, utmost caution must be exercised to avoid any sort of service degradation due to irregularity of settlement mechanisms.

4.5.6 Faults, Overloads, and Disaster Management of NG Services

The distributed mode of operation of emerging digital packet-based networks makes networks less vulnerable (an advantage) and at the same time less manageable (a drawback) in a centralized manner. However, certain technologies and their advancement are simplifying faults, disaster, and overload (FDO) management. These technologies include software as a service (SaaS) and platform or hardware as a service (P/HaaS), driven by virtualization and the emerging grid-computing methodologies. These enablers allow service providers to have a tool-kit based approach to end-to-end health and welfare monitoring, including analyzing the collected data. The objective is to provide service-level reporting, and management of the network, making the assurance of service delivery a minuscule derivative of the entire scheme.

4.6 NEXT GENERATION SOCIETY

With the advancement of next generation technologies and their deployment in networks, services will be more sensitive [8] to the media, context, personality, and location of the user, in addition to being more automated in operations. These are expected to contribute to developing a more relaxed and enjoyable lifestyle irrespective of whether the computing and communicating services are being used for work or play or entertainment. Societies must pay close attention: these capabilities must add to human development, not subtract.

Both technologist and policy maker will have roles to play in this new technology-dominated society and environment. For example, employees of the NG service providers must learn and demonstrate both depth and breadth of the subject matter of concern. A router engineer must also have working knowledge of the optical transport layer and storage or data center requirements. In addition, the technology life cycle is also shrinking every year, and sometimes the first-generation technologies (ISDN, analog hi-definition video, ATM, etc.) are nothing but false starts! Technology transformation drives both the communications and entertainment transformations, and these in turn make the entire society more vulnerable, unless these are transformations implemented/executed cost-effectively and responsively.

4.6.1 NG Technology-Based Humane Services

As NG technologies mature, we will see increasing use of automation. Application of these new technologies in services that humans touch and expression of and reaction to feelings, are of utmost importance. These include patient care in hospitals and elderly care in nursing homes, hospitality services' attendants, and first responders in harsh environments. One feasible option would be to continuously augment—not completely replace—these service attendants' jobs by automated NG technology based gadgets. Of course, the human attendants must be continuously trained to keep themselves up-to-date to compliment their NG-technology based counterparts—the machines and gadgets.

4.6.2 Ethical and Moral Issues in Technology Usage

It is often said that conscientious minds will always develop technologies and utilize them for harmonized evolution of civilization [9], maintaining proper sustainability of the society. Since peace and harmony really matter in this globalized era, societal progress must happen through harmonized use of technologies and services in a peaceful manner. For every technology, and the system based on that technology, there must be a set of methods and mechanisms to trace and prevent abuse and misuse of the NG technology-based services. NG digital copying technologies must have built-in copyright protection and audit trail support mechanisms to prevent theft, and so on.

Yes, of course, the technologist must stay a few steps ahead of social progress but not at the cost of social discontentment. For example, would not it be great to invest in developing a workforce that is optimized to satisfy evolving societal requirements? Why not invest in educating the emerging workforce in areas in which there will be demand for jobs and services in the society. A statement of caution is that cross-pollination of ideas and techniques from one discipline to others is acceptable only when it is executed with moderation. Recent development and subsequent smashing of the economic bubbles in Internet working, real estate, and financial sectors should be treated as painful reminders and lessons from irresponsible behaviors and irrational expectations.

To that end, development of technologies that reduce waste and utilize more of (a) recyclable materials and (b) renewable energy sources should be encouraged. This may call for development of (a) recyclable components based network elements and (b) wind-, solar-, and other ambient-source-based energy cells which are more efficient for use in network elements and communication/entertainment devices. It will also be extraordinarily helpful to encourage the use of more of the earth-friendly, i.e., free from harmful toxic supplements and materials in manufacturing network elements, connection servicing devices (human-friendly wireless links may be the best), customer premises equipment, and user (e.g., hand-held) devices.

4.7 CONCLUSIONS AND FUTURE WORKS/TRENDS

Telecommunications is over a trillion-dollar industry today. The growth of ubiquitous high-speed wired and wireless broadband access coupled with high-definition audio, video, and other digital multimedia services across a multitude of devices will cause meteoric rise of traffic in the network over the next five years. A tremendous socio-economic opportunity therefore lies ahead, and everyone involved must act *responsively*.

Communications being done via texting today will be achieved using variable resolution video and multimedia messages over adjustable (flexible) extended-size screens. Consumers will increasingly play the roles of producers of information and entertainment, and these activities will make the traffic flow across the access lines more symmetrical. In addition, many users may be uploading and downloading contents at the same times, and that will significantly reduce the bandwidth utilization gains from traditional statistical multiplexing. This mode of operation may call for a revised design of the traditional capacity planning methods and tools.

Next generation networks, therefore, must be capable of supporting features and functions beyond supporting merely the emerging IP networking technologies and offering the legacy PSTN and TV services. Of particular interest are the following: (a) reduction of cost and complexity and cost via unified support of self-managed (or autonomous) scalability of network and services, (b) graceful deployment of open, ubiquitous, and transparent networking with high degree of support for privacy and security, (c) seamless support of virtualization to offer location- and identification-independent on-demand applications, (grid computing) services, and entertainment—including gaming, super-high-definition, and three-dimensional movies, using the already deployed and available networked resources, and (d) increasing the use of earth-friendly materials in communications and entertainment devices, and simultaneously making these devices and the supporting elements more energy-efficient.

The emerging next generation of businesses and services will be based on secure personalized broadband social networking applications supporting human-to-machine and machine-to-machine communications in context-sensitive fashion. The personalized communication devices will be media and service aware, just like the network elements will be, and will continuously adapt the configuration (via virtualization) and communication bandwidth in order to maintain the quality of experience for the service being used. The service can range from monitoring premises to the health of a patient, sharing stored and live video sessions with friends remotely, taking the personalized network and service to anywhere in the world using, e.g., the evolved DLNA (Digital Living Network Alliance, www.dlna.org) specifications, and so on.

4.8 REFERENCES

1. ITU-T Recommendation Y.2010: Functional Requirements and Architecture of the NGN, 2006. Geneva, Switzerland: International Telecommunication Union Telecommunication Standardization Bureau.
2. MultiService Forum, "Global MSF Operability 2008," http://www.msforum.org/interoperability/GMI.shtml (accessed December 2008).
3. Khasnabish B, Shukla V. Next Generation Network and Service Interoperability. (Submitted for publication).
4. ATIS, "Next Generation Networks," http://www.atis.org/TOPSC/archive.shtml#NGN (accessed December 2009).
5. Khasnabish B. 2003. *Implementing Voice over IP*. Hoboken, NJ: Wiley-Interscience.
6. Tatipamula M, Khasnabish B, eds. 1998. *Multimedia Communications Networks Technologies and Services*. Norwood, MA: Artech House.
7. Khasnabish B. 2007. Converged Services and NGN: Implementation and Implications. IEEE Distinguished Lecture, Bengaluru and Chennai, India, December 2007.
8. Maeda Y, Sherif MH. 2009. The first ITU-T kaleidoscope event: "Innovations in NGN." *Communications Magazine, IEEE* 47(5):80–81.
9. Khasnabish A. 2009. *Humanitarian Identity and the Political Sublime: Intervention of a Postcolonial Feminist*. Lanham, MD: Lexington Books.
10. ITU-T Draft Recommendation Y.2205: Next Generation Networks—Emergency Telecommunications–Technical Considerations, 2009. Geneva, Switzerland: International Telecommunication Union Telecommunication Standardization Bureau.

CHAPTER 5

IMS AND CONVERGENCE MANAGEMENT

Keizo Kawakami, Kaoru Kenyoshi, and Toshiyuki Misu

5.1 IMS ARCHITECTURE

Service providers (SPs) recently have gained momentum with the migration to the Next Generation Networks (NGN) and the move towards full IP-based networks. This trend is based on the intensifying competition environment among the carriers and the long-term decrease of the incomes from the telephone communication fees. Both fixed and mobile carriers are accelerating the migration to full IP-based network in order to reduce the network operation cost and to provide new value-added services to make difference from other competitors. In particular, fixed mobile convergence (FMC) enables the integration of both fixed and mobile phones facilitating the creation of new value-added services. FMC is expected to create new market and new source of revenue for fixed and mobile service providers. In NGN, various services are provided by the service control function called IP Multimedia Subsystem (IMS), which is located in the service stratum of NGN. By introducing IMS, IP-based multimedia services are provided to various terminals (mobile terminal, wireless LAN (WLAN) terminals, etc.) independently of the access networks. Thus, in NGN, IMS is expected to be the common service control architecture applied to both fixed and mobile. Figure 5-1 shows the NGN architecture overview, which is defined in the ITU-T Recommendation Y.2012 [4].

IMS standardization was set by the 3rd Generation Partnership Projects (3GPP/3GPP2). IMS Phase 1 was completed by Release 5 issued by the 3GPP in 2003. Only the basic call connection was defined in Phase 1, however as a result of the following continuous functional extensions such as provisioning of multimedia services, Release 8 specification is currently being defined. IMS has been initially defined for mobile network and in later years IMS has been expanded by ETSI TISPAN and ITU-T as the NGN service control function for both fixed and mobile networks. Currently, 3GPP is also studying and developing common IMS standards for the fixed and mobile networks.

IMS comprises CSCF (Call Session Control Function), which controls sessions and services, MRF (Multimedia Resource Function), which controls multimedia

Next Generation Telecommunications Networks, Services, and Management, Edited by Thomas Plevyak and Veli Sahin

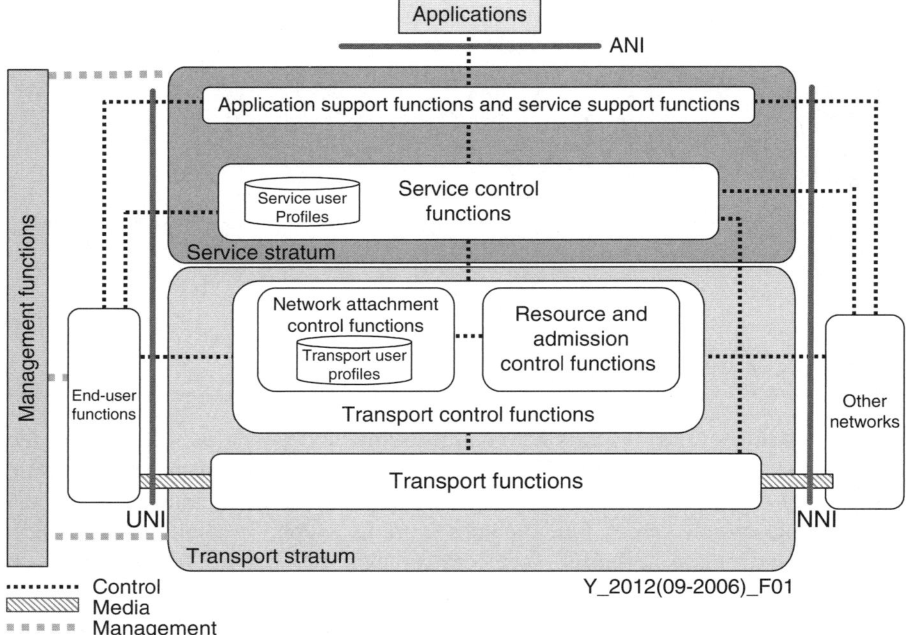

Figure 5-1. NGN architecture overview (ITU-T Rec.Y.2012 [4])

resources, HSS/SLF (Home Subscriber Server/Subscription Locator Function), which manages user profiles, media gateway MGW/MGCF (Media Gateway/Media Gateway Control Function), which performs the inter-working with the existing networks, and the application server (AS), which provides services and applications to be used in IMS. CSCF is a SIP server that controls sessions and services using the SIP protocol, and realizes the core features of the IMS architecture including user terminal access control, roaming control, and activation of the services that are provided by AS. Figure 5-2 shows the reference architecture of the IP Multimedia Core Network Subsystem, which is specified in 3GPP TS 23.228 [7].

5.1.1 Serving CSCF (S-CSCF)

S-CSCF performs session routing based on the destination address (SIP-URI, telephone number, etc.) specified by the initiating user and sets, manages, and releases the session between the initiating and terminating users by using user profile information and user location information managed by HSS. S-CSCF uses SIP-based standard interface called ISC (IMS service control) to connect with the common enablers (general-purpose function realizing individual services such as presence, messaging, etc.) shared by different services and the application servers controlling each service. S-CSCF can provide various services through the connection with multiple application servers. Open service environment (OSE) is introduced to

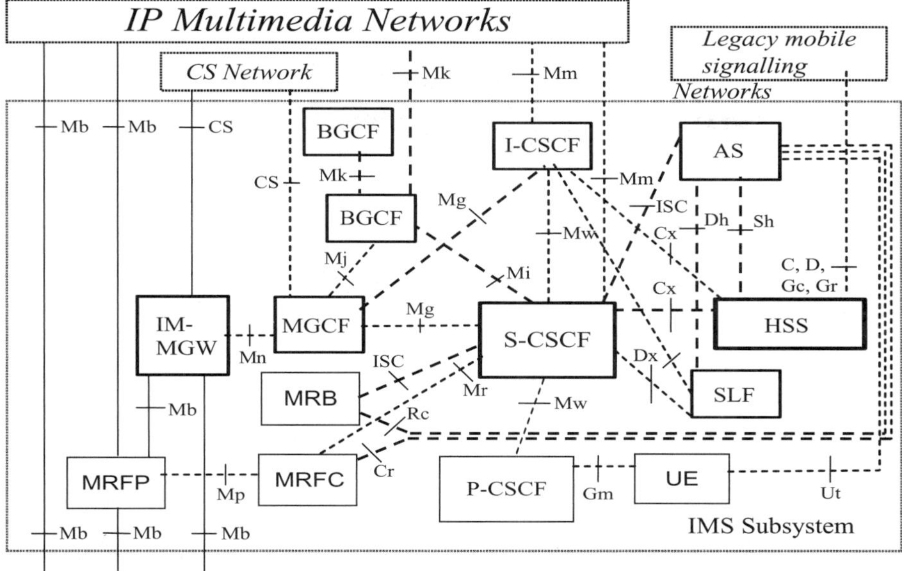

Figure 5-2. Reference architecture of the IP Multimedia Core Network Subsystem (3GPP TS 23.228 [7]). AS, Application Server; BGCF, Breakout Gateway Control Function; CS, Circuit Switched; HSS, Home Subscriber Server; I-CSCF, Interrogating-CSCF; MGCF, Media Gateway Control Function; MGF, Media Gateway Function; MRB, Media Resource Broker; MRFC, Multimedia Resource Function Controller; MRFP, Multimedia Resource Function Processor; P-CSCF, Proxy-CSCF; S-CSCF, Serving-CSCF; SLF, Subscription Locator Function; UE, User Equipment

facilitate the integration between IMS and various applications by converting the IMS service control (ISC) interface provided by S-CSCF into a logical easy-to-use API for the applications. As a result, it becomes possible to use the functions for setting and releasing sessions provided by IMS and acquisition of user location information from various applications and servers, thereby enabling flexible development and deployment of new services. In NGN, the application support function allocated in the service stratum realizes service reusability and service portability, and achieves connection with external applications provided by third parties through application network interface (ANI). Figure 5-3 shows the extended NGN architecture positioning the OSE which is specified in ITU-T Y.2234 [5].

5.1.2 Proxy CSCF (P-CSCF)

P-CSCF is connected to a user terminal through the access network and it is allocated during the connection of the user terminal. The network connection with a user terminal is realized through a packet switch called Gateway GPRS Support Node (GGSN) in the W-CDMA network, and through a relay router in the access type IP network for fixed broadband access and access from wireless LAN. For the user

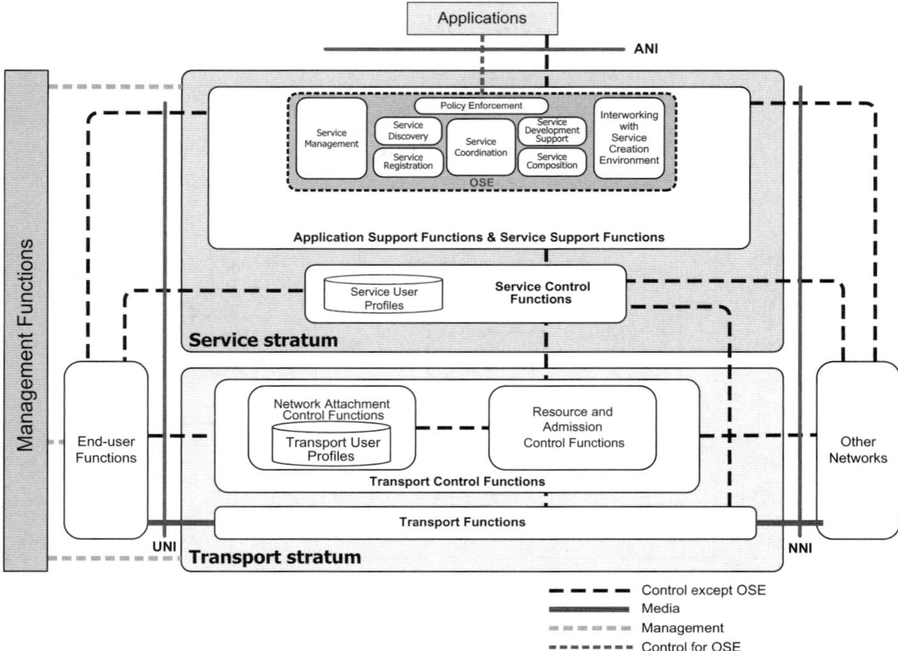

Figure 5-3. Extended NGN architecture positioning the OSE with expanded view of the OSE functional group

terminal connected to the access network, an IPsec tunnel is established with the user terminal after the authentication is performed at connection, and SIP messages from the user are transferred through IPsec, and the integrity of the messages is checked by P-CSCF. For efficient transfer of SIP messages between users, P-CSCF compresses and encodes messages as required. This function is particularly useful in the case of wireless access with limited signal channel bandwidth. Users can use either P-CSCF of the visited network or P-CSCF of the home network at roaming. When P-CSCF of the home network is used if the IMS function is not available in the visiting network, user traffic such as VoIP will also be routed through the home network. By contrast, if P-CSCF of the visited network is used, the traffic between users on the visited network is efficiently transferred locally within the same network. However, P-CSCF of the home network is always used when the user is in the home network.

5.1.3 Interrogating CSCF (I-CSCF)

I-CSCF, located in the home network, accesses the SLF when required at user registration and identifies the HSS that contains the subscriber information. Based on the instructions from HSS, I-CSCF passes registration processing to the S-CSCF that serves the user. When there are multiple HSS in the network, I-CSCF performs the load balancing smoothly and hides the network configuration from external resources.

5.2 IMS SERVICES

5.2.1 Push to Talk over Cellular (PoC) Service

The PoC service realizes 1-to-1 or 1-to-n voice connection between mobile phones. The first commercial service started by Nextel (currently known as Sprint Nextel) of the United States in 2002, this service helped to make difference from other competitors and attracted much attention due to its success related to its high average revenue per user (ARPU) and low customer churn. In Japan, the service started in November 2005 as a new way of communication, different from the traditional voice and mail services. The PoC service platform consists of the IMS platform (handling SIP session control (S-CSCF/P-CSCF) and managing subscriber data (HSS)) and the application server (PoC (AS)), which provides "session initiating/terminating control," "voice initiation control," and "floor re-entry control." Flexible service expandability is also achieved by concentrating the service-dependent features in the application server (PoC (AS)). The main role of PoC (AS) in the session initiating/terminating control function is as follows:

5.2.1.1 Service Authentication This function allows the verification and the authentication of the user before starting the PoC connection; when a mobile terminal requests PoC connection, validation of the terminal and check of the maximum allowed number of members is performed before starting the connection.

5.2.1.2 Floor Information Management A floor is created when a PoC service call occurs as a unit to manage the session. The floor information (identifier, establishment time, number of members, etc.) and information of members joining the floor (member names, status (calling, online, offline, etc.)) can be managed in an integrated manner. By managing the status of each member, the floor information management function notifies the user about the current status of opponent user.

5.2.1.3 Message Duplication and Transmission in 1-to-n Communication
1-to-n communication, which is a feature of the PoC service, is achieved by installing the function that duplicates and distributes the messages from an initiating mobile unit to all the receiving mobile units.

The PoC service requires voice initiation control since it applies a half-duplex communication mode. Voice initiation control is performed through RTP Control Protocol (RTCP) after receiving a voice initiation request from a mobile unit. By managing which user requests the voice initiation, the function enables deterring the voice initiation from other users. 1-to-n communication is enabled by duplicating voice data from the mobile unit with voice initiation and distributing it to other users using Real-time Transport Protocol (RTP). The service provides a re-entry function to enable the user who once left the floor to re-participate in the floor that is continuing. This function is necessary in the mobile environment, since nodes can be disconnected due to the absence of network signals or failures. When receiving a request from a user, the function checks whether the floor that the user wishes to join still exists. If the floor exists, the function connects the mobile unit to the floor. In addition to the presence of the floor as a condition to approve re-entry, adding extra conditions for re-entry based on the service specification of the carrier is also possible.

5.2.2 IMS-Based FMC Service

IMS based FMC service commenced in July 2008 in Japan. In this service, when indoor, 3G/WLAN dual-mode terminal can use the home antenna (Wireless LAN Router) to get connected to a broadband environment (FTTH etc.) through the wireless LAN and then access the internet. with high speed packet transmission/reception up to 54Mbps. IMS based FMC service allows to use 3G services in the 3G service area (outdoors etc.) and when in the WLAN service area (indoors: inside the house), it allows to use VoIP and data access services provided by the IMS core functions; Call Session Control Function (CSCF) and Packet Data Gateway (PDG).

The function of each node that provides the IMS based FMC service is as follows.

5.2.2.1 CSCF As the core function of IMS, CSCF provides session (connection) control, management, authentication, routing, etc. through the SIP based call control. In the IMS-based FMC service, CSCF provides the VoIP service to 3G/WLAN dual terminal using SIP.

5.2.2.2 PDG PDG is a gateway that provides the functions required for the interconnection between the wireless LAN network inside the house and the 3G network. PDG ensures the security when the user connects to a 3G service from the wireless LAN network. Through this gateway, PDG provides users with the services difficult to be achieved by the 3G network only, such as large contents distribution service and low-cost IP telephone service.

In the IMS-based FMC service, initiating/receiving processing is performed using the number starting with 050 in the WLAN area and the number starting 080/090 in the 3G area at the initial commencement of the service. However, currently, one number is provided (number starting with 090/080) for both the 3G service area and the WLAN area for initiating/receiving processing. By using a single-number service, 3G/WLAN dual-mode terminal users can perform initiating/receiving processing with a single number (number starting with 090/080) in either the 3G or WLAN area. The single number service is realized by the interconnection between CSCF, which provides WLAN, and the 3G core network that provides the 3G services. The initiating user can transmit messages using a number starting with 090/080 without being conscious of the location of the receiving IMS based FMC service user, and receiving messages from the IMS based FMC service user is displayed with a number starting with 080/090 regardless of the location of the IMS based FMC service user. Thus, the usability of the FMC service has been enhanced.

5.2.3 IMS-Based IPTV Service

The standardization of IPTV provided by NGN is being promoted by various standardization organizations such as Open IPTV Forum, ATIS, ETSI TISPAN, and

ITU-T. In ITU-T, since 2006, three types of architectures of IPTV in Focus Group-IPTV have been discussed, namely non-NGN, non IMS based NGN, and IMS based NGN architectures. From 2008, the related SG such as SG9, SG11, SG13, and SG16 hold joint meetings as IPTV-GSI, and the formulation of recommendation is still under work. In this discussion, IMS is expected to be the service control platform which provides IPTV. In IPTV-GSI, formulation of recommendation for network architecture Y.1910 [2] was approved in May 2008 and at the same time, the service use case, Supplement 5 [3] to Y-series Recommendations, was approved as the supplementary document. In September 2008, the formulation of recommendation for the IPTV requirement condition Y.1901 [1] was approved, completing the basic recommendations for providing IPTV through IMS-based NGN. QoS (Bandwidth/Priority Control) Guarantee, user authentication, billing, profile management, mobility, FMC realization, provision of services for users in different networks during roaming etc. are the key merits of selecting the IMS as a service control platform for IPTV services. By integrating services provided by IMS such as VoIP, realization of blend services such as display of the call ID on the screen while receiving a call or messages during viewing of IPTV. Realization and control of seamless contents delivery to the terminal by integrating IPTV with presence information is being studied.

5.3 QoS CONTROL AND AUTHENTICATION

This section describes QoS Control function and authentication function required for NGN. These functions are provided respectively by Resource and Admission Control Subsystem (RACS) and Network Attachment Subsystem (NASS) in the transport stratum. Figure 5-4 shows the system configuration including RACS and NASS.

RACS and NASS, cooperating with the Service Control Function in the service stratum, perform QoS control and authentication. The details are described in the following sections.

5.3.1 QoS Control in NGN

Different from the traditional networks individually assuring the service quality, NGN, which is fully IP-based and handles a wide variety of media and services in an integrated manner, is required to provide these services without failure or downgrading the quality level. In an IP-based environment, it becomes easier to implement on-demand broadcasting and services cooperating with Web applications, however, the quality of existing services (e.g., telephone services) must be guaranteed. Specifically, assurance must cover voice service without delay or loss, priority control of emergency calls, quality of calls from start to end, etc., all of which require QoS control at end-to-end level.

NGN performs packet transfer according to the priority when service traffic passes through a network. In this case, since the priority control can be performed separately to each call session based on the authentication information of the relevant

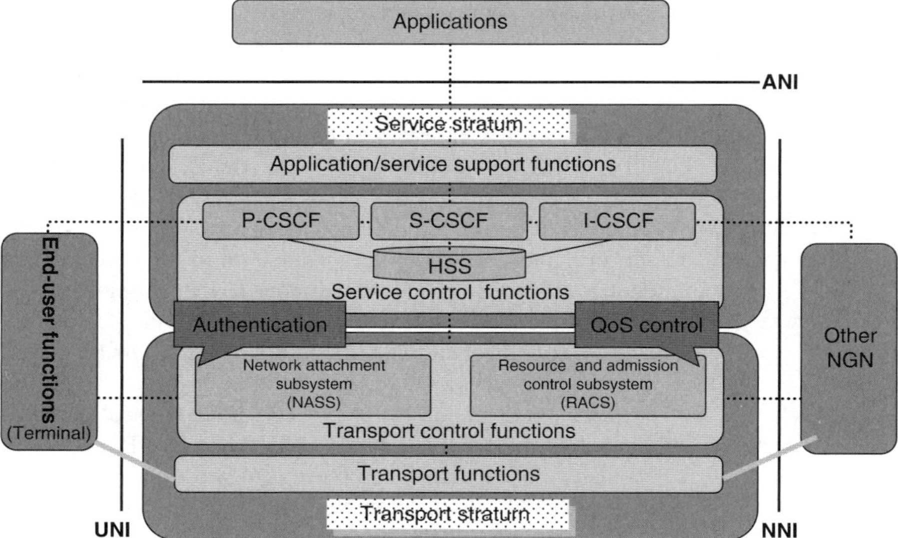

Figure 5-4. Overview of NGN architecture

session, the service levels can be provided by classifying the voice calls into emergency calls and ordinary calls, or per-user charging basis.

NGN also performs admission control for QoS assurance. Specifically, by managing the network bandwidth, NGN secures the bandwidth used by priority traffic for which communication is permitted through acceptance control, and controls malicious/congested traffic. Since the edge router opens and closes the RTP port interlocking with call control through the control from RACS, calls during congestion and the spoofing traffic not compliant to the formal SIP negotiation can be restricted. As a result, the legacy call quality and security are ensured.

5.3.2 RACS

RACS is a subsystem that provides admission control function and gate control function defined by the ETSI TISPAN standard. QoS control in NGN is implemented by RACS, thereby added values such as service levels can be provided.

5.3.2.1 Functions Provided by RACS RACS provides the following functions:

1. *Admission control.* Performs admission of QoS resource requests based on the user profile, operator-specific policy rules, and resource reservation provided by NASS.
2. *Resource reservation.* Verifies if the QoS resource request is within the permitted bandwidth in the access network, and reserves the resource.
3. *Gate control.* Provides NAPT (Network Address Port Translation) control and priority traffic control, and performs gate control of the edge router based on the approved QoS resource request.

5.3.2.2 Function Blocks Comprising RACS RACS consists of the following function blocks:

1. *A-RACF (Access-Resource Admission Control Function).* Performs QoS control to ensure the communication quality for the access network.
2. *SPDF (Service Policy Decision Function).* Receives a resource reservation request, and determines the policy.

Figure 5-5 shows the network components associated with QoS control and the control procedure. The control procedure is as follows.

<1> In the access authentication when starting the access to the network from a terminal, RACS receives and retains the QoS profile from NASS.

<2> The originating terminal sends a session initiation request (SIP message) to CSCF.

<3> CSCF sends a QoS resource request to RACS together with the information such as the bandwidth required for service execution, service class, and reservation priority.

<4> After receiving the QoS resource request, RACS collates the requested conditions with the retained user profile and edge router information, and performs admission control. RACS also calculates the bandwidth value required for guaranteeing the service quality.

<5> RACS performs gate control for the edge router based on the calculated bandwidth value.

<6> RACS responds about the securing of the QoS resource to CSCF.

<7> CSCF sends a session initiation request to the terminating terminal.

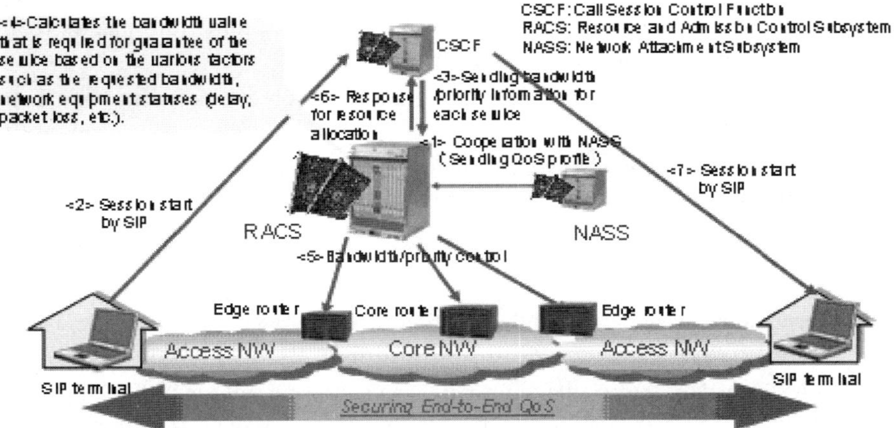

Figure 5-5. Network components and control flow for QoS

5.3.3 Authentication in NGN

In NGN, authentication is defined in two layers: authentication at the access level for allocating IP addresses and authentication at the service application level in order to use SIP. NASS performs authentication at the network access level based on the user profile.

Roles and functions are specified as follows:

1. *Profile management.* Manages user profiles (subscriber ID, access ID, location information, etc.).
2. *Access authentication.* Receives a connection request from a terminal and performs authentication at the access level based on the user profile.
3. *DHCP (Dynamic Host Configuration Protocol) server function.* Assigns an IP address to the terminal.
4. *Connection information management.* Manages the connection information and the IP addresses of the currently connected terminals.
5. *RACS Interaction.* Manages QoS profile information required for RACS bandwidth management, and provides QoS profile information to RACS.

5.3.4 NASS

NASS is a function to perform IP address distribution and authentication defined in the ETSI TISPAN standard. NASS consists of NACF, CLF, UAAF, PDBF, AMF, and CNGCF functions blocks.

1. *NACF (Network Access Configuration Function).* It assigns IP addresses.
2. *CLF (Connectivity Session Location Function).* Manages the assigned IP address information, and transfers QoS profile information containing the IP address and QoS information to RACS.
3. *UAAF (User Access Authorization Function).* Performs authentication based on the user profile information.
4. *PDBF (Profile Database Function).* Manages user profile information (e.g., subscriber ID, subscribed services etc.).
5. *AMF (Access Management Function).* Distributes requests from terminals to NACF/UAAF.
6. *CNGCF (Customer Network Gateway Configuration Function).* Performs additional terminal settings not implemented by NACF.

Figure 5-6 shows the authentication procedure and related network components. The authentication procedure is described as follows:

<1> Terminal requests access authentication to NASS when initiating the access to the network.

<2> NASS verifies the profile information such as the subscriber ID, physical access ID, and logical access ID for the terminal connection request, and performs authentication.

■Authenticates the network access level based on the user profile.
■Manages IP addresses of terminals and network information to be set in the terminals, and provides the information to the terminal whose connection is permitted (DHCP server function).
■Manages QoS profile information that is required for RACS bandwidth management and provides the information to RACS.

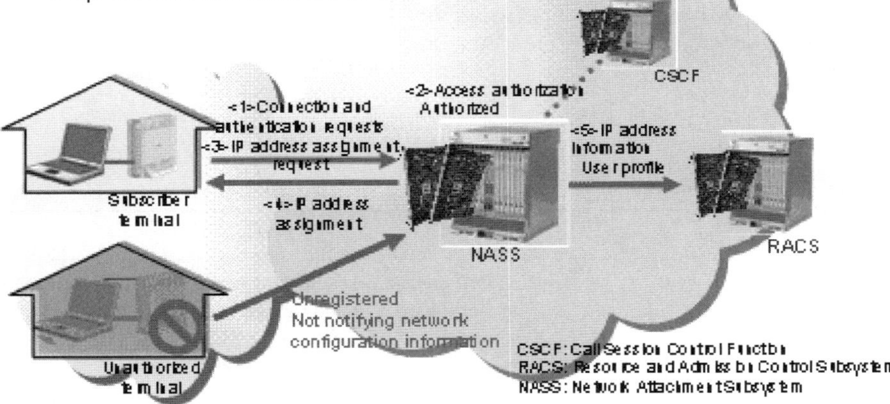

Figure 5-6. Network components and procedure for authentication

<3> The authenticated terminal requests IP address assignment to NASS by sending a DHCP request.

<4> NASS assigns an IP address to the terminal through DHCP.

<5> When interacting with RACS, NASS sends the QoS profile information to RACS together with the subscriber ID and the IP address. (See <1> of the QoS control procedure in Section 5.3.2.2)

5.4 NETWORK AND SERVICE MANAGEMENT FOR NGN

5.4.1 Introduction

Considering the diversity of services provided in NGN, demands for high quality, evolution of network technologies, and diversity of providers conducting businesses on the NGN environment, the NGN management functions must meet not only the requirements for traditional existing networks but also those from multifaceted aspects.

This chapter describes the requirements of NGN management by classifying them into the following categories:

- Network management operation requirements
- Service management operation requirements
- Service enhancement requirements

- B2B realization requirements
- Compliance with legal regulations requirements

In each category, the requirements for FCAPS are listed and described in Table 5.1. FCAPS refers to the Fault management, Configuration management, Accounting management, Performance management, and Security management. They are represented as the general management functionalities in the ITU-T Recommendation M3400 [6].

TABLE 5-1. Requirements for FCAPS

No.	Function	Description
1	Fault management	The fault management allows detecting a fault event in the network, identifying the cause of fault, locating and minimizing the impacts of the fault on services, arranging operations (issuing a trouble ticket), analyzing faults, and providing fault recovery. The fault management supports the following features: • Fault monitoring • Fault isolation and identification of the major causes • Fault diagnosis • Fault recovery • Trouble management
2	Configuration management	The configuration management provides NE (Network Element)/Path Route configuration management, NE parameter management, synchronization and modification management of configuration information with current network, status data collection, configuration information setting to NE, and NE control. The configuration management supports the following features: • Network planning and designing • Network construction • Service planning • Service provisioning • Status management and control
3	Accounting management	The accounting management allows measuring the network services usage amount per user, and notifying the amount to the service provider (service department). The account management also allows sending notifications to the users about their network services usage fees. The performance management supports the following features: • Measurement of the services used amount • Usage fee management • Billing, payment, and credit management • Settlement among service providers

TABLE 5-1. Continued

No.	Function	Description
4	Performance management	The performance management allows measuring, analyzing, and reporting the performance and the quality of the NE and the network. The performance management also allows collecting of NE and Network statistic information, either periodically or at ad-hoc, comparing the threshold values, analyzing the data, and confirming the existence or absence of problems. NE performance data, network traffic data, and QoS data are object to analysis. The performance management supports the following features: • Performance data collection/traffic data collection • Performance analysis/traffic analysis • Network capacity analysis • QoS data collection • SLA (Service Level Agreement) determination QoS data includes, for example, packet loss, delay, and jitter as flow quality for VoIP and video services. If these are defined as the SLA conditions, the performance management compares the measured values with the conditions, and determines the quality of service.
5	Security management	The security management provides functions to detect security abnormal events on networks (e.g., security violation event) and to audit security. The security management also provides management functions such as user authentication among systems and among user systems in the Network Management system, access control, and operation log history. The security management supports the following features: • Security abnormality monitoring • Security abnormality prevention • Security recovery • Security auditing

5.4.2 Network Management Operation Requirements

From the viewpoint of network operation management, reduction of network operating cost must be compatible with high-quality network maintenance. The major requirements include efficiency in providing network resources satisfying the demands, assurance for service provisioning quality, prompt and immediate interventions during failures, proactive handling against failure warnings, and real-time accounting. These requirements are described below:

Configuration Management:

- Providing the ability to manage NGN system resources, both physical and logical (including resources in the core network, access networks, interconnect components, and customer networks and their terminals)
- Integrating an abstracted view on Resources, which is hiding complexity and multiplicity of technologies and domains in the resource layer
- Efficient network expansion according to the network utilization status
- Provision of efficient test tools during network expansion

Fault Management, Performance Management, and Security Management:

- Provision of integrated monitoring
- Monitoring of network service quality (QoS)
- Early detection of network failures and QoS deterioration and identification and isolation of failure causes at early point.
- Advance detection of failure indicators
- The ability to have proactive trend monitoring

Accounting Management:

- Collection and accounting (rating) of service usage information in real time
- Supporting the availability of management services any place any time to any authorized organization or individual (e.g., access to billing records)

Common Functions:

- Automation and acceleration of end-to-end operation process

5.4.3 Service Management Operation Requirements

From the viewpoint of service management operation, service provisioning quality maintenance must be compatible with operating cost reduction. Since the quality of services has a direct relation with customer satisfaction, addressing the requirements from customers is quite important. The requirements from the viewpoint of service management operation are as follows.

Configuration Management:

- Automation and acceleration of service provisioning
- Automation of provisioning from service level to network level
- Automatic testing linked with service orders
- Customers' self-service provisioning

Fault Management, Performance Management, and Security Management:

- Monitoring at SLA provisioning level
- Early detection of SLA violation and early identification of failure causes
- Analysis of affected range at failure occurrence or QoS deterioration, and early notification to customer

- Prompt notification to customer at SLA violation
- Service quality confirmation test at customer inquiry, dispatch of maintenance operation, and operation progress status notification

Accounting Management:

- SLA-based accounting adjustment
- Service usage logs and provided quality tracing
- Real time information on accounting and settlement status for users

5.4.4 Service Enhancement Requirements

Increase in highly sophisticated services due to the high-functionality equipments and the advancement of video services increases the needs for device remote control, video quality of experience (QoE) measuring, and other customer support requirements. In addition, it is also required to cope with service personalization based on the user access method and context information (location and presence). Since the advancement of services involves in the reduction of services lifecycle, lifecycle management has become necessary as a service management platform.

Based on this background, the following requirements need to be considered:

Configuration Management:

- Remote settings for terminals (mobile terminals and CPE) and firmware update
- Service provisioning function corresponding to service lifecycle reduction

Fault Management, Performance Management, and Security Management:

- Collection of quality information from terminals (mobile terminals and CPE)
- Monitoring quality of experience (QoE)

Accounting Management:

- Accounting function corresponding to service lifecycle reduction
- Accounting model corresponding to user segmentation and access method

Common Functions:

- Collection of user context information (presence, location, preference, etc.) from terminals
- Providing user context information to service functions
- Protection of user context information

5.4.5 B2B Realization Requirements

In NGN, it is necessary to realize B2B process between the service providers, as well as between NGN providers. The requirements for B2B realizations are as follows:

Configuration Management:

- Service provisioning functions covering multiple providers

Fault Management, Performance Management, and Security Management:

- SLA-based SLA monitoring among multiple providers, and dispatching of SLA violation notification to the provider

Accounting Management:

- Inter-provider settlement function for distributing income among multiple providers

Common functions:

- Disclosure of user context information to 3rd parties

5.4.6 Compliance with Legal Restrictions Requirements

In NGN, the legal regulations imposed on the existing communication services must be observed continuously and the operation functions must be provided for the following requirements:

- Emergency communication
- Lawful interception
- Confidentiality of communication contents
- Storage of communication records
- Protection of personal information

5.5 IMS ADVANTAGES

This section describes various advantages of the IMS platform to be constructed under a multi-vendor environment according to the international standards. Cost reduction for operation and maintenance, and quick provisioning of a wide variety of services using open interface installed in the upper level of the Service Delivery Platform (SDP) over IMS, will be described particularly in this section.

5.5.1 Reduction of Maintenance and Operating Cost

The IP-based network enabled the integration of fixed and mobile networks allowing not only the migration of telephone networks but also the diversity of communication services such as FMC and rich multimedia services and the shifting of businesses to the non-traffic field. Under such circumstances, communication carriers find themselves required to shift their investments to the service and IT fields more than ever by improving the operation efficiency while ensuring safe and secure networks.

On the other hand, IMS, whose specifications are defined by 3GPP, is a system that provides IP-based multimedia services to mobile phones and WLAN terminals independently of the access network. Rather than providing only IP-based services, IMS can also provide services by interacting with PSTN and the Internet. IMS employs SIP as the core control protocol. In the traditional architecture, since SIP depends a lot on the implementation specifications, the interconnectivity needs to be thoroughly examined. However, by employing SIP as IMS, interconnectivity among communication carriers has become simple.

In these circumstances, the following advantages can be listed for communication carriers to introduce IMS from the viewpoint of business investments.

5.5.1.1 Reduction of Time Required for Introducing New Services (Time to Market) By providing an open interface through employment of the open service architecture, service development and deployment become faster. Using the open interfaces, personalized services can be created and provided by service operators and third-party application providers.

5.5.1.2 Cost Merits Introduction of IMS enables effective use of facilities, for example, voice and data that have been handled separately in the traditional method can be integrated. In the traditional architecture, a system has been individually constructed for each service; however, in IMS a variety of services can be implemented in the common service platform, which facilitates inter-working among these services and leads accordingly the reduction of costs for development, maintenance, and operation. By realizing a multi-vendor environment based on the standard architecture, optimum components can be installed with less cost. Interconnections among communication carriers for basic services become easy, enabling communication carriers to provide differentiating services and make investments by introducing services at higher levels.

5.5.2 Roles of SDP and Development and Introduction of New Services

In NGN, the key issues for communication carriers and service providers are how to develop and deploy efficiently new nontraditional differentiating services. The Service Delivery Platform (hereinafter referred to as SDP) responds to the above demands. SDP is attracting much attention as a common platform to provide various functions ranging from service development, operation and maintenance, and accounting to user management functions.

5.5.2.1 Positioning of SDP in NGN Figure 5-7 shows the positioning of SDP in the NGN architecture. SDP is a platform belonging to the application layer positioned in the upper level of IMS. SDP consists of a set of service execution functions called service enablers and various service control/management functions. The following are examples of service enablers (Application Servers (AS)):

- Call Control AS
- Presence AS
- Media Resource Control AS

These application servers use SIP and other protocols to inter-work with the nodes in the control layer. The application servers also provide open APIs as easy-to-use programming and interfaces to realize the interaction with external systems. These open APIs allow developers to develop and deploy services easily without requiring knowledge of the network control layer configuration and protocols.

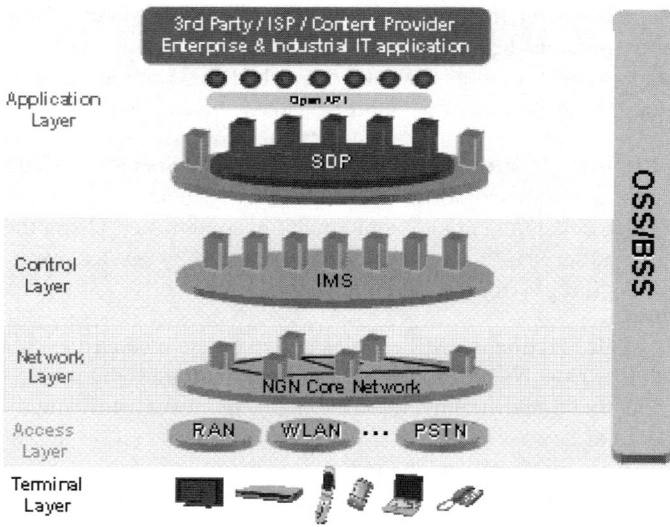

Figure 5-7. Service delivery architecture for NGN

SDP realizes the integration of IT applications (e.g., business systems) and Internet-based Web services with the SIP-based communication platform, enabling quick development and fast delivery of new services as well as expansion of service area.

5.5.2.2 Features of SDP This section describes the features of the SDP.

1. *Fast service development/deployment.* Java/SOAP (Simple Object Access Protocol)/REST (Representational State Transfer) and others provide easy-to-use Open APIs to improve the development time-frame and allow fast deployment of new services.
2. *User-friendly programming.* The SIP support interface verifications and connection tests required during each services development in the traditional structure are no longer necessary; SDP hides the complicated SIP call control procedures.
3. *Flexible system configuration.* In the SDP configuration, the application servers providing individual services are located on top of the commonly used application servers providing the control of services. This configuration allows easy deployment of new services and flexible scalability of the system depending on the type and scale of the services.

5.5.2.3 Examples of Application Servers This section describes the key Application Servers of the common service execution layer.

Call Control. Call control AS realizes various connection controls as value-added services. Call control provides simple programming APIs to be easily handled by the application programmers, and allows the control of SIP-based services (e.g., two-party call, voice distribution etc.) from the application.

The call control performed based on program logic, on manual operation, or on event occurrence is referred to as 3PCC (3rd Party Call Control). The call control AS is considered as the SIP application server realizing the 3PCC function. In addition to the basic call control, the APIs provided by the call control AS enable also transferring and combining of the calls in progress. For example, it is possible to play voice guidance during the call, transfer the call, or switch from two-party call into a conference call.

Presence. Presence AS manages the presence status of either a person or a machine, allows the search and the update of the presence information from terminals and applications, and notifies terminals and applications about the new presence information when updated. The contents of presence information could be an "activity"; Available/Busy/At Lunch, "location"; Home/Office/Café, "mood"; Happy/Angry/Puzzled, "privacy"; Public/Private, or "detailed location"; Latitude/Longitude and Tokyo/New York/London.

In the presence service, the party that references or views the presence information is called "watcher" and the party whose presence information is referenced is called "presentity." Presence AS provides access control function to prevent the "watcher" from referencing the presence information without the permission of the "presentity" or of the presence AS administrator. Both the "watcher" and the "presentity" can be either a terminal or an application. For example, the "watcher" can be a terminal while the "presentity" can be another terminal, or the "watcher" can be an application while the "presentity" can be a terminal etc.

The presence information is used in various scenes. For instance, the presence information is very helpful in selecting the convenient communication media when trying to contact another party. If the opponent is in the office at work for example, then a phone call may be appropriate, but if he or she is in the middle of a conference, then the caller can either choose to call in a later time or send an instant message. The Presence AS also allows the calls to subscriber's home phone to be forwarded to his or her mobile phone when the subscriber's presence status is set to "Outside." This kind of call forwarding is called "presence-based call forwarding."

Media Resource Control. Media Resource Control AS provides the audio and video control functions. It is used for example to transmit voice messages or to set a video conference etc. The following are examples of provided media control functions:

Media transmission: This is used for example to send a voice message to a terminal, or to play selected calling tone (CRBT (customized ring back tone)) to the caller while waiting for the Callee to answer the phone, or also to play IVR (interactive voice response) synthetic speech, etc.

Media mixing and media distribution: This is used to mix the media from multiple members participating in a voice or video conference and then distribute it to the conference participants.

Media recognition: This function includes DTMF detection and speech recognition. The application is notified about the detection and recognition results.

Media storage: This function enables the user to store and play of recorded voices similarly to the answering machine (voice mail) service.

Messaging. By the interaction with the applications, messaging AS performs transmission/reception of SMS (short message service) and MMS (multimedia message service). Message types include texts, images, videos, and ringtones. Messages can be sent from an application by specifying the message type and the recipient address (terminal URI). The message transmission result notification can either be received automatically or stored to be checked later.

One of the service examples is a simultaneous broadcasting service used in emergency cases, such as transmission of evacuation messages to the residents of a specific area during Tsunami forecast.

Location. Location AS manages the location information (latitude, longitude, precision, time stamp, etc.) of the terminal and provides this information as a response to the request from the application. Several methods are used to determine the terminal location; the method of using a GPS device in the terminal, the method of determining the location information of the base station to which the mobile terminal is connected, and the method of determining the location information of the nearest wireless LAN access point to the terminal.

The PULL type and the PUSH type are available for obtaining location information of a terminal from an application. In the PULL type, the application specifies an URI (such as SIP-URI and TEL-URI) used to obtain the location information, of a specific terminal, stored in the server. The PUSH type includes the periodic notification of the location information of a specific terminal, and the notification of the URI of all terminals that entered or exited a specific area to the application.

As examples of services using location AS, sale/discount information provided by the shops can be transmitted to the subscribers passing by the shop, and also warning messages can be sent to the person who intruded into a prohibited area.

Device Management. Device management AS provides remotely and in a centralized manner the management of devices; mobile phone/PDA/PC/car navigator/game equipment/etc. Management functions include updating of the firmware, collecting of devices power status (ON/OFF), remotely locking the devices, and restrict/release of specific device functions.

For example, Device Management AS can provide a "device fault remote diagnosis service" where device failures are analyzed remotely through the collection of the device logs and status, and repairs are performed promptly and automatically. The log data collected from devices used to play games for long time can be used for marketing. If a device is lost or stolen, the device can be locked and the data can be remotely wiped from the device (Lock & Wipe) increasing user security.

AAA. AAA AS is a server that performs authentication, authorization, and accounting. Authentication is to authenticate an individual or service provider based on the ID and the password. Authorization is to check the utilization authorization or access authorization before approving the access. Accounting is to record approved conducts. AAA AS can perform authentication based on the NGN authentication or IMS-AKA authentication. Interaction with OpenID enables using IDs simultaneously in multiple websites. Authentication adopting the single login feature allows the user to automatically log-into the other services once he logged into one service. Online settlement feature can also be realized through the interaction with settlement systems (e.g., electronic money, credit card).

User Profile Management. User profile management AS manages personal information and preference information of end users. Personal information includes the name, date of birth, address, telephone number etc., and preference information refers to the field of interest of each user (baseball, travel, jewel, etc.). As a service example, the "recommendation service" that allows providing suitable goods information to the user based on the user profile and user's purchase history. Another example is the "target advertisement system" that based on the user preference, user activity history, and time, selects appropriate persons from those who are near the restaurant, and sends an advertisement or information about the restaurant.

5.5.2.4 API Parlay is one of the telecommunication industry standard APIs (Application Programming Interface). Parlay is developed and standardized by the Parlay group formed by members from telecommunication related companies and IT companies, Parlay is also defined as part of OSA (Open Service Architecture) API in the 3GPP standard.

Parlay API is designed by abstracting the section related to the hardware, enabling the common use of the API without being conscious of the differences between fixed and mobile network, current network, and NGN. It also aims at enabling various applications to access the network resources of carriers from remote sites.

In addition to Parlay, Parlay-X is released to be used in telecommunication field. While Parlay is a set of APIs intended to be used by application developers familiar with Java and other programming technologies enabling detail control, Parlay X is designed to enable the application developers to access from external sites and make use of the functionalities, assets and resources of the carriers' network. Parlay X enables the use of communication services from the applications via Web Service messages. Parlay X can be used via SOAP making application development easier and simpler than Parlay.

Considering the recent rise of Web services in the communication market, the future prospect of Parlay X is promising as the API uses SOAP, which is highly compatible with Web services.

Parlay X APIs enable the control and management of services including but not limited to Call Control, Messaging, Presence, Location, Account, Terminal Status, etc.

5.5.3 Services Implemented on NGN

With the popularization of IP-based networks and diversification of services, the PoC (push to talk over cellular) service described in Section 5.2.1.1 has been extended from a voice-based solution to a multimedia solution (from push to talk to push to X), the development of presence and location based FMC services, and the integration of communication and broadcasting services (IPTV) are expected to significantly increase. In addition, the communication ways are also expanding from a simple communication services to a variety of services more related to the actual daily life and from person to person communication to person to machine, and then machine to machine communications.

The following sections give examples of prospective services to be implemented on NGN and discuss the relation with IMS and SDP.

5.5.3.1 Push to X PoC is a half-duplex call service that allows the user to transmit a voice by pressing a button on the terminal, similar to the transceivers and business radio equipments. By making use of the NGN and SDP features, Push to X (where X refers to "any") was developed as an enhanced solution of PoC, handling other types of multimedia.

Push to X is a communication service realized by making use of the Enablers provided in the SDP, and NGN capabilities; QoS, real-time communication, etc. Figure 5-8 shows an implementation example. Push to X includes call control (CC), presence (PR), messaging (MS), and media resource control (MR) application servers. The functions provided by these application servers are used by the Push to X application through APIs.

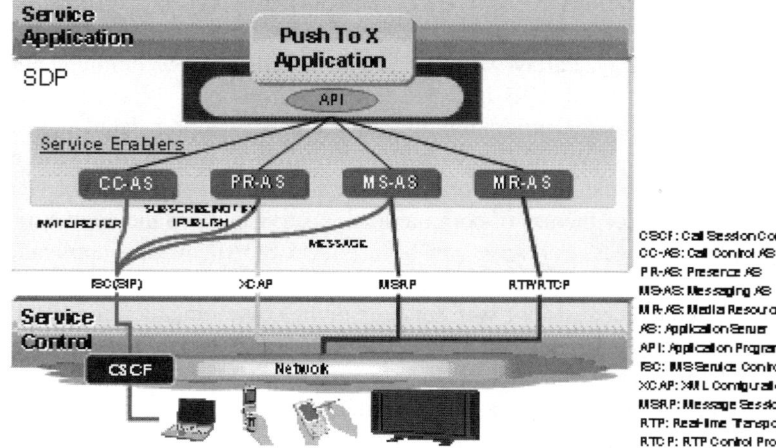

Figure 5-8. "Push to X" on SDP architecture

In addition to the transceiver-like voice solution, Push to X provides end users with more realistic and diversified ways of communication. Push to X provides, image sharing, video sharing, whiteboard sharing, Web sharing, and text chatting. Push to X solution is expected to be a rich communication tool within family or community members, or to be also used as a groupware.

5.5.3.2 IPTV IPTV is considered one of the promising services that can be realized on top of NGN.

There are various definitions for IPTV, however, this document follows the definition given by international standardization organization, ITU-T. IPTV is defined by ITU-T as follows:

"IPTV is defined as multimedia services such as television/video/audio/text/graphics/data delivered over IP based networks managed to provide the required level of QoS/QoE, security, interactivity and reliability." Basically, IPTV is not an NGN-unique solution. Services can also be provided to users in the existing networks as long as these networks are IP-based.

5.5.3.3 IPTV Architectures Three different architectures are defined for IPTV. Figure 5-9 shows three IPTV architectures.

Non-NGN. In this case NGN is not used, services are provided by using the Internet or the operator's closed network. Since the existing network is used in this case, the initial cost is generally low. However, in many cases, vendor-unique specifications are applied. Most of these specifications are not standardized, resulting in the remaining of various issues on interoperability or interconnectivity.

Non-IMS (NGN). In this case part of NGN is used, however, IMS is not used (non-IMS). Although the NGN functions are used for the security and bandwidth control, IMS is not used for the control. For Non-IMS IPTV, in most cases

Mode		Non-NGN	NGN	
			non-IMS based IPTV (Dedicated Subsystem)	IMS-based IPTV
Network outline diagram		 Closed Network	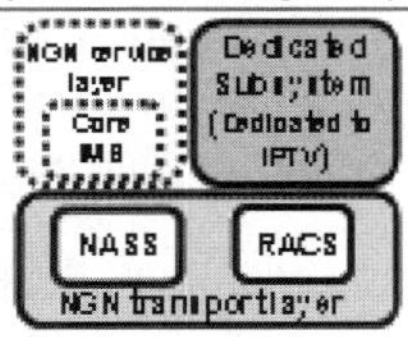 NGN service layer Core IMS Dedicated Subsystem (Dedicated to IPTV) NASS RACS NGN transport layer	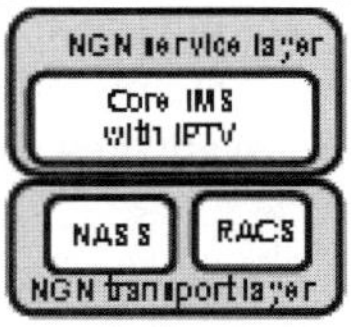 NGN service layer Core IMS with IPTV NASS RACS NGN transport layer
Overview		Mode that provides IPTV in a closed network	Mode that installs IPTV functions in the subsystem different from Core IMS	Mode that installs IPTV functions together with Core IMS
Standard-ization group	TISPAN	X	O	O
	ITU-T	X	O	O
	ATIS	X	O	O

Figure 5-9. Comparison of IPTV architectures

special systems are required in order to integrate IPTV with the existing IMS services such as telephone, presence, and PoC.

IMS-based (NGN). This is an IMS-based case, where IMS is used for the security and bandwidth control. In this case, the interaction of IMS IPTV with the existing IMS services (e.g., telephone, presence, and PoC) can be easily performed, and the management and accounting systems of IMS can also be reused.

5.5.3.4 Advantages of NGN (IMS-based) IPTV As described in the previous section, several implementation modes are available for IPTV. This section describes the advantages to implement IPTV in NGN (IMS-based) mode.

Enhanced Video and Sound Quality. At present, key applications using the network include Web browsing and message transmission/reception. The control protocol provides the re-transmission of the data when packet loss occurs within the network, ensuring the browsing and reception of correct data by the end users. It also minimizes the delays during the network packet delays. Therefore, even in an Internet-based network where the bandwidth is not guaranteed, users do not encounter serious problems.

However, in IPTV, the end user receives the transmitted videos and sounds in real-time. Therefore, when packet loss or delay occurs within the network, the problem cannot be solved by re-transmitting the packet, causing fuzzy reception of videos and sounds troubles to the end user.

Compared with this, introducing the bandwidth guarantee mechanism of IMS, the bandwidth required by the user application is secured on demand, thus occurrence of packet loss and delay can be controlled. This enhances the quality of videos and sounds, and improves the user experience quality.

Control of CAPEX/OPEX. When IPTV is implemented in IMS, the accounting and bandwidth guarantee mechanisms can be standardized with that of the telephone service. Therefore, the providers who have already been implementing telephone services on IMS can reduce the operating cost for service maintenance and the initial investment cost during the implementation of IPTV service by sharing part of the systems of the telephone service.

Taking the above into consideration, contrarily to the Non-IMS which requires special systems for bandwidth guarantee and accounting, the IMS-based system has better advantages and can be easily realized as a business model.

Improvement of Usability. The standardization of the existing services such as telephone, presence, and PoC in IMS, facilitates the interoperability among multiple vendor equipments. When IPTV is implemented on IMS, other IMS services can be easily integrated with IPTV. For example, the caller's phone number and name can be displayed on the IPTV terminal screen when the user receives a call while watching TV. It is also possible to record the IPTV program while talking on the phone. The mash-up of multiple services improves the convenience and user-friendliness of solutions.

Quicker Service Delivery. The IPTV services are not limited to the videos and sound transmission. The IPTV service includes distribution of advertisements according to the users' preferences and also control of the contents addressed to children. By implementing such services on SDP as described above, multiple types of advertisement distribution engines can be selected and new services can be created easily and quickly.

As described above, implementing IPTV on top of IMS has better advantages compared to the other modes of implementation. IMS-based IPTV is regarded as one of the most important items by the NGN standardization organizations (ITU-T, TISPAN etc.), and is formulated as part of the NGN Release 2 (NGN Release 1 mainly handles the telephone service). In addition, the IMS-based IPTV interoperability tests between multiple vendor equipments in GMI (Global MSF Interoperability) clearly indicate that IMS-based IPTV is attracting a lot of attention. From these reasons, the IMS-based IPTV service is expected to be one of the promising key services that utilize the features of NGN.

5.6 REFERENCES

1. ITU-T Recommendation Y. 1901. Requirements for the Support of IPTV services, 2008. Geneva, Switzerland: International Telecommunication Union Telecommunication Standardization Bureau.
2. ITU-T Recommendation Y. 1910. IPTV Functional Architecture, 2008. Geneva, Switzerland: International Telecommunication Union Telecommunication Standardization Bureau.
3. ITU-T Y. 1900 Series, Supplement 5. IPTV Service Use Cases, 2008. Geneva, Switzerland: International Telecommunication Union Telecommunication Standardization Bureau.
4. ITU-T Recommendation Y. 2012. Functional requirements and architecture of the NGN release 1, 2006. Geneva, Switzerland: International Telecommunication Union Telecommunication Standardization Bureau.
5. ITU-T Recommendation Y. 2234. Open service environment capabilities for NGN, 2008. Geneva, Switzerland: International Telecommunication Union Telecommunication Standardization Bureau.
6. ITU-T Recommendation M. 3400. TMN management functions, 2000. Geneva, Switzerland: International Telecommunication Union Telecommunication Standardization Bureau.
7. 3GPP. 2008. IP Multimedia Subsystem (IMS); Stage 2 (Release 8). 3GPP TS 23.228 V8.7.0. Sophia-Antipolis, France: 3GPP Mobile Competence Centre.

5.7 SUGGESTED FURTHER READING

1. ITU-T Recommendation M. 3060. Principles for the Management of Next Generation Networks, 2006. Geneva, Switzerland: International Telecommunication Union Telecommunication Standardization Bureau.
2. TM Forum. 2008. NGN Management Strategy. TR133. Morristown, NJ: TM Forum.
3. ETSI. 2005. ETSI TR 188 004 Telecommunications and Internet converged Services and Protocols for Advanced Networking (TISPAN); NGN Management; OSS vision. Sophia Antipolis, France: European Telecommunications Standards Institute (ETSI).
4. ETSI. "TISPAN." http://www.etsi.org/tispan/ (accessed October 1, 2009).
5. ETSI. "ETSI Standards." http://www.etsi.org/WebSite/Standards/Standard.aspx (accessed October 1, 2009).
6. 3GPP. 2003. Open Service Access (OSA). Sophia-Antipolis, France: 3GPP Mobile Competence Centre.

7. Open Mobile Alliance. "Presence and Availability Working Group." http://www.openmobilealliance.org/technical/PAG.aspx (accessed October 1, 2009).
8. Open Mobile Alliance. "Messaging Working Group." http://www.openmobilealliance.org/Technical/MWG.aspx (accessed October 1, 2009).
9. Open Mobile Alliance. "Location Working Group." http://www.openmobilealliance.org/Technical/LOC.aspx (accessed October 1, 2009).
10. Open Mobile Alliance. "Device Management Working Group." http://www.openmobilealliance.org/Technical/DM.aspx (accessed October 1, 2009).
11. Open Mobile Alliance. "OMA Push to talk over Cellular V1.0.2" http://www.openmobilealliance.org/technical/release_program/poc_v1_0.aspx (accessed October 1, 2009).

CHAPTER 6

NEXT GENERATION OSS ARCHITECTURE

Steve Orobec

6.1 INTRODUCTION

The intention of this chapter is to give the reader an introduction to the world of Operations Support Systems (OSS) architecture. The focus is on areas that are being implemented in Service Provider (SP) OSS and vendor commercial off the shelf (COTS) applications. It is restricted to service management systems that manage SP networks. Hence coverage of Network Element (NE) manager interfaces, from the Simple Network Management Protocol to Transaction Language 1 (SNMP or TL1) are not in its scope. It is important that, if Next Generation Networks (NGN) are an enabler for efficient network management, the prime driver for SP profitability is the ability to leverage these NGN for building productized services. NGN management systems cannot be considered in isolation; they must be considered in terms of a wider OSS enterprise perspective.

For the SP, standards are not merely about intellectual debate and academic idealism, rather they represent real requirements to reduce complexity and cost and increase flexibility, reuse, and agility. These standards should support the customer (or end user) wherever they are and over any communications device, as well as providing the user with good quality-of-service (QoS).

It is impossible to cover every single standards body or initiative, therefore, I have attempted to identify trends for convergence and show how technology, eclecticism, and cutting edge standards work can filter through into these other areas in the future. Typically, among SPs and vendors there is a movement toward utilization of standard data taxonomies (such as SID). It is necessary to consider how components that are described by a general OSS taxonomy may be integrated with a more insular data taxonomy used by an NGN management system. Ultimately it is the data impedance (different data definitions on either side of OSS component boundaries) and discrepancy that complicates and adds cost to any OSS integration (so-called integration tax).

Traditionally, the ITU-T Telecommunications Management Network (TMN) triangle has a focus of the network characterized by its large base representing the

Next Generation Telecommunications Networks, Services, and Management, Edited by Thomas Plevyak and Veli Sahin

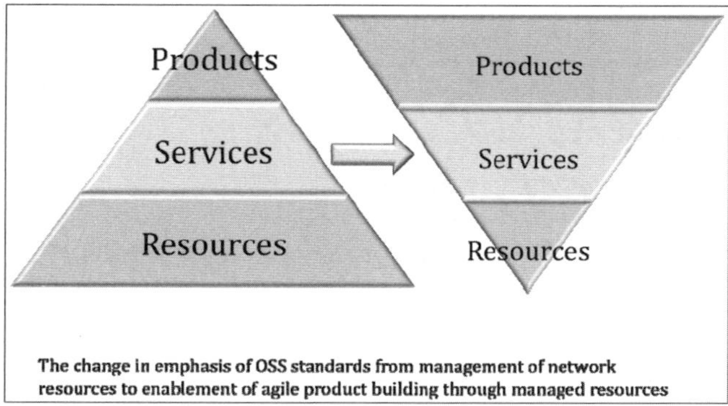

The change in emphasis of OSS standards from management of network resources to enablement of agile product building through managed resources

Figure 6-1. Transition of TMN layered model showing transition from historic relative importance of network to modern focus on products and services

relative importance of the network in comparison. In today's environment and tomorrow's, it will be argued that the triangle should be turned upside down, with the focus on management of products and services (Figure 6-1). Management of the network is important because it provides the foundation for all service management. The ability to rapidly create new services and products relies on good network modeling and management, but ultimately it is the business and service layers that generate revenue.

This chapter will focus on convergent efforts to provide this agile business/service capability and touch on how these capabilities may be implemented.

Compared to network standards, OSS standardization is in relative infancy after a few false starts in the previous decade (the TMN Q3 interface and Common Management Interface Protocol (CMIP) only achieved partial penetration into Network Management).

From an SP point-of-view, the end goal is not merely the specification of working standards but the adoption and productization of these standards such that they become off-the-shelf commodities. Marketing and politics are now as much a part of the standardization process as technical aspects, so there is plenty of opportunity for the reader to undertake interesting and ground-breaking work.

6.2 WHY ARE STANDARDS IMPORTANT TO OSS ARCHITECTURE?

Early in this decade, following the dot.com bubble, many SPs were facing the prospect of impending obsolescence of existing network equipment. They were seeking to replace much of it in the near future to avoid future rising costs. IT budgets were spiraling due to the increase in integration costs of mixing tactical legacy systems and strategic COTS platforms. Vendors were also in a position of trying to reduce development and testing costs, following the cost-cutting period of the post dot.com bubble. Previously, a vendor may have provided different SPs with bespoke network

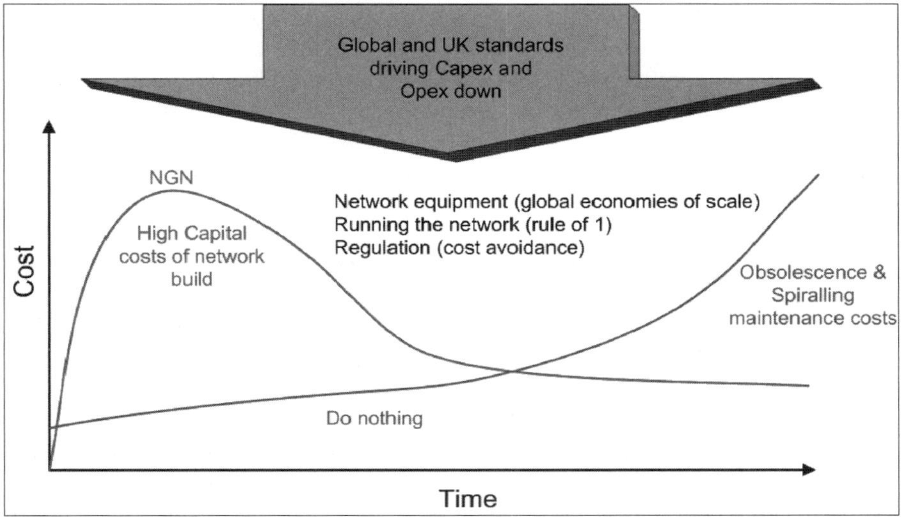

Figure 6-2. Cost of high initial Capex spending versus future high Opex spending

and OSS management platforms. Due to the need to reduce development, and testing labor costs this was no longer feasible.

Figure 6-2 shows projected capital expenditure (Capex) versus long-term operational expenditure (Opex). At some point in time, doing nothing and maintaining obsolete assets was going to cost more than investing in all new agile network and management systems. Additionally, SPs were seeing further requirements from regulators and competition from companies outside the traditional fixed line operators. Standards were seen as a way of reducing the so-called integration tax and improving agility in getting products to market.

Traditionally, standards have focused on the management of network nodes through vendor-provided element managers. However, due to the changing environment in which SPs now operate with voice-over-IP (VOIP) services provided free or at very low cost to customers by third parties, and the growth of the Internet (i.e., the so-called Web 2.0), the focus is now very much on the idea of SPs transitioning to the so-called "Soft Telco." There has been a major push in CTOs (Chief Technology Offices) to develop strategies for the development products (or services) as rapidly as possible, e.g., the concept of product factories and active product and service catalogs. Even more recently, other technologies such as cloud/grid computing are being researched and developed. In the context of previous chapters, NGN become an enabler and foundation for the rapid development of aservice provider's products that are the main key to future profitability.

Unlike traditional network standards, which focus on the definition of concrete attributes and parameters pertaining to networks or devices, the focus on OSS standards is different. In developing OSS standards we define a relatively abstracted definition of a network or piece of hardware and utilize this as a kind of hardware abstraction layer (as found in many computers which the complexity of the underlying hardware is hidden from the user by a layer of adaptive software). Furthermore, unlike network management standards, OSS standards seek to increase the level of abstraction moving up the OSS from abstracted network layer to product management. Up to the early part of the 21st century much of the focus of standards development was on the network and its abstraction. Now the focus is on using this abstracted capability in the OSS to develop products and services.

6.3 THE TELEMANAGEMENT FORUM (TM FORUM) FOR OSS ARCHITECTURE

Many SPs chose TM Forum as part of a collaborative choice because it hosted three main pieces of implementation based work, eTOM, SID, and MTNM, coupled with a large vendor presence. Hence it was the place to develop future NGN standards in one location. The first two specifications gave SPs a means to create a rational and organized OSS architecture. The third specification gave SPs a means to manage their layer 2 networks (e.g., SDH, ATM, DSL, etc.). The ongoing intention is to absorb and develop technology-specific work done elsewhere in future specifications.

More importantly, TM Forum had a membership that included nearly all OSS/BSS and NE vendors and SPs from around the globe. Many of the programs were actively delivering concrete outputs such as implementable data models and application programming interfaces (APIs). Many organizations have limited resources in terms of budget and people; hence it has made sense to focus on converging the number of different standards group activities across groups. There has been a desire to restrict the work that gets undertaken to a manageable number of groups. In terms of OSS management, this has been predominantly, but not uniquely, TM Forum. (Note that TM Forum is not an SDO; it develops consensus specifications. Many refer to the ITU-T for standardization and many SDOs take TM Forum documents for input into their standardization initiatives.)

Examples of collaboration agreements with other bodies all given in Table 6-1. Standards are often managed strategically by CTOs in organizations, through liaisons between SP CTOs over periods of time as well as via personal efforts.

In addition to growing organically, TM Forum has grown by merging with other groups such as Global Billing Alliance, IPSphere Forum, OSS/J, Internet Protocol Detail Record Organization (IPDR), acting as a single focal point for coordination. It now covers a wide range of OSS areas from IP management to billing to digital content management.

The TM Forum as a specifications body has been further boosted by the recent influx of industry sectors, which have not been traditionally thought of as telecoms, for example,

TABLE 6-1. Other standards bodies that have liaisons with the TM Forum

ATIS/TMOC	Cable Labs	DMTF	ITU	ItSMF	ETSI	OMG	OMA
Rosetta Net	The Open Group	IEEE	Oasis	HGI	3GPP	MSF	MEF

- Defense: The TM Forum now hosts regular Defense Interest Group meetings and recently the U.S. Department of Defense stated that it had mandated its suppliers to use TMForum standards where they exist. This should add momentum to further standards development in future.
- Content Encounter: A technology incubator for companies and organizations interested in creation, delivery and monetization of digital media services, attended by major names in media content and its distribution.

6.4 OTHER STANDARDS BODIES

Other important areas of interest for SPs have been the management of IPTV and mobile fixed line convergence. ITU-T, ATIS, ETSI, and 3GPP hold leadership roles, hosting many areas of collaborative work. Mobile-fixed line work has been undertaken between ETSI TISPAN and 3GPP, with its experience in mobile networks. SDOs have come together to develop the concept of IP Multimedia Subsystem (IMS). Recently, ATIS has led initiatives to circulate its work to 3GPP and ITU-T.

ITU-T is the world's primary *de jure* telecommunications standards body. The requirement of many governments and regulators is to mandate the use of *de jure* standards, as they can be legally enforceable in procurements where *de jure* standards are offered. ITU-T also attracts much attention, with diverse areas of study such as numbering and naming, ID management, IP management, and data management.

6.5 TM FORUM'S ENHANCED TELECOMMUNICATIONS OPERATIONS MAP (eTOM)

The objective of OSS development in tier 1 service providers is to move from a purely IT nuts and bolts approach of NGN management to a business process orchestration of abstracted capabilities. TM Forum's eTOM is a Business Process Framework, a blueprint for process definition and a component of their New Generation Operations and Software Systems (NGOSS) set of frameworks. SPs may use eTOM to define their internal process engineering relationships, partnerships, alliances and general working agreements with other providers. Vendors may use eTOM to outline potential boundaries of software components as well as the required functions, inputs, and outputs that must be supported by products. It will be later shown that a further role of the eTOM is the structuring of data that is used in enterprise architecture into domains of similar related data; this in turn has consequences for OSS integration with NGN devices.

In a simplistic way, it can be said that eTOM has mappings to certain aspects of the Zachman Framework (Figure 6-3) which defines the what, how, where, who, when, and why of an IT infrastructure. The Zachman Framework is structured horizontally over 6 levels, with level 1 being the high level contextual aspects down to the runtime deployment at level 6.

	DATA *What*	FUNCTION *How*	NETWORK *Where*	PEOPLE *Who*	TIME *When*	MOTIVATION *Why*
Objective/Scope (*Contextual*) → *Role: Planner*	List of Things important in the Business	List of Core Business Processes	List of Business Locations	List of important Organizations	List of Events	List of Business Goals/Strategies
Enterprise Model (*Conceptual*) → *Role: Owner*	Conceptual Data/ Object Model	Business Process Model	Business Logistics System	Work Flow Model	Master Schedule	Business Plan
System Model (*Logical*) → *Role: Designer*	Logical Data Model	System Architecture Model	Distributed Systems Architecture	Human Interface Architecture	Processing Structure	Business Rule Model
Technology Model (*Physical*) → *Role: Builder*	Physical Data/ Class Model	Technology Design Model	Technology Architecture	Presentation Architecture	Control Structure	Rule Design
Detailed Representations (*Out of Context*) → *Role: Programmer*	Data Definitions	Program	Network Architecture	Security Architecture	Timing Definition	Rule Specification
Functioning Enterprise → *Role: User*	Usable Data	Working Function	Usable Network	Functioning Organization	Implemented Schedule	Working Strategy

Figure 6-3. Zachman Framework. (http://commons.wikimedia.org/) ref 1

The six Zachman viewpoints are:

1. Scope (contextual) viewpoint—aimed at the planner
2. Business model (conceptual) viewpoint—aimed at the business owner
3. System (logical) viewpoint—aimed at the designer
4. (Implementation) technology (physical) viewpoint—aimed at the developer
5. Detailed representations (out-of-context) viewpoint—aimed at the subcontractor
6. Functioning enterprise viewpoint

The six aspects—and the interrogatives to which they correspond—are:

1. Data aspect—What?
2. Function aspect—How?
3. Network aspect—Where?
4. People aspect—Who?
5. Time aspect—When?
6. Motivation aspect—Why?

Similarly, eTOM (Figure 6-4) is split over several horizontal layers of increasing details; currently levels 1 to 3 have been defined for most areas and level 4 decompositions exits for some areas. (Domains such as Resource Management and Operations, Service Management and Operations, etc., are sometimes referred to as level 0.)

Figure 6-5 shows how decomposition of use cases exists at different layers. Figure 6-6 represents the decomposition of problem handling use cases at level 3 system level. By drilling down and decomposing each leaf at level 3 use case we get level 4, which is synonymous with the implementation view. Decomposition of the level 3 leaf "Isolate Customer Problem" is shown in Figure 6-7.

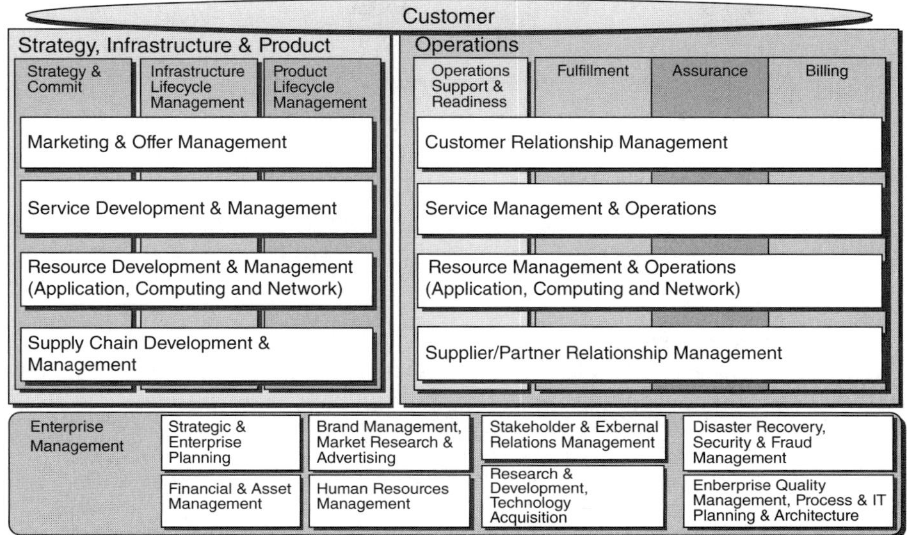

Figure 6-4. eTOM levels 0 and 1. GB921, The Business Process Framework (eTOM) (http://www.tmforum.org/) ref 2

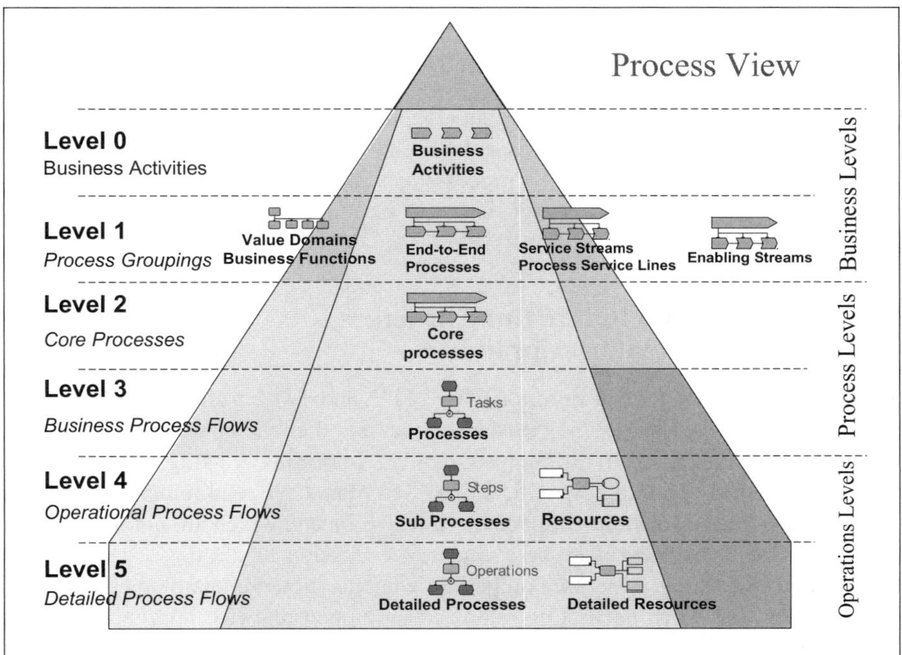

Figure 6-5. eTOM Process levels. GB921, The Business Process Framework (eTOM) (http://www.tmforum.org/) ref 3

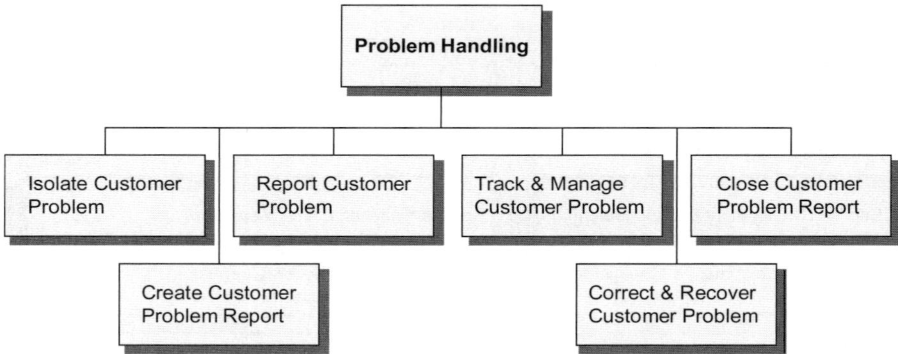

Figure 6-6. Example of eTOM level 3 process. GB921, The Business Process Framework (eTOM) (http://www.tmforum.org/) ref 4

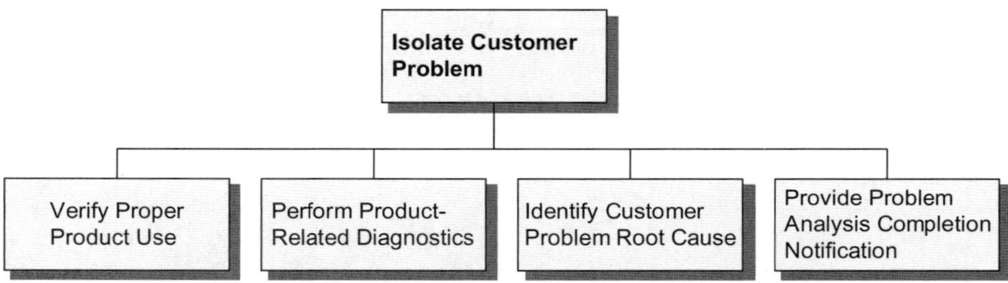

Figure 6-7. Example eTOM level 3 Process decomposed to level 4 GB921. The Business Process Framework (eTOM) (http://www.tmforum.org/) ref 5

Each of these lower-level components maybe decomposed into more detailed level 4 and level 5 definitions, respectively. Additionally, a SP may extend the eTOM and create their own level 4 processes and below when required, although it is requested that any such decompositions to be made as written contributions to future eTOM releases.

6.5.1 Relationship to ITIL (Infrastructure Technology Information Library)

The IT Service Management Forum (itSMF) ITIL and TM Forum eTOM both offer separate perspectives for defining business processes that can be used in OSS enterprises. They originated from different sources (ITIL, originally being a UK Treasury set of information technology specifications). SPs often have requirements to support one or both of these standards, especially in ICT projects, which are not specifically telecom-based. This may often be a cause of confusion to the unwary, who may believe that eTOM and ITIL are competing, whereas they are complementary. This is not to say that work needs to be done to harmonize the two approaches. They attempt to achieve similar goals and there is some overlap; there are differences in terminology and definition. Following are basic definitions, that, while similar, are subtly different.

For example, eTOM defines a "service" as follows:

> Services are developed by a Service Provider for sale within products. The same service may be included in multiple products; packaged differently, with different pricing, etc. A Telecommunications Service is a set of independent functions that are an integral part of one or more business processes. This functional set consists of the hardware and software components as well as the underlying communications medium. The Customer sees all of these components as an amalgamated unit.

However, ITIL does not have a concept of a product, which is fundamental in eTOM. ITIL defines a service as follows:

> A means of delivering value to customers by facilitating outcomes customers want to achieve without the ownership of specific costs and risks. IT organizations provide IT services that bring value to IT customers. Products are not provided.

For reference, here is the eTOM definition of a product:

> Product is what a supplier offers or provides to a customer

Adding complexity, SPs may also have their own legacy definitions in different aspects of the organization.Therefore, caution should be used when referring to commonly used terms such as service, product, etc. in order to determine which definition is being used in the discussion.

The TM Forum, itSMF, and ITIL communities have been working for some time to analyze and define an integrated framework that leverages both areas. This working party has delivered an initial report, which is a joint study by TMF and itSMF on integrating ITIL and eTOM. The TM Forum continually works on new releases of eTOM utilizing ITIL concepts.

6.6 INFORMATION FRAMEWORK

Historically, SID has been a cornerstone of TM Forum's NGOSS Framework, representing what is known as the Business and System view of NGOSS (now called Business Services). Additional NGOSS views were originally envisaged, called the Implementation and Deployment views, respectively. SID is defined in UML (Unified Modeling Language, predominantly a UML class model), which transitions from one UML release to the next. The OSS enterprise data is structured into specific domains, according to the eTOM horizontal layers, with resource management at the bottom and customer and product definitions at the top. There is a general movement of increasing data abstraction moving up from the network to the product layer. Abstraction is a key goal in defining data since it allows an architect to have maximum flexibility in creating interworking models. The Business view captures taxonomy of interrelated and decomposed business concepts mapped from eTOM levels 1 to 2 with additional supporting concepts such as time, money, etc.

As the software lifecycle moves through from analysis through design and development, the Business view entities are enriched with additional attributes; entities existing in the Business view do not disappear in subsequent views. Referring back to the earlier eTOM definition of a Service as an example, the SID shows the related entities to Service and any associated attributes.

The System view can thus be considered to add extra detail useful to an OSS designer to the Business view as in the following example taken from the SID Primer. Note that in both Business and System view the Card class is abstract (as denoted by its name in italics) and must be sub-classed as a new specific type to create a concrete class for instantiation.

An example of a Business view of a card enriched in the system view is shown in Figure 6-8.

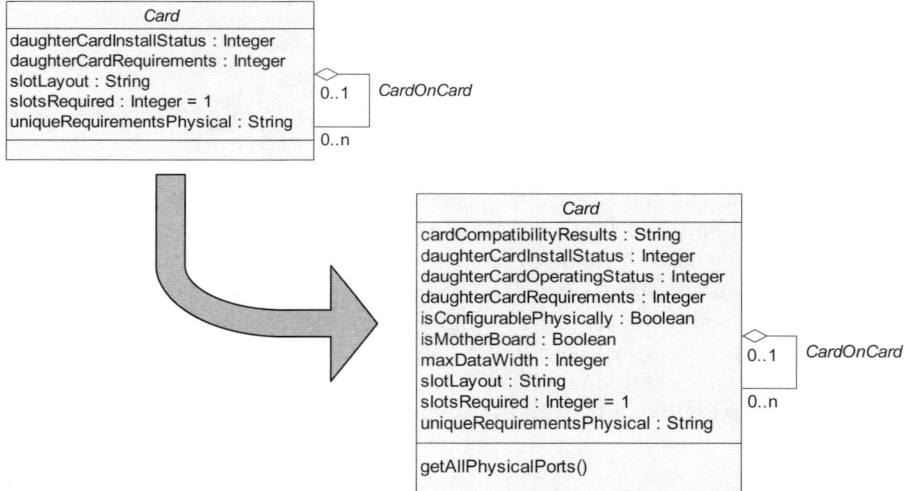

Figure 6-8. Example additional meta data enrichment in the transitioning to different view—Business view enriched with new System view attributes. GB922 Information Framework (SID) Solution Suite Release 8.0 (SID Primer) (http://www.tmforum.org/) ref 6

The Implementation view can be considered to be the developer view, detailing the software and hardware implementation. The Deployment view concerns the management of deployable runtime and supporting policy.

When involved in procurement, a further ambiguity seems to arise in that several companies claim to implement SID. Given that SID is currently defined as a Business view model, what does it mean to implement SID? It turns out that there are two main ways that SID may be implemented. One may literally create an object model that exactly mirrors the Business view UML model. This is the approach that has been used in the past when implementing the SID in research projects. Alternatively, SID may be used as a guide, but the actual object model or database implementation may be refactored subtly for more efficient runtime efficiency. This is an issue since UML associations in the model which make sense at the Business/System view do not necessarily reflect best efficiency in database or object implementation; (i.e. use of design patterns as used in SID versus use of implementation patterns, the two types of pattern are not always complementary).

This duality of approaches (direct mapping or refactored runtime) arises because the Implementation view was never officially defined. While this may have been a problem in the past, a NGOSS approach may mean that a single technology specific view representation is now unnecessary. This is because the NGOSS framework relies on automated tool driven development and therefore does not need to be defined manually on a class-by-class basis. A technology specific implementation, Java, C#, or CORBA, may be generated from a technology neutral model according to the rules defined in a technology specific template at design time. (See Section 6.24 on tool driven development.)

Regardless of how the SID is implemented, it is often necessary to extend the SID via UML sub-classing to meet implementation requirements, for example, when modeling a new network card. This is how the SID was intended to be used, it is said to be an extensible model.

In a Service Provider OSS this can lead to some interesting integration challenges as it is possible to have platform A have one interpretation of the SID and platform B have a second slightly different interpretation of the SID. This may be tackled in several ways;

1. Expose API capabilities via very simplistic CRUD (Create, Read, Update, Delete) interfaces using a minimum of standard attributes.
2. By using traditional J2EE and Gang of Four (GOF) type patterns such as façade adapters.
3. Using commercial model mapping tools that produce bespoke runtime mapping components to translate between models at either side of the interface.

Recently, the TM Forum through liaisons with the ITU has had the SID Product and Service domains adopted by the ITU as recommendation ITU M.3190. Further down-streaming is planned for the future. Liaison work with TISPAN and 3GPP is also underway to develop a common Subscription Management model (as used in IMS, an issue here being the mapping of data structures defined by SID for the wider OSS, and data structures identified by IMS which are not structured explicitly to eTOM format).

The SID also has a Resource domain, which is a feature it shares with TIP; these models are not static and are evolving to meet the challenges of new technologies (NGN, Web 2.0 etc), processes, and methodologies (tool-driven automation).

6.7 DMTF CIM (DISTRIBUTED TASK FORCE MANAGEMENT)

Although the CIM is not part of the TM Forum it is an important model to consider in that it has overlaps plus additional focus on areas that the SID does not cover, e.g., IT infrastructure such as data centers and IT hardware. Increasingly there is collaborative effort on producing a harmonized approach between the two models. Collaborative catalysts between the two groups, such as the Harmony Catalyst series, have demonstrated this. DMTFs have also adopted the concept of ITIL's CMDB (configuration management database). At some stage a grand unified data model will need to embrace SID, CIM, and ITIL. Fortunately, this is recognized and work to do so is in progress.

6.8 TIP (TM FORUM'S INTERFACE PROGRAM)

The interfaces for connecting all of the inter- and intra-Service Provider components have recently been unified under TM Forum's Interface Program. These interface specifications have historically evolved from several different teams working in different areas at different times:

- MTOSI (Web-Service XML interfaces for network and service management)
- MTNM (CORBA management interfaces that model multi-technology layer 2 networks)
- OSS/J (EJB/Web-Service based APIs for component-based OSS systems)
- IPDR (interfaces used for usage data management and accounting)
- CO-OP (this is an initiative to promote unified standards for network element management and user management).

Originally, the MTNM team produced fully specified standards consisting of

1. A UML data model (TMF608) based on ITU G.805, which modeled connection oriented layer 2 networks.
2. A set of business requirements and solution specification documents plus supporting documents.
3. A CORBA IDL (TMF814) specification based on the data model.

The TMF608 model is an implementation optimized, but abstracted interpretation of ITU G805 and G809 allowing a network element management system (EMS) to offer a simpler representation of the network to higher layer service platforms. In effect, the complexity of the network is hidden or abstracted by the EMS.

MTOSI originally started as a sub-team of MTNM to deliver Web-service representations of CORBA interfaces. Both MTOSI and MTNM originally defined APIs based in a common data model (TMF814). After the initial successful release of MTOSI v1 the scope of the MTOSI became a team in its own right and the MTOSI model (TMF854) expanded to cover a wider area than the MTNM model, taking in the management of Services and componentized data OSS data transfer. Typically, MTOSI interfaces are XML/SOAP implemented over HTTP or JMS.

MTNM has expanded its scope to include the management of connectionless Ethernet based networks by expanding the UML model to move to the more elegant ITU G.800, superseding the previous G.805 and G.809. Both the MTNM and MTOSI core projects cover what is considered the eTOM Resource domain.

At this time, another external group called OSS/J, which was working under the auspices of the Java JCP program, affiliated with the TMForum. Originally OSS/J APIs were predominantly based on the SID and occasionally 3GPP data models.

After a period of study that was required to quantify all the interdependencies (MTOSS/J), it was determined that OSS/J plus MTNM/MTOSI were complimentary and future specifications would be a unified output with existing OSS/J APIs supported in maintenance mode or used as seeding specifications for new TIP specifications where no MTOSI implementations existed.

The TIP team have been bolstered by the arrival of two other teams that originally existed outside the TMF, namely IPDR and CO-OP. The experience of the TIP team, while not always a smooth transition to harmonization, is an example of convergence of standards bodies, which is something most major tier, 1 Service Providers are encouraging through their lobby groups such as Fireworks and the TM Forum's SPLC (Service Provider Leadership Council).

It will be recalled that in the previous section it was stated that the SID has a Resource domain data representation, and also we have just stated the TIP team have a highly detailed implementation model covering the Resource domain. Unfortunately, there have been some uncomfortable differences between the way common entities are modeled in the SID and TIP in the past. A specific example of such is the Physical Termination Point (PTP).These are model representations of the physical ports on network cards. Fortunately, after some years of debate, both the SID team and the TIP team are now in a position to start work on a common harmonized view of the Resource domain. This is currently an ongoing activity, which will greatly benefit all teams involved.

Given that service providers have deployed aspects of both SID and MTOSI models already, the reader may question how this may be so given the apparent discrepancies in certain domain areas. The answer is that the Service Provider has undertaken a proprietary local harmonization within their Enterprise data model. As standards can take time to develop from requirement to release, the Service Provider may also need to work with draft versions and align later. This is made easier if the Service Provider is involved in the development of that standard. Sometimes it is necessary for short-term pragmatism to apply.

6.9 NGOSS CONTRACTS (AKA BUSINESS SERVICES)

NGOSS is the cornerstone of the TM Forum's standards frameworks. It defines the concept of a Contract (by convention, Contract is capitalized) between parties, similar to a legal contract in that each party has certain expectations and obligations.

Original work on Contracts dates back to the start of the TM Forum with the SID as the heart of the Contract definition. At the time the SID was held in a Rational Rose UML 1.4 model, which was intended to hold four views.

- Business
- Service
- Implementation
- Deployment

The idea of a Contract was that it would utilize the SID and define an electronic representation of everything a service provider or vendor needed to know in order to build or procure in order to achieve what would now be called a SOA implementation.

In addition to enterprise data models in the SID the original definition of NGOSS had extensions for items such as Policy, Security, and what was then referred to as a meta-model. The meta-model was intended to be a semantic blueprint detailing the components of the Contract and how each component of the Contract fitted together and constructed.

At the time of its inception OMG's Model Driven Architecture was gaining attention, along with Bertrand Meyer's programming by contract. It was intended to utilize these methodologies in the NGOSS Contract specification.

The Contract methodology introduced the idea that each NGOSS Component, which was the fundamental atomic unit of NGOSS, could have several contract interfaces associated with it (Tasks). Each Contract and Task would support the use of

1. *Pre-condition*: A condition that must exist or be established before something can occur or be considered.
2. *Post-condition*: A condition that must always be true just after the execution of some section of code or after an operation.
3. *Invariant*: An expression whose value doesn't change during program execution.

The Contract contains similar concepts such as conditions from Meyer's work, but they are based on a more pragmatic interpretation of these terms, as the original conditions were quite draconian on the failure of a pre-condition. Conditions were also part of the NICC standards (Network Interoperability Consultative Committee) specifications, which were injected into Contract design along with requirements for service level agreements and support for complex process flows. (NICC is a technical forum for the UK communications sector that develops interoperability standards for public communications networks and services in the UK. Historically NICC has been formally constituted as a committee reporting to the communications sector regulator Ofcom on standards for NGN networks.)

At runtime, the deployment aspect of the Contract would be available in a repository such that any client application or component wishing to undertake a defined action of work could search for a suitable Contract deployed in the repository. Once found, the client could bind to Contract provider and, provided the pre- and post-conditions were met, then this would result in a guaranteed outcome for the client. This is typically referred to as a "find-bind and execute" paradigm. It was anticipated at the time that technologies such as Universal Description, Discovery and Integration (UDDI) might have been suitable for such a repository. Unfortunately, due to limitations of then current supporting software, progress in Contract development gradually slowed.

In the summer of 2007 development of the NGOSS Contract was re-activated. By late 2007 the open source software group Eclipse was offering the ability to developers to build their own software tools. For the first time it was possible to create a true NGOSS meta-model that would create both a fully electronic definition of a Contract and support the generation of runtime code from a model. In June 2008 the Architecture team released its first guidebook, GB942, which is the reference specification of this project.

The benefit of the Contract is that encompasses the full software lifecycle from analysis to deployment in a self-consistent and packaged way, providing trustworthy electronic documentation of all these stages. While it is possible for applications in a single organization to implement what is called "SOA," by itself it is a meaningless statement since it doesn't explain to an outsider what "SOA" (service-oriental architecture) is or how it is built. Moving to Contracts removes the concept of integration being purely an IT exercise with several possible vendor specific styles of SOA to choose from and puts the problem directly in a business context with a clearly defined blueprint for integration. Hence NGOSS can be considered at runtime to be a standard SOA blueprint, the objective being to build on, and extend existing interface specifications and data models. Incorporating existing standard APIs, creating new APIs where necessary, and re-using working legacy and manual solutions where appropriate, all in a business-oriented runtime software wrapper.

Hence, in the context of a B2B (business to business) API, each business may implement different SOA processes, which may not be fully understood or sufficiently documented for both parties to interpret unambiguously. Contracts have sufficient richness to provide a standard SOA blueprint such that both client and provider know exactly the obligations on either parties or what form the delivered result should be, i.e., a contract. Contracts maybe used anywhere inside an organization, or between organizations, nor are they intrinsically telecom specific.

Contracts have beneficial consequences for the integration of NGN management layers into a wider OSS environment. Currently the data associated standard enterprise wide models such SID differ, and are structured differently from that found in common NGN models such as IMS; as such data impedance occurs. By using commercial adapter technology, coupled with a standardized common data model for API data transfer and standard eTOM based process flows, we may minimize any integration tax. (The amount of data discrepancy determines the amount of integration work required; if both platforms on either side of an API share same standard data definitions then this is minimized. This is why enterprise and NGN standards convergence is such a desirable goal.)

Architecture team members are collaborating with Oasis (Organization for the Advancement of Structured Information Standards) to formalize our specifications in the SOA for Telecoms working group.

A reference implementation of the Contract software has been built and uploaded to the TM Forum SourceForge, which is available to TM Forum member organizations

The goal of this software is to automate future design and development of OSS standards.

The current vision of the team is to be agile in delivery, with regular milestone deliverables in tooling, specifications, and reference implementations. The specification documents detail the content of the electronic tooling, which would generate the reference implementation. The reference implementation would be used for evaluation and any changes would be fed back to the next iteration of specification and tooling.

Recently the term "Business Services" has been officially defined as an alias for the NGOSS Contract; the terms may still be officially used interchangeably. In order to avoid confusion later on, the term NGOSS Contract will be used to denote the older specification TMF53-based Contracts; Business Services will be used to denote Contracts based on the newer GB942 specification. The specification of the Business Service has evolved since TMF53; all current work in the TM Forum is based on the Business Service model and GB942.

6.10 MTOSI CASE STUDY

MTOSI can be considered the state of the current art in terms of a Web service-based OSS standard. In terms of strategic standards development it can be considered a transition to a future NGOSS Contract-based standard, providing the necessary foundations for future work as any standard API could be wrapped in a Contract, the effort in doing so being proportional to the alignment with SID as the API lingua franca.

6.10.1 Will Web Services and MTOSI Scale?

Going back to the first release of an XML-based standard for telecoms, there was some concern from other standards groups (especially those having promoted CORBA) as well as traditional SNMP-based vendors over the scale-ability of Web services. Typically in the past network APIs had ranged from basic SNMP command line, RPC, to CMIP and CORBA. The reason for the concern being the potential high volume of traffic from a telecom network. As an example, a quoted validated figure being 3.5 million alarms per day from the legacy network was used as reference, i.e., 40.5 alarms per second. XML is a text-based payload, which requires serialization and de-serialization at either end of an API, possible processing and persisting to a database. It was often stated as fact that Web services did not scale to real-world problems, despite after research finding no examples of anyone actually testing this assertion!

Thus the test assertions were:

- XML/SOAP is too slow to process.
- DCN routers will have insufficient bandwidth to cope with the XML volumes.
- MTOSI (i.e., standards in general) are too complex and too hard to understand.

In order to determine the factual validity of the above assertions, a test rig was created to test this. The test consisted of two basic low specification PCs, one running a simple open-source database and the Java Reference application server (JRI), the other running a Java Swing-based client generating alarms according to MTOSI format. The JRI was chosen because it was the reference standard and to demonstrate that no special vendor features were being used. A basic 802.11g (54 Mbit/s) wi-fi router was used to simulate the network DCN routers, which were also stated to be incapable of dealing with that volume of XML traffic. The test environment simulated an SDH element manager processing network alarms and sending them to an OSS alarm management system.

A Web service was generated directly from the TM Forum alarm specification WSDL downloaded from the TM Forum Web site. Typically Web services can be generated on the command line by running an interpreter (wscompile) provided with the application server package (JRI) with the WSDL file as the input. The result of this is to produce three sets of outputs:

1. A server side java interface and a java class "stub" that implements the interface, the stub may have one or more operations, see figure 6-10.
2. A client side "proxy" class that mirrors the service operations on the remote server (service endpoints) plus the necessary java functionality to bind to the server, see figure 6-9.
3. A set of Java files which are generated from the XSD files corresponding to the data in the XML payload.

The programmer fills in the business logic for the operations on the generated server side stub such that they write the output to a database running on the server PC.

The programmer also writes code to first serialize the alarms on the client using the generated XSD files. Secondly, they write a class, which accepts the serialized data and calls the proxy (or local dummy service endpoint), which now exists on the client machine. The proxy then accesses the real service endpoint via generic library code running on the client machine.

Figure 6-11 shows the opened SOAP packet and XML payload for each alarm. Data is first serialized (converted to a more compact binary format) by JAX-WS protocols on the client before being passed to the WSDL-generated client side proxy. The proxy acts as a black box in sending this data to the appropriate server, as defined in the client-side configuration.

Figure 6-11 represents the data, which is carried in the XML data payload between a client and a server (aka provider). The application server that runs on the provider will strip out the content prefixed with SOAP tags in the background. The stripped-down payload is then passed to a receiving JEE5 (aka J2EE for Java 5) POJO (Plain Old Java Object or a Servlet) where it is de-serialized. From there, data can be sent to any appropriate object running in memory for further processing or persisted to a database.

The swing-based client was set up to generate configurable alarms, display them in a window and batch fire them to a web service running on a second machine. For the test, 2,000 alarms were generated and sent to the server in quick succession. Figure 6-12 shows a screenshot of the Swing GUI client.

The results were as follows:

```
packageMtosiDemo;

importcom.bt.server.*;

importcom.bt.server.NotificationService;
importjavax.xml.namespace.QName;
importjavax.xml.soap.SOAPMessage;
importjavax.xml.ws.Dispatch;
importjavax.xml.ws.WebEndpoint;
importjavax.xml.ws.WebServiceRef;


/**
 *
 * @author 802998277

name=<wsdl:portType name="NotificationConsumerInterface">
targetNamespace=tmf854WS="tmf854.v1.ws"
wsdlLocation=Not currently used by JAX-WS 2.0
serviceName=<wsdl:service name="NotificationService">
endpointInterface=The qualified name of the service endpoint interface
portName=<wsdl:port name="NotificationConsumerInterface"
binding="tmf854WS:NotificationConsumerSoapBinding">
see https://jax-ws.dev.java.net/jax-ws-ea3/docs/annotations.html
*/

public class CallNotificationService {
@WebServiceRef(
wsdlLocation="http://localhost:8080/mtosi/v1/NotificationService?wsdl"
   )

private final String NAMESPACEURI = "tmf854.v1.ws";
private static final String SERVICE_NAME = "NotificationService";
private static final String PORT_NAME = "NotificationConsumerInterface";
private static final String OPERATION_NAME = "Notify";
privateQName SERVICE_QNAME = new QName(NAMESPACEURI, SERVICE_NAME);
privateQName PORT_QNAME = new QName(NAMESPACEURI, PORT_NAME);


staticNotificationServicenService = new NotificationService();


Dispatch<SOAPMessage>dispatchMsg = null;

@WebEndpoint(name = "NotificationConsumerInterface")
public  void invokeOneWay(SOAPMessagereqMsg) {
try
{
dispatchMsg = nService.createDispatch(PORT_QNAME, SOAPMessage.class,
javax.xml.ws.Service.Mode.MESSAGE);
dispatchMsg.invokeOneWay(reqMsg);
      }
catch (Exception e)
      {
System.out.println("Exception Thrown in invokeOneWay");
e.printStackTrace();
      }
   }
}
```

Figure 6-9. Example of code of client calling the server using generated proxy. NotificationService is the generated server endpoint

```
//imports omitted to save space
importjavax.jws.WebService;

@WebService(serviceName = "NotificationService",
portName = "NotificationConsumerInterface",
endpointInterface = "ws.v1.tmf854.NotificationConsumerInterface",
targetNamespace = "tmf854.v1.ws",
wsdlLocation = "WEB-INF/wsdl/AlarmService_1/NotificationService.wsdl")
@ServiceMode(value=Service.Mode.PAYLOAD)
@PersistenceContext(name="jdbc/Alarm",unitName="AlarmService")
public class AlarmService implements ws.v1.tmf854.NotificationConsumerInterface {
   @Resource private UserTransactionutx;
privateEntityManagerem;
private Context initCtx;
private Context envCtx;


@WebMethod(operationName = "Notify", action = "http://localhost:8080/mtosi/v1/NotificationConsumer")
   @Oneway
public void notify(ws.v1.tmf854.MTOSIHeaderT mtosiHeader, ws.v1.tmf854.NotifyT mtosiBody) {
// the following UnsupportedOperationException is a core java class
//throw new UnsupportedOperationException("Not yet implemented");
     // Start of User written code
EventT event;
AlarmT alarm;
AlarmWrapalarmw = new AlarmWrap();

event = mtosiBody.getMessage();
alarm = event.getAlarm();
alarmw.setAlarm(alarm);

try {
initCtx = new InitialContext();
envCtx = (Context) initCtx.lookup("java:comp/env");
em = (EntityManager)envCtx.lookup("jdbc/Alarm");
// End of User written code
     } catch (NamingException ex) {
System.out.println("NamingException");
ex.printStackTrace();
     }

try {
utx.begin();
em.persist(alarmw);

try {
utx.commit();
       } catch (IllegalStateException ex) {
System.out.println("IllegalStateException");
ex.printStackTrace();
       } catch (SecurityException ex) {
System.out.println("SecurityException");
ex.printStackTrace();
       } catch (SystemException ex) {
System.out.println("SystemException");
ex.printStackTrace();
       } catch (javax.transaction.RollbackException ex) {
System.out.println("javax.transaction.RollbackException");
ex.printStackTrace();
       } catch (HeuristicRollbackException ex) {
System.out.println("HeuristicRollbackException");
ex.printStackTrace();
       } catch (HeuristicMixedException ex) {
System.out.println("HeuristicMixedException");
ex.printStackTrace();
       }
     } catch (SystemException ex) {
System.out.println("SystemException");
ex.printStackTrace();
     } catch (NotSupportedException ex) {
System.out.println("NotSupportedException");
ex.printStackTrace();
     }
}
```

Figure 6-10. Example of server side code, most is generated, some configured, and the business logic used for database persistence is marked as “User written code”

```
<SOAP-ENV:Envelopexmlns:SOAP-ENV="http://schemas.xmlsoap.org/soap/envelope/">
<SOAP-ENV:Header>
<ns2:header xmlns:ns2="tmf854.v1" extAuthor="Steve" extVersion="1.0" tmf854Version="1.0">
<ns2:domain>CuMSAN</ns2:domain>
<ns2:activityName>notification</ns2:activityName>
<ns2:msgName>Alarm</ns2:msgName>
<ns2:msgType>NOTIFICATION</ns2:msgType>
<ns2:payloadVersion>1.0</ns2:payloadVersion>
<ns2:senderURI>http://senderEndpoint</ns2:senderURI>
<ns2:destinationURI>http://replytoEndpoint</ns2:destinationURI>
<ns2:correlationId>0001</ns2:correlationId>
<ns2:communicationPattern>Notification</ns2:communicationPattern>
<ns2:communicationStyle>MSG</ns2:communicationStyle>
<ns2:timestamp>20060517230336.36+0100</ns2:timestamp>
</ns2:header>
</SOAP-ENV:Header>
<SOAP-ENV:Body>
<ns2:Notify
xmlns:ns2="tmf854.v1"><ns2:topic>HTTP</ns2:topic><ns2:message><ns2:Alarm><ns2:eventInfo><ns2:notificationId>1786917863</ns2:notifi
cationId><ns2:objectName><ns2:mdNm>CuMSAN</ns2:mdNm><ns2:meNm>Felixstowe/01</ns2:meNm><ns2:eqNm>Some
equipment</ns2:eqNm><ns2:ptpNm>some ptp</ns2:ptpNm><ns2:ctpNm>some
ctp</ns2:ctpNm></ns2:objectName><ns2:objectType>OT_CONNECTION_TERMINATION_POINT</ns2:objectType><ns2:osTime>20060517
230336.36+0100</ns2:osTime><ns2:neTime>20060517230336.36+0100</ns2:neTime><ns2:edgePointRelated>false</ns2:edgePointRelated></
ns2:eventInfo><ns2:isClearable>true</ns2:isClearable><ns2:aliasNameList><ns2:alias><ns2:aliasName>nativeName</ns2:aliasName><ns2:alia
sValue>nativeValue</ns2:aliasValue></ns2:alias></ns2:aliasNameList><ns2:layerRate>LR_Line_OC768_STS768_and_MS_STM256</ns2:laye
rRate><ns2:probableCause><ns2:type>AIS</ns2:type></ns2:probableCause><ns2:nativeProbableCause>nativeProbableCause</ns2:nativeProba
bleCause><ns2:additionalText>Additional
Text</ns2:additionalText><ns2:perceivedSeverity>PS_CRITICAL</ns2:perceivedSeverity><ns2:affectedTPList><ns2:name><ns2:mdNm>CuM
SAN</ns2:mdNm><ns2:meNm>Felixstowe</ns2:meNm><ns2:eqNm>Equipment</ns2:eqNm><ns2:ptpNm>ptp</ns2:ptpNm><ns2:ctpNm>ctp
1</ns2:ctpNm></ns2:name></ns2:affectedTPList><ns2:serviceAffecting>SA_NON_SERVICE_AFFECTING</ns2:serviceAffecting><ns2:rcaiI
ndicator>true</ns2:rcaiIndicator><ns2:acknowledgeIndication>AI_EVENT_ACKNOWLEDGED</ns2:acknowledgeIndication><ns2:X733_Eve
ntType>X733_eventType</ns2:X733_EventType><ns2:X733_SpecificProblems><ns2:specificProblem>Specific problem
1</ns2:specificProblem><ns2:specificProblem>Specific problem 2</ns2:specificProblem><ns2:specificProblem>Specific problem
3</ns2:specificProblem></ns2:X733_SpecificProblems><ns2:X733_BackedUpStatus>X733_BackedUpStatus</ns2:X733_BackedUpStatus><ns
2:X733_BackUpObject><ns2:mdNm>CuMSAN</ns2:mdNm><ns2:meNm>Felixstowe</ns2:meNm><ns2:eqNm>Equipment</ns2:eqNm><ns2:
ptpNm>ptp</ns2:ptpNm><ns2:ctpNm>ctp
1</ns2:ctpNm></ns2:X733_BackUpObject><ns2:X733_TrendIndication>X733_TrendIndication</ns2:X733_TrendIndication><ns2:X733_Corre
latedNotifications><ns2:correlatedNotifications><ns2:name><ns2:mdNm>CuMSAN</ns2:mdNm><ns2:meNm>Felixstowe</ns2:meNm><ns2:e
qNm>Equipment</ns2:eqNm><ns2:ptpNm>ptp</ns2:ptpNm><ns2:ctpNm>ctp
1</ns2:ctpNm></ns2:name><ns2:notifIDs><ns2:notificationId>Notification ID
1</ns2:notificationId></ns2:notifIDs></ns2:correlatedNotifications></ns2:X733_CorrelatedNotifications><ns2:X733_MonitoredAttributes><ns2
:SpecificProblemList><ns2:specificProblem>Specific problem 1</ns2:specificProblem><ns2:specificProblem>Specific problem
2</ns2:specificProblem><ns2:specificProblem>Specific problem
3</ns2:specificProblem></ns2:SpecificProblemList></ns2:X733_MonitoredAttributes><ns2:X733_ProposedRepairActions><ns2:proposedRepai
rAction>Replace card</ns2:proposedRepairAction></ns2:X733_ProposedRepairActions></ns2:Alarm></ns2:message></ns2:Notify>
</SOAP-ENV:Body>
</SOAP-ENV:Envelope>
```


Figure 6-11. SOAP packet and contents

- Over a run of 11,000 tests, an average of over 52 transactions per second was achieved, persisted to database, higher than the alarm rate of the legacy network, see figure 6-13.
- Load on the server stayed between 2% and 4%.
- Utilization of the "DCN" substitute router averaged 0.08%, with an alarm volume 25% greater than the mean daily alarm volume for a typical real SDH network.

It was noted that not persisting the alarms resulted in much faster processing, and that the most significant overhead involved writing the alarm to a database, as opposed to processing the SOAP and XML. Using a higher specified commercial database is likely to have increased performance much more.

As a result of empirical testing it was proven and accepted by all parties that XML-based Web Services are not an issue for telecommunications OSS management. The moral of this story is to never simply accept an unproven argument or assertion. If in doubt, do the test!

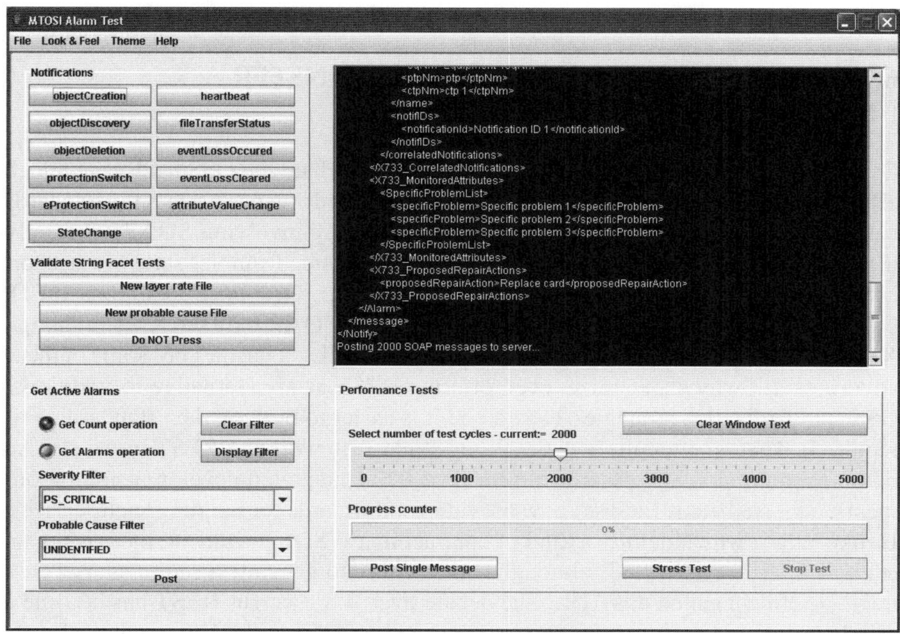

Figure 6-12. Screenshot of Java Swing client

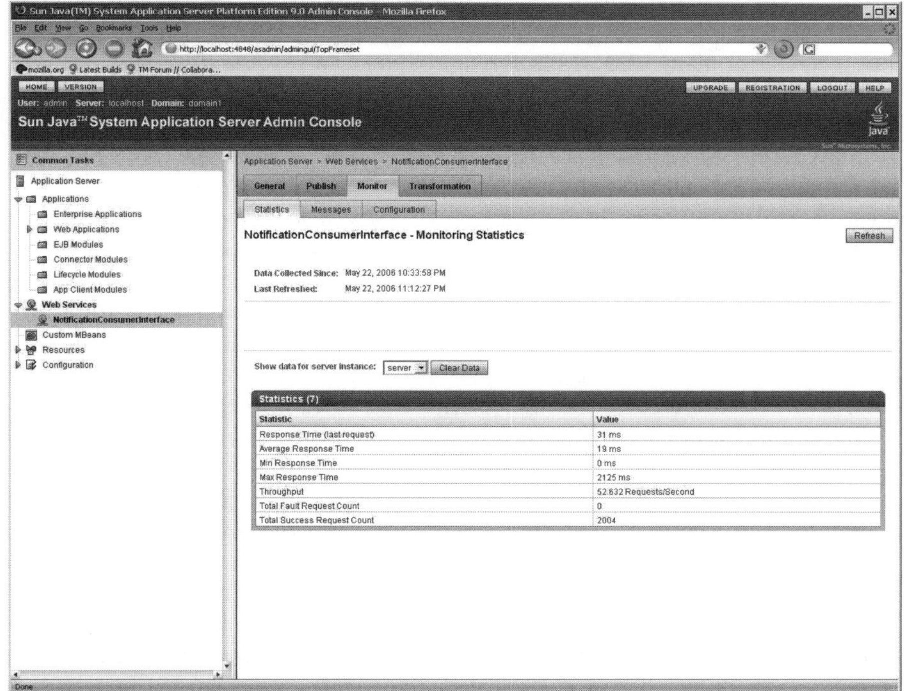

Figure 6-13. Screenshot of server side Application Server (Java Reference Implementation showing achieved throughput of over 52 transactions per second)

6.11 REPRESENTATIONAL STATE TRANSFER (REST)—A SILVER BULLET?

Over the past years, as Web Services promised to overhaul the OSS landscape, much effort was put into the adoption of WS-* based standards (generic name for standards based on W3C and Oasis specified extensions usually involving SOAP). Recently there has been much effort by some evangelists for a much simpler style of transport called REST, involving HTTP using the defined Put, Get, Post operations HTTP offers. REST also does away with the requirement for SOAP and even XML, although XML may be a possible payload (POX—plain old XML), it needn't be. Some of these advocates ask why standardize, even offering REST as an alternative to traditional OSS standardization strategies.They argue it is easier to build it, then standardize (if required). That is, after all, how the World Wide Web works. A REST-based service doesn't require a WSDL, many current Java IDE (integrated development environments, e.g., Netbeans, Eclipse) now provide good wizards for building such services. Unlike WS-*, which defines a kind of contract in a WSDL document, there is not an equivalent artifact in REST, although it is possible to generate a rudimentary "contract" document called a WADL. The debate over WS-* versus REST has become a new holy war in IT. Does REST negate the need for OSS and NGN standards?

At first glance it seems a righteous struggle by the new evangelists. REST promises agility and simplicity. In terms of business opportunities, an organization need not wait for some external standards organization to spend years debating what is required; nor does it need to wait for the necessary tool support. Simplicity means a developer may just build as many REST APIs as required and meet all their customer deadlines. A similar argument may even be used for other agile frameworks, such as Ruby on Rails. This is a seductive but dangerous argument. Although this is how much of the World Wide Web works, there is still a need for a contract of some kind. Unfortunately, the nature of this contract is often overlooked by virtue of the simplicity of a transaction and often is not known, nor is the required data well documented. Typically, when a customer orders a product from a Web retailer there is still a need for some kind of pre- and post-condition. The retailer imposes the pre-condition that if you want to buy their product the customer must fill in a order form (often Web browser-based), which is defined by the retailer. Hence the data for the order is implicitly packaged in some defined way prior to being posted to a server URI (Uniform Resource Identifier) or URL (Uniform Resource Locator). In return the server fulfills its post-condition obligation to fulfill the order and the data is processed by some proprietary back-end business logic. This approach works because the retailer constrains the way data maybe sent to server by the client. If, however, the back-end logic changes but the form doesn't (or vice versa) then the order request is not guaranteed to work.

On the Web, this approach is often adequate as each retailer doesn't need to deal with the quantity and volume of complex data that exists in a telco, nor do they need to participate in many like-minded business-to-business transactions. The fundamental aspect is that the transport technology or middleware is not the key issue here; it is the data and how it packaged. Badly managed data and poor design amounts to the same problem regardless of whether the middleware is CORBA, WS-*, or REST-based. In addition, there is also a danger that the feeling of freedom engendered by REST will result in numerous unmanaged and undocumented REST APIs, as opposed to the similar problem in other middleware. The simplicity of a REST-based solution should not, therefore, be a substitute to a proper standards-based solution. The world is not black and white and there is still a place in certain mission-critical areas for other transports such as JMS and even WS-*, so future developments may require both SOAP and REST. Contracts themselves could be implemented in either technology, as they are designed to be technology neutral.

HTTP is not natively a guaranteed delivery protocol; messages are normally delivered on a best-effort basis. Other technologies such as JMS and MQ can guarantee resilience, and to address this WS-Reliability was created. REST does not offer such reliability guarantees for mission-critical work.

6.12 REAL NETWORK IMPLEMENTATION OF A STANDARD

Following the release of MTOSI v1, a UK SP had a requirement to provide management of MSANs (Multi-Service Access Nodes, next generation DSLAMs with VOIP capability). This would replace existing deployments of legacy DSLAMs with legacy EMS-OSS interfaces. The initial requirement was for inventory management. A standard was chosen for the following business reasons:

1. *Improved customer experience*: Simple, straightforward and complete, it was better specified than existing legacy API and the prospect of creating a new proprietary API.
2. *Regulatory change*: SPs' undertakings, including equivalence of input, transparency, separation of Access Services, etc. (Equivalence of input is a UK regulatory term which means that all internal but unregulated parts of a service provider can not have preferential OSS access over an external third party service provider requiring use of our network).
3. *Increased market responsiveness*: Reduced concept to market cycle time. The API was already defined as an off-the-shelf multi-technology standard. Why develop proprietary API's from scratch?
4. *Reduction in cost*: Legacy OSS had hundreds of systems incurring many hundreds of millions of Euro in annual IT budget s., the move to standardized OSS was seen as a way to reduce overall costs.

There were also technical challenges to meet:

1. *Enhance customer experience*: User stories were focused on customer need. The vision was to prove full OSS automation from customer relationship management (CRM) down to the network.
2. *Deliver the OSS support of the operational requirements for NGN deployment and services.*
3. *Deliver OSS support for a major restructuring of business responsibilities, accountability, systems ownership and data ownership.* The legacy OSS has

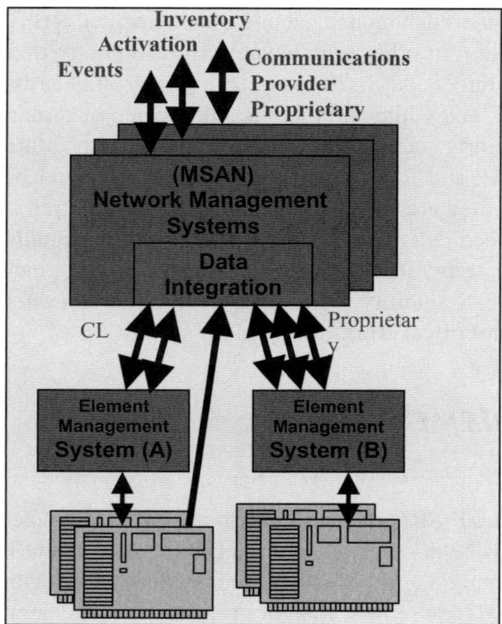

Figure 6-14. Example legacy Element Manager and NE environment

evolved into a series of technology-specific stovepipes with their own systems data and processes.

4. *Achieve major reduction in OSS development and support costs.*
5. *Achieve major reduction in number of Operational Support Systems (OSS).* A mixture of over 1400 legacy, tactical, and COTS platforms was becoming unwieldy.

Figure 6-14 shows a typical legacy or proprietary multiple vendor interface implementation. Each system is likely to hold a proprietary data model, each of which will require integration with other OSS systems.

Typically, costs will be very high to support additional new suppliers' interfaces as vendors are added to the network.

There is minimal reuse; each system would require its own test specifications on integration. Due to interface and data diversity there would be significant per-EMS integration time, costs and especially increased project management risk associated with each new vendor integration.

The goal was to migrate to a structured architecture connected by standardized APIs (MTOSI) and using a common data model (TMF608). We are not concerned about the EMS to managed network element API that is typically SNMP or TL1. The vendor-provided element manager (EMS) provides a standardized data mapping from the EMS's own internal model, which maybe any format so long as it is mapped to a common format when exposed to the OSS. Obviously, the closer the EMS data model is to the data standard the less work the vendor has to

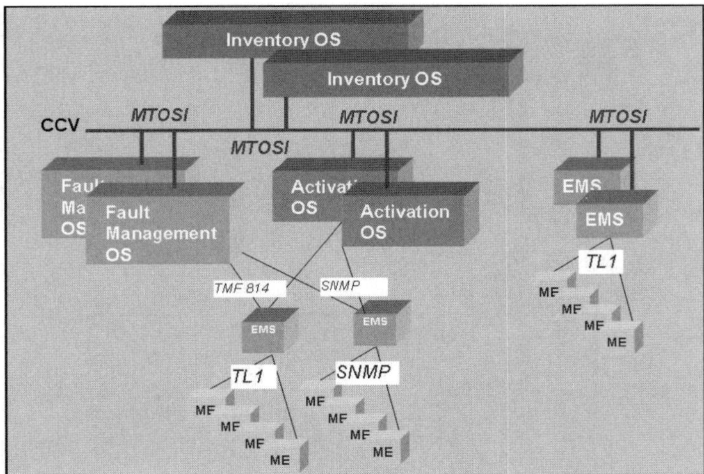

Figure 6-15. High-level MTOSI integration. TMF517 Business Agreement for MTOSI Release 1.1 (http://www.tmforum.org/) ref 7

do. This is a macroscopic analogy to the smaller-scale concept of the hardware abstraction layer (HAL), as found in many computer architectures for peripheral management.

Idealized OSS network management connected by a logical bus of Web service APIs is shown in Figure 6-15.

The initial development took place with one OSS COTS vendor and two network vendors. The whole process went relatively smoothly in comparison to traditional legacy developments, and additional systems were added incrementally including WDM and SDH network management systems and other OSS systems to build the deployment shown in Figure 6-16.

The first system deployment represents some overhead in getting the infrastructure and the human knowledge base set up, but subsequent additional integrations can be done relatively quickly and cheaply.

6.13 BUSINESS BENEFIT

Following the above deployment it was decided to evaluate what cost savings, if any, had been made. It is possible to show that for N systems without a common data model the cost of integration is proportional to N * (N – 1), while in a standard data environment the cost is proportional to 2 * N (in the above implementation a common data model based on TMF608 was used). In Figure 6-17, the point-to-point architecture requires an explicit data translation on each API, while the systems in the bus architecture share a common data model and need only load data on and off a middleware bus.

Table 6-2 gives examples of benefits of a single activation system evolving to SOA (where we want to get to).

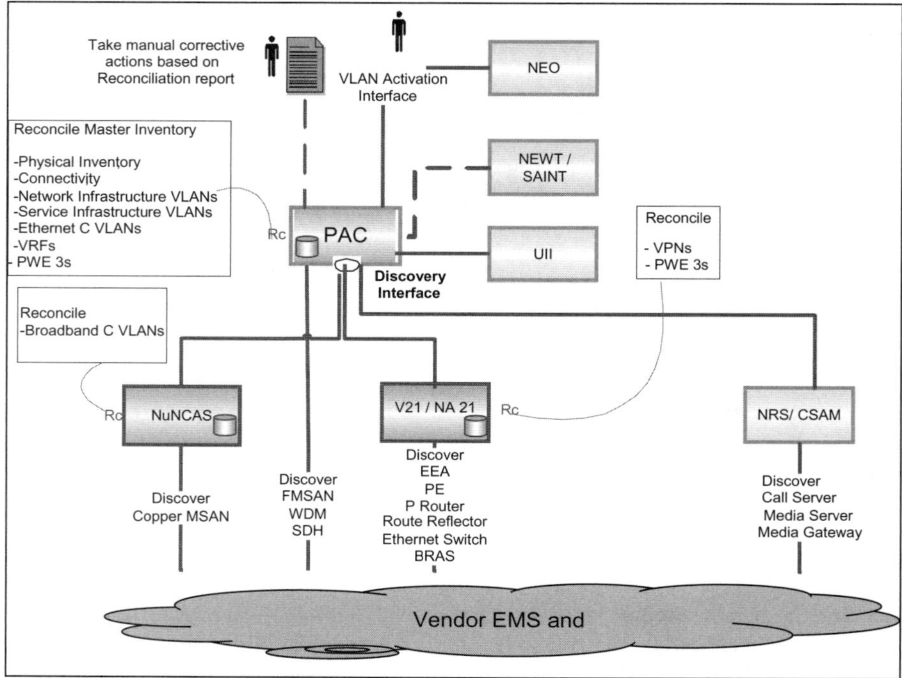

Figure 6-16. Representation of MTOSI API's in SP environment

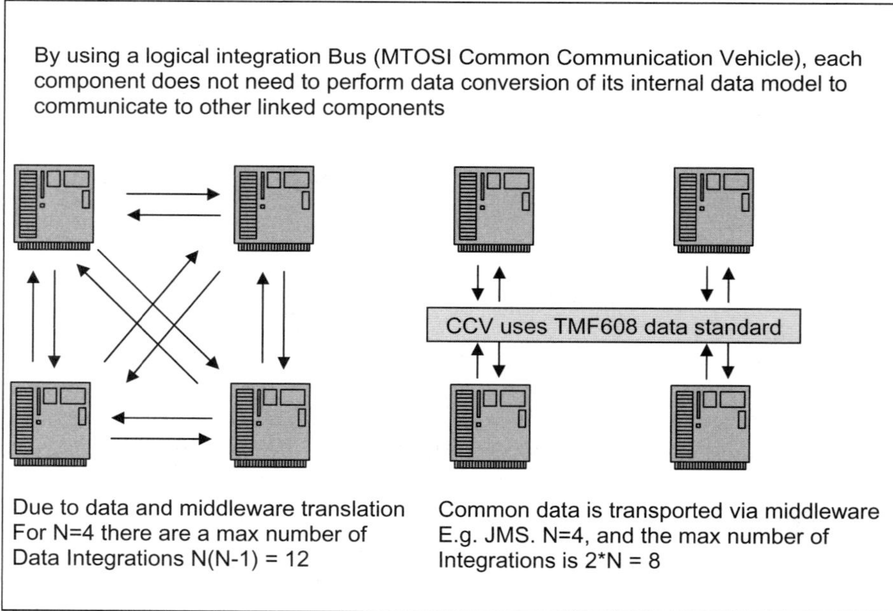

Figure 6-17. Diagram showing scalability problem of point-to-point versus data bus architecture

TABLE 6-2. Benefits of CapEx vs OpEx

Cost item (across all parties)	Cost Reduction
CapEx (capital expenditures, which include procurement of new equipment—Estimates)	
EMS Interface specification cost	80%
EMS Interface specification elapsed development time	50%
EMS Interface development cost	75–80%
EMS Interface development time	50%
Integration testing cost	80%
Integration testing time	90%
Data Integration Software	100%
Middleware costs	–10%
OpEx (On-going cost for running a product—Estimates)	
One off cost for introducing new supplier	90%
Support cost saving from using non custom EMS adaptors	100%
Conformance testing tools reference implementations	80%
Specification maintenance	95%
Support costs on systems per supplier	70%

6.14 OSS TRANSITION STRATEGIES

The solution may sound simple to replace existing legacy systems with new standards-based OSS platforms. Unfortunately there are several issues that need considering. Typically SP OSS environments are not green field sites; many platforms are well designed doing a perfectly adequate job, and these may need to integrate with new platforms. While a standards-based solution may be strategically ideal, financial pressures, the need to deliver key products to a customer by a certain date, and numerous other inter-dependencies also affect implementation decisions. Ultimately, SPs would like to see standards delivered "out-of-the-box" by vendors. This is going to take time as vendors are subject to similar internal pressures between strategic delivery and the need to sell products in a financial year. In all of this a healthy degree of pragmatism and patience is required.

In a typical SP environment, many platforms have grown up, some strategic, some tactical. In our example SP environment these numbered approximately 1,400. It is essential that these platforms be cataloged in order that sound judgments may be made. One way of organizing this catalog is by grouping the applications into platforms of similar functionality, see fig 6-18. We have seen this pattern in the TM Forum Application Map. Each platform becomes a collection of interconnected applications organized by function. An easier way of supporting this transition than standardizing everything at once is to try to "standardize" the interaction of the platforms rather than the platform components. This could involve international standards, if they exist (unfortunately they don't always); more often it involves creating an in-house standard or capability. Recall, connecting process-oriented components is one of the goals of the Business Service Framework. As the platform components become defined by their functional capability the tactical systems and strategic systems may be identified.

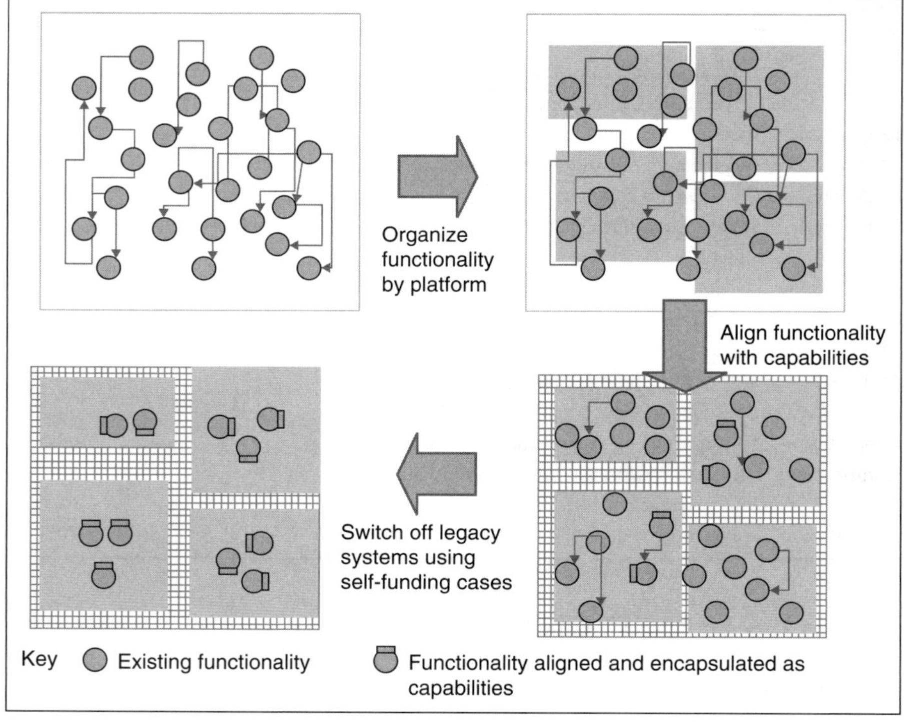

Figure 6-18. Gradual migration from a legacy OSS of mixed applications to a capability based collection of platforms. BTs Matrix Architecture W G Glass (http://www.btplc.com/Innovation/Journal/BTTJ/current/HTMLArticles/Volume26/08Matrix/Default.aspx) ref 8

Each platform has a specific owner; over time the decision can be made to switch off tactical applications, thus streamlining the OSS. The goal of such a transition should be, at the minimum, sharing standardized data (extended SID based) across platform boundaries; if this can be done inside a platform, this can be considered a bonus. Typically a SP provider may find transition to standard APIs easiest at the Resource Domain closest to the network. This is because layer 2 networks are already well defined and come with supporting standards. There may be instances where a specific network technology or feature is not explicitly modeled by the standard, but in this case a successful strategy is to work with vendors within the SP space and to support a parallel standardization effort and have a strategy of gradual convergence on a new standard. (Typically this will involve extending existing standards' data models in line with existing design patterns of the standard.)

6.15 ETSI TISPAN AND 3GPP IMS

IMS is an industry initiative based on fixed-mobile line convergence. In order to ensure that divergent standards for IMS didn't converge between the fixed line operators and mobile operators ETSI TISPAN (Telecommunications and Internet converged Services and Protocols for Advanced Networking) have a working liaison with 3GPP, a federation of primarily mobile telecoms vendors and mobile operators who developed the mobile 3G standard. They agreed to work jointly on a harmonized IMS network standard and OSS management capability. TISPAN would embrace much of the 3GPP model for network however 3GPP and TISPAN WG-8 would work jointly on the OSS management aspects.

The focus of this activity was creation of a service-oriented architecture (SOA) model underpinned by a TISPAN data model extended from the 3GPP model. It was desired that the TISPAN information model would eventually be harmonized with the SID using multiple NOSI interfaces. This was harder than first envisaged in that, since the origin of the 3GPP model differed from the SID, the terminology was different, the class representations were different, even some basic concepts differed. And finally, since people are human, there were lots of differing opinions on how this complex and world-leading project should be delivered.

There are not yet any normative implementations of SID/3GPP/TISPAN Subscriber Management as the model harmonization is still a work in progress. The relationship between 3GPP and TISPAN is interesting. TISPAN is a group of teams that works under the auspices of ETSI (European Telecommunications Standards Institute, part of the EEC), which together with the ITU (part of the United Nations) are *de jure* standards bodies whose standards or recommendations can have legal authority. 3GPP, DMTF, and the TM Forum are *de facto* standards bodies, having no legal authority. However, these and several other organizations often downstream their specifications to the ITU or ETSI (as an example, the SID Product and Service domain is also ITU-T M.3190).

A further connection, which is no accident, is that TISPAN WG-8 which is predominantly the part of TISPAN dealing with OSS, defined a SOA framework based around IMS Subscriber Management called Business Services and NOSIs (NGN OSS Service Interfaces). We will see later that we have a convergence point here with the TM Forum NGOSS model—a NGOSS Contract is related to a Business Service and a NOSI is an example of a NGOSS Task.

After several years of vigorous debate, ironically the current world financial crisis may have helped slightly in bringing convergence closer. The desire to save money and to consolidate standards bodies to avoid duplication of effort is bringing about a new sense of compromise and pragmatism to the benefit of those who continue to invest in standards development.

6.16 OSS INTERACTION WITH IMS AND SUBSCRIBER MANAGEMENT (SuM)

In the 3GPP specification for IMS, the HSS database (Home Subscriber Server) has similar behavioral characteristics to a traditional Element Manager (although this is a very loose analogy) (Figure 6-19).

In terms of OSS effort in TISPAN, Working Group 8 has focused on interfacing the data in the HSS with that in the OSS as defined by eTOM. This data has

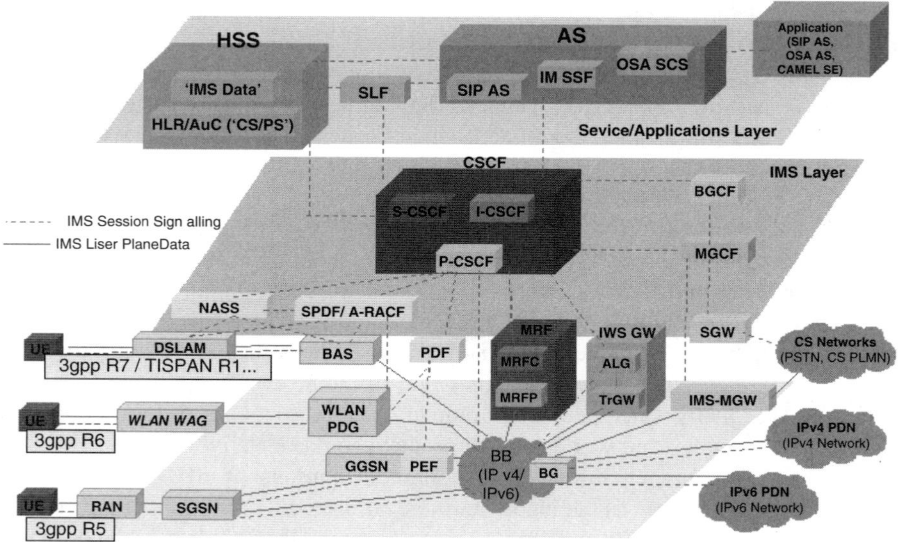

Figure 6-19. 3GPP IMS model. (http://commons.wikimedia.org/) ref 9

focused on the management of TISPAN Customers (3GPP Subscribers) and TISPAN Users. The other seven TISPAN groups are predominantly focused on intra-IMS and network management standards, which are not covered here, as they are not directly involved in OSS service management.

The definition of simple terms like Customer and User are not as simple as first imagined. In 3GPP, which is a mobile standard, a user is synonymous with a customer and 3GPP uses the term Subscriber to identify them. This is because a 3G-phone user typically has a contract with a service provider, the service may be shared with other family members but the only identification is via the SIM details on the mobile device, i.e., subscriber John Smith has a registered service against device X.

A Subscriber is an entity (associated with one or more users) that is engaged in a Subscription with a service provider. The subscriber is allowed to subscribe and unsubscribe services, to register a user or a list of users authorized to enjoy these services, and also to set the limits relative to the use that associated Users make of these services.

The goal of TISPAN, see reference architecture figure 6-20 is to have any service over any technology on any device, anywhere. This means, for example, the ability to watch an IPTV service via broadband on the main home TV, or remotely via hand-held device. In a future release this was intended to include nomadicy or the ability to have this flexibility over a network other than the contracting service provider (via B2B agreements with other service providers, similar to the ability to roam by voice internationally on mobile networks). In order to support this ability, TISPAN stated its OSS vision thus:

> The "OSS Vision" document presents business, regulatory and operational requirements, utilizing the eTOM as reference business process framework,

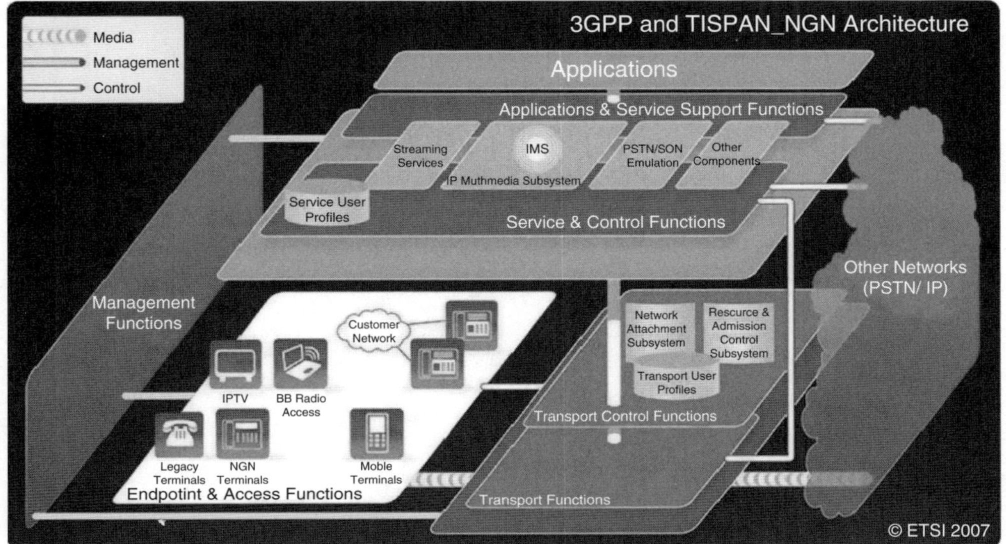

Figure 6-20. TISPAN high-level model. (http://portal.etsi.org/) ref 10

and referring to the TM Forum NGOSS (New Generation Operation Systems and Software) program as reference industry approach for development of OSS.

There is a contractual relationship between a person (or organization) called a Customer, the Customer is allowed to order, modify, and delete services according to contractual agreement with the service provider. The Customer may elect to have one or more Users associated with these services, the customer may nominate himself or herself as a User, and the Customer may wish to define the services associated with each User on an individual basis.

An example of this would be a family; a parent may have a legal contract with a service provider. They may wish to grant full access to all subscribed services to themselves and their spouse but prohibit access to certain services, e.g., international calls, to their children. In order to achieve this functionality it is necessary to modify the 3GPP model slightly by creating new design entities of Customer and User to replace Subscriber plus provide additional model attributes and associations.

The 3GPP information model, see figure 6-21, does contain some additional challenges when attempting a direct mapping to TM Forum SID classes in that the naming convention for 3GPP IOCs is slightly unusual. Based on the SuM NRM (TS 32.172 v7.0) the 3GPP classes for User, Subscriber, and Subscription are not obviously identified, for example:

- SuMSubscriberProfile can be viewed as representing the subscriber.
- SuMSubscribedService can be viewed as the class by which the subscriber has a subscription to a service, having an offer/contract agreement relationship to SuMService.

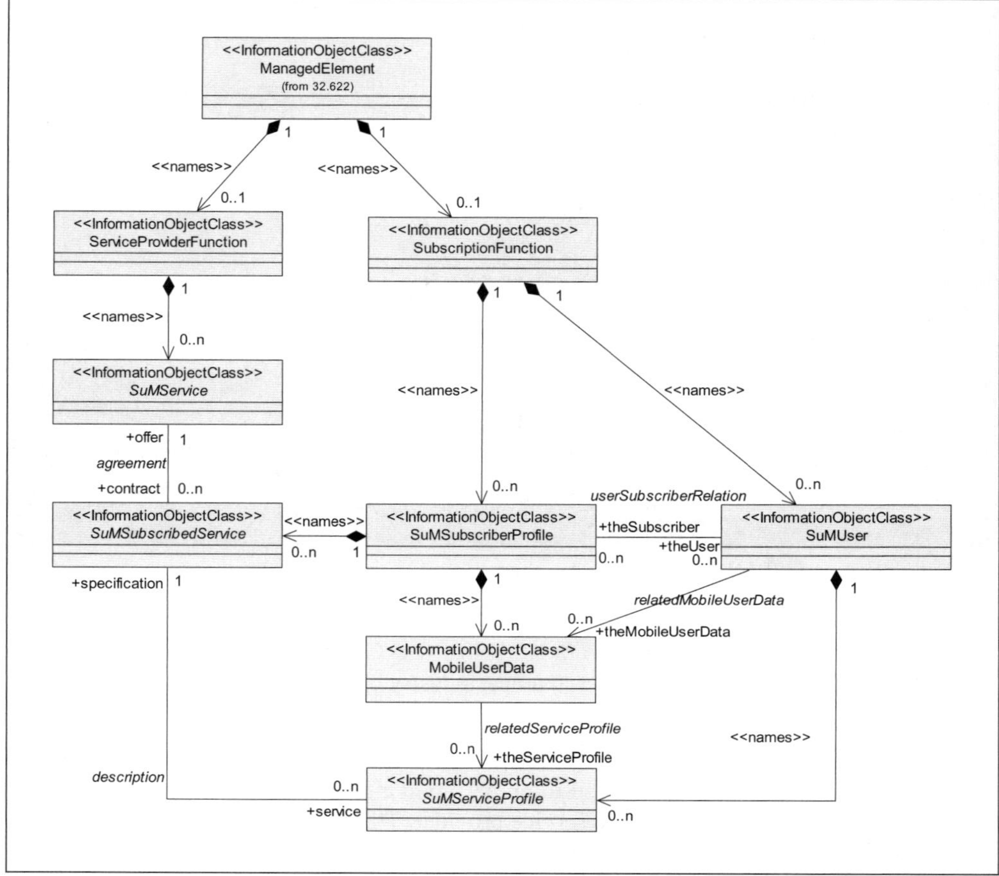

Figure 6-21. Example of 3GPP relationships from TS32.172 v8.0. Telecommunication management; Subscription Management (SuM) Network Resource Model (NRM) Integration Reference Point (IRP): Information Service (IS) (http://www.3gpp.org/ftp/Specs/html-info/32-series.htm) ref 11

- SuMUser (old name SuMSubscriptionProfile) can be viewed as representing the User.
- SuMServiceProfile can be viewed as a User's profile for using a service, or a user's subscription to a service.

The rules of the 3GPP NRM place certain requirements for classes to inherit from a global super-class and for the above IOCs to be "owned" in a specific containment relationship by a Managed Element. Typically in the eTOM and SID a Managed Element would be a Resource Domain artifact and would not have containments of people such as Subscribers and Users. This is a major difference from the SID definition of a Customer and rationalization is a work in progress.

Care also needs to be taken with inheritance relationships in models since not all languages contain support for multiple-inheritance, e.g., XML, and Java also poses some limitations. This is an important challenge because IMS platforms are likely to be built, specified, and procured from network vendors using a 3GPP/TISPAN model, but service providers need to integrate the IMS, including this data model into a SID-based OSS model, possibly using IT technology like Java and XML.

6.17 NGN OSS FUNCTION/INFORMATION VIEW REFERENCE MODEL

Let us examine how the IMS domain may interact with our SID-based OSS. The TISPAN OSS Vision document [TS188001—2005] introduced the definition a NOSI as part of a distributed SOA environment.

Figure 6-22 represents an NGN OSS Service object (NOSI). An object is a runtime instantiation of an Object-Oriented class (OO class). This object exposes two interfaces as depicted by the "lollipops"; these are equivalent to the UML 2.0 ball and socket notation.

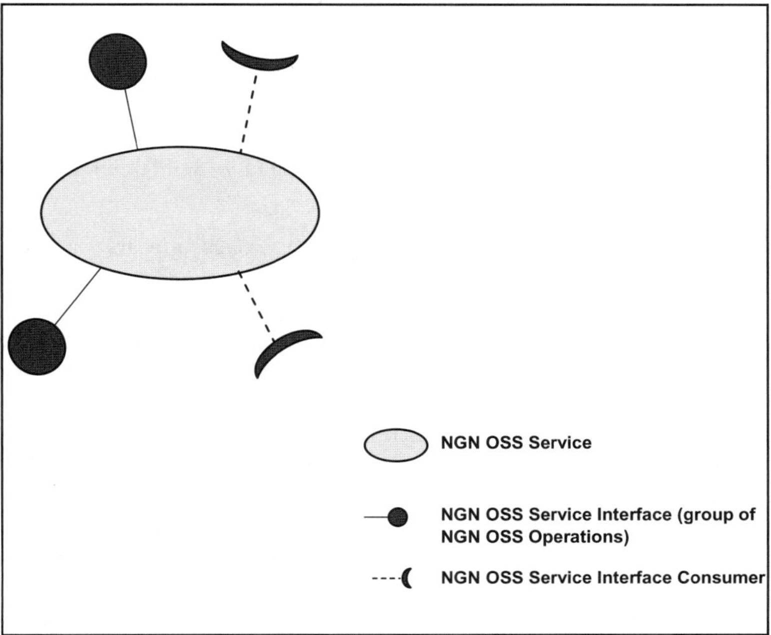

Figure 6-22. NOSI concept (note similarity to SDF model). Telecommunications and Internet converged Services and Protocols for Advanced Networking (TISPAN); Network and Service Management; Subscription Management; Part 3: Functional Architecture TS 188 002-3. (http://portal.etsi.org/) ref 12

New functionality and capability can be created by composing these NGN services via interactions of their NOSI interfaces, rather like children's building blocks. The big question is, therefore, what does a NOSI look like and how does one build it?

At its creation, the TISPAN vision document stated,

> TMF NGOSS (TMF 053D) is the only known source of metamodels to specify SOA.
>
> NOTE: In the present document the definition of SOA is taken from OASIS. NGOSS is based upon the use of the eTOM, SID, and NGOSS platform specifications including the key concept of an NGOSS Contract to specify the services exposed by NGOSS Components. The critical features of a SOA are captured in the NGOSS principles:
>
> - Common Communications Vehicle—Reliable distributed communications infrastructure e.g. Software bus integrating NGOSS components and workflow;
> - Externalize Process Control—Separation of End to End Business Process Workflow from NGOSS Component functionality;
> - Shared Information Data Model—NGOSS Component uses /implements a defined part of the SID model;
> - Business Aware NGOSS Components—where component services/functionality are defined by NGOSS Contracts;
> - Contract trading and registration using NGOSS Framework Components (covering things like directories, transactions, HMI, security, etc.).
>
> There is a need to converge TMF NGOSS, ITU-T, 3GPP, and TMF MTNM and IPNM specifications.

Hence it is no accident that the TISPAN approach appears similar to that of the TM Forum's NGOSS. From the onset it was planned to aim for convergence of OSS standards, the rationale for this apparent duplication of work was as much to do with the political and financial situation at the time as the technical. Indeed, many contributors attended both TISPAN and NGOSS/Business Services meetings.

In the intervening 3-year period, IPNM has become affiliated with the TM Forum as part of TIP; TISPAN/3GPP are liaising with the SID team on SuM; eTOM, SID, and TIP standards are down-streamed to the ITU, and certain other ITU recommendations such as M.1400 and Z.601 may eventually find their way into the Business Services Framework. We are making positive headway in the goal of convergence among standards bodies.

The future OSS to IMS integration may be tied to the creation of new NGOSS or Business Service Contracts and a way to physically build such a Business Service. As we will later see, a NOSI may be considered as being a slimmed down Business Service. (Later work in TISPAN aligns the NOSI with a Business Service Task.)

6.18 DESIGNING TECHNOLOGY-NEUTRAL ARCHITECTURES

As explained previously, transition strategies are useful but they do not tell the whole story. Service providers will need to find a way to:

- Reduce costs
- Speed up delivery times
- Cope with fewer skilled human resources
- Manage their ever-growing or changing OSS real estate while coping with the above requirements
- Develop new consistent standards models quicker and achieve a measure of testing vendor conformance to these standards
- Manage inventories of new and existing OSS services

Thankfully, new open-source and commercial tooling technologies may help with these requirements. Sometimes even old ideas, which were ahead of their time but failed to gain popularity due to politics or technology support issues, may be revisited.

A famous quote of David Wheeler goes: "*All problems in computer science can be solved by another level of indirection.*" This is often deliberately misquoted with "abstraction" substituted for "indirection." We shall see how this may help bring around a converged standardized solution through machine automation.

6.19 UML AND DOMAIN SPECIFIC LANGUAGES (DSLs)

Before covering the current work on Business Services it is necessary to explain some concepts and issues with traditional approaches. This section is complex and may be omitted, however, knowing the background will help the reader understand the how and why of our approach to standards design. It is not intended to be a UML, MDA, or DSL reference. The goal is to show that OSS standards do not stand in isolation in terms of APIs, process, and data, but rather can be linked by a common blueprint that tells a user how to configure these parts together. Ultimately, these tools may be configured as factories to turn out standards artifacts automatically.

All the current work in standards involves creating a model. This model is just an abstract representation of reality. It is possible that the same real-world phenomena may have several alternative but consistent representations.

In order to build a model, the actual framework or tool containing the model must have knowledge of the rules (or semantics) for building models. The Object Management Group (OMG) has created the UML specification for the purpose of building models. These could be models of telecom products, services, and resources, or they could be models on how to classify financial products. UML is generic enough to be capable of creating models of diverse physical phenomena.

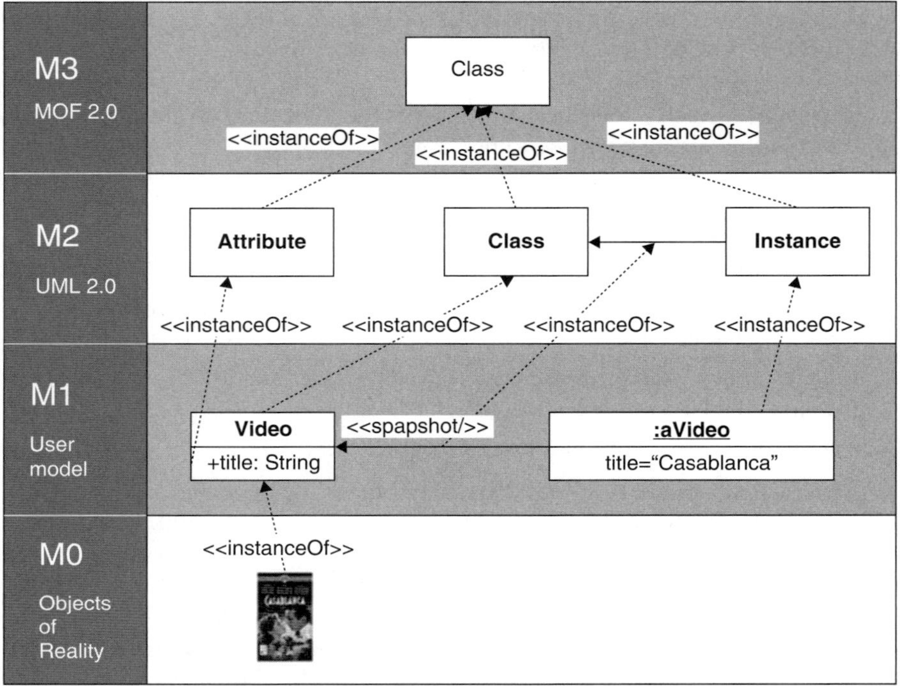

Figure 6-23. OMG MOF-UML and explanation of meta-levels (http://commons.wikimedia.org) ref 13

Underlying the ability of UML is a kernel, which embedded in a commercial or open-source tool; this kernel is the UML meta-model. The meta-model (M2 in OMG shorthand terminology) defines the set of semantic rules that allows a **tool user** to create a model (M1 in OMG terminology) in a certain way (e.g., in a UML Class diagram, a class node may have several attributes but an association link may not).

The meta-model itself is constrained by a further set of rules defined by a meta-meta-model (M3). Hence, one meta-meta-model may semantically define many meta-models. A specific meta-model may support the creation of a virtually unlimited number of models, see figure 6-23. A few notes:

- There is no M4 meta-meta-meta-model; the MOF meta-meta-model is able to describe itself unambiguously and consistently using its own semantic rules.
- The term meta-model is synonymous with the term "abstract syntax," which is used in certain older texts and recent advanced ITU specifications such as ITU-T M.1400 series and Z.601.
- M3 and M2 workings are not usually of interest to tool users, but they are of interest to the tool builders.

The OMG specification for a meta-meta-model is called the MOF (Meta-Object Facility). In implementation terms it was historically hard coded into the UML tool by a tool builder. From there, the UML tool had a hard coded meta-model hidden behind the scenes in the tool. Users do not commonly interact with the meta-model. This meta-model or kernel supported the creation of numerous user-created UML models.

The SID is an example of one of these M1 level models, as is the TIP model. Both models contain a representation of a physical and logical resource domain; they have several similarities but they are different (another model that overlaps here is the DMTF CIM). As they grew up from differing starting points you can not say that either is better than the other, but it is likely that a harmonized model containing aspects of both is a better representation of reality than the SID and TIP models alone. However, they are all still constrained by the same UML meta-model. This is one reason why there is much activity in the TM Forum in working toward this SID-TIP resource domain harmonization.

There are two main issues with this approach. First, due to the way that standards are sometimes created, there is enough slack in the standard to allow different vendors to interpret things differently (i.e., two vendors see different interpretations from viewing the same slightly ambiguous text—implementation detail!). Hence, as a result, different vendor products may claim to implement the same OMG UML specification but there is enough divergence in the implementation to prevent interoperability (the ability to share model information between differing tools via XML is an example).

This may not seem a major problem for an architect working in a single company environment, but for a collaborative environment containing several organizations, the inability to exchange conceptual model data is a drag on standard development. (This is ultimately tied to the commercial realities of product development and deployment in the Service Provider or Vendor. Service Providers want solutions for managing their OSS now; the tradition multi-year development cycles of standards is too slow.)

Secondly, due to the desired scope of UML as a general-purpose modeling language, a commercial UML tool must enforce some generic but restrictive rules that limit the freedom of the architect when creating models. That is, the meta-model constrains what the architect may build (in terms of models). The lack of a physical reference implementation when specifying a standard means that it is unlikely a paper specification will be right first time. Hence, real-world phenomena may not be modeled as elegantly as the architect envisaged or desired. The tool over-constrains what a user can create in a model.

It is possible to slightly enrich the semantic capability of the UML meta-model by using UML 2.0 extensions called profiles. Here the tool builder (also called toolsmith in Eclipse parlance) may access a read-only definition of the UML meta-model and add extensions called stereotypes to it. These stereotypes may have additional properties called tagged values. The result is a more specialized version of UML that may provide a closer fit to the desired problem space being modeled.

The military MOD Architecture Framework, DoD Architecture Framework (MoDAF and DoDAF) UML models are examples of such extended models, as is

The Open Group Architecture Framework (ToGAF). There is another problem here, going back to issue 1: these models have been developed in one tool and it is difficult to import them into other tools and use them.

Note: there are two xml-based interchange format called XMI, one XMI representation for data and one for UML diagrams. Unfortunately, XMI data are not truly interoperable, and currently no one implements the diagram exchange version. Hence, XMI import/export is not always guaranteed to work between different tools, but the situation is improving.

6.20 AN EMERGING SOLUTION: THE DOMAIN SPECIFIC LANGUAGE

Given what has been said so far, things may appear pretty hopeless. Fortunately, technology has come to the rescue. Over the past several years the open-source platform called Eclipse has been maturing projects based on what are called Domain Specific Languages or DSLs.

Eclipse provides a tool builder with access to a drawing tool that allows us to define our own tool kernel or meta-model, which may be "compiled." We can then use our created kernel to define our own models built according to our own rules and semantics. These rules can be tailored to be as close as possible to a customer's problem space. The beauty of this approach being two-fold:

1. We no longer need compromise ourselves by forcing a domain-specific solution from a generic commercial tool. We can have the correct bespoke tool, which isn't limited by a one set of rules fits all, made for a specific job.
2. Commercial tool builders can create these new tools; equally skilled architects (tool-smiths) with some programming ability may create tools. Hence, other than learning the language you have little or no external dependencies.

The major difference between DSLs and the generic approach of UML is that, unlike UML, DSLs do not attempt to capture a generic set of semantic rules for the set of all possible models in every problem space. Rather, a DSL focuses on the semantic rules for an individual problem space and does so as closely as is humanly possible. Hence, a DSL potentially creates a 100% perfect fit or representation for a problem space. A problem space or domain could be a telecoms resource model or the rules for building an electronic circuit, for example (the clue is in the name, they are domain specific!).

In addition to defining model rules, DSLs have the ability to be coupled with Eclipse-based code generators. This enables a software architect to define a specific model definition in a DSL editor, and then create a runtime implementation in a technology of choice via the code generator at the push of a button.

These runtimes code generations are also semantically constrained since they are created from models which are constrained by the rules of the meta-model (aka DSL). We have the possibility of creating a true software factory from a semantically constrained blueprint for any problem space.

6.21 FROM MODEL-DRIVEN ARCHITECTURE TO MODEL-DRIVEN SOFTWARE DESIGN

Behind the scenes, these Eclipse tools are based on a meta-model (M2) called Ecore (Figure 6-24), which is considered by many to be a pragmatic reference implementation of OMG Essential MOF. (Ecore is self-reflective so can be considered its own meta-meta-model—M3.) Tool builders may use this in the Eclipse EMF project (Eclipse Modeling Framework) to create their own meta-model or DSL. The DSL is "compiled" and a domain-specific editor is generated. This editor is visually generated by another Eclipse project called GMF (Graphical Modeling Framework) and may be further semantically enhanced by OCL.

Note: Although Ecore and GMF editors are written in Java, this doesn't mean that the output of such editors is tied to Java implementations or problem spaces.

The concept isn't new; it originally stems from a methodology called MDA, which was developed by the OMG from 2001. Prior to this, there were also CASE (Computer Aided Software Engineering) tools in the 1990s. MDA originally had a goal of starting with a technology neutral representation of a model or PIM (Platform Independent Model based on UML) and using a particular technology, e.g., CORBA

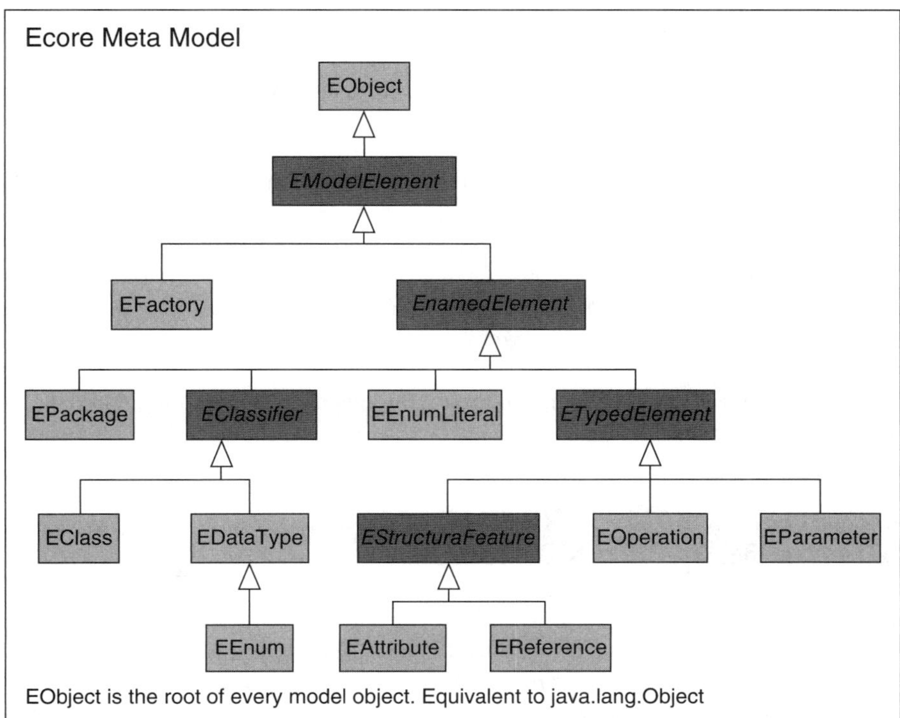

Figure 6-24. Ecore meta-meta-model representation. Simplified representation of Ecore Class Hierarchy (http://www.devx.com/Java/Article/29093/1954) ref 14

mapping it onto a technology specific representation or runtime PSM (Platform Specific Model in Java or C++, etc.).

As stated earlier, UML is an example of a meta-model that is defined by the MOF, however, it isn't the only meta-model MOF constrains. OCL (Object Constraint Language) is an example of a semantic constraint language based on pi-calculus and can be used to remove ambiguities that may arise in UML. Further discussion of OCL is out of scope; the reader should refer to *The Object Constraint Language*, by Warmer and Kleppe. Additionally, OMG defined a model-to-model transformation language called QVT (Query View Transform language). QVT is a script-based language in which a user uses OCL-like constructs and user-written procedures to take one model as input and output to another model. While in practice these could be any models, they could equally be PIM-PSM transforms.

Unfortunately, QVT is complex and is implemented from three different specifications, part of the reason for this being a desire to support model round tripping. The price for this ability was complexity and a long gestation time for the development of QVT. In theory, it should be possible to modify either PIM or PSM and feed the changes bi-directionally via QVT. The complexity and long gestation of QVT and MDA as a whole was a primary reason why those of us working on OSS standards never really warmed to MDA despite liking the concept.

The new approach of MDSD differs from MDA in that it has simpler and more pragmatic requirements. For example, the strict PIM-PSM round tripping concept has been replaced by DSL to model to text generation. Other tools such as JET and Xpand set of languages have removed a reliance on the relatively more complex QVT for textual code generation. There are now growing numbers of companies researching and using DSL-based development. The reason for this is not just the ability to build "correct models" but also the potential cost benefits. DSL done correctly offers the prospect of ultra-agile development times. Some verified productivity increases range from 1,000 to 2,000% over traditional hand-crafted methods. Certainly once the original model has been created (which isn't necessarily very long), new runtime implementations can be generated very rapidly, in minutes.

Although Eclipse is referenced throughout this chapter, other commercial and academic MDA platforms exist.

6.22 OTHER STANDARDS MODELS (DMTF CIM, 3GPP, AND TISPAN)

The SID is based on UML 2 but there are other information frameworks belonging to other standards bodies that are also based on UML (not all UML 2). For example, another body that is very important in defining IT infrastructure (e.g., call centers) is the DMTF CIM or Distributed Management Task Force Common Information Model. The scope of the DMTF is actually much wider but the IT infrastructure is an important touch point for the SID. The 3GPP and ETSI TISPAN models aim to model the management capabilities of IMS (IP Multimedia Subsystem) networks and their management of customers and users. The main focus is currently on Subscription Management or the ability to model the interaction between Service Providers, Services, Subscribers (the people who pay for Services) and Users (people who use the Service that may or may not include the Subscriber).

As all these bodies are important and expert in their areas it is important that the SID, CIM, and 3GPP/TISPAN models are harmonized to the extent that interoperability exists between them. Currently this is a work in progress, but important strides forward in demonstrating interoperability have been achieved through close liaisons between these bodies.Many people are members of both groups and several catalysts have been developed showing how differing data models may be utilized and bridged in a large OSS space.

A catalyst is not technically a normative standard. It may be pretty close to a standard implementation but it is more focused on providing a working demonstration based on interpretations of current standards work with a view to speeding up further standards development. Typically, they are implemented by a collaboration of like-minded SPs and vendors.

6.23 PUTTING THINGS TOGETHER: BUSINESS SERVICES IN DEPTH

In a typical SP OSS we may have a vision of an agile Service-Oriented Architecture (SOA) that supports the rapid creation and deployment of new products. The challenge is how to achieve this; a good reliable architecture needs good architectural design rather than ad hoc organic growth. This requires an architecture that is aligned to SP best practice in terms of process. This is the goal that Business Service Contracts aim to fulfill.

This vision is built around having our processes defined down to at least eTOM level 3 and 4. Level 3 provides fully articulated and specific business processes with a real-world business goal. Many processes already exist and are defined within eTOM, however, it is possible that the SP provider may wish to create new Tasks, as they become part of their business model. These processes define significant and meaningful business activities, which can be used by a business analyst in designing new configurations for reuse.

These processes are composed of Tasks and can be represented by Use Cases. Drilling down to Level 4, the expectation is that software, or manually operated systems (Tasks may be a manual activity such as a technician replacing a card in a NE) will be able to deliver the function without further decomposition to an *exposed and explicit business process.* In this sense they are considered to be atomic Tasks that are combined into business processes at the higher level.

From a software implementation aspect, these Tasks could be SOA services provided by a system. Tasks are grouped together into Business Service Contracts (Contracts may support one or more Tasks) in order to support a specific business purpose between a client and provider application as defined by the Application Map. This may differ from definitions of SOA that may be defined in other books or on the World Wide Web. The key difference in a SP provider OSS is that the SP knows exactly what applications exist in their OSS, and solutions are designed with this in mind rather than attempting to discover the best provider to bind to in order to deliver a service. This is not to say that there is no role for policy, for example, deploy a service via vendor "A" in geographic location "X" or deploy using vendor "B" in location "Y."

Below Level 4, further decomposition is considered to be an internal function of the system, which may be custom or standardized. A Task may be implemented by one or more Operations of the system, but this is an IT implementation issue, not a business process issue. In this aspect a Task may be considered analogous to the familiar Web Services API that people are familiar with; indeed, there is scope for re-using good existing legacy APIs within the Business Service Framework. Each Task may have its own data associated with it. This data that is exposed on "the wire" is expected to be SID based, but if it is not then some kind of adaptation software will be required to map from the API on to a form that is understood by the back-end application database.

A Contract specifies the requirements for a business interaction between two systems. It does this by considering a managed entity, or entities, over an extended time period, e.g., the lifetime of that entity. It brings together the set of tasks (services) that the provider system is required to perform, such as "create SNC across device," or "insert card into device." In this way, Business Service Contracts are the evolution in Service Orientated Architecture (SOA), transitioning SOA from IT architecture to a business-oriented architecture. The word "Task" is taken from eTOM, rather than IT "Services," and is used in the Contract specification to represent this transition.

The sequence of Tasks and their outcomes represent the lifecycle history of the Contract. This is the Business view. The System view adds extra detail commensurate with a solution designer's worldview. These Tasks may require some choreography in that in order to fulfill a specific business purpose "Task 3" may need to complete before "Task 2" may start. The Contract moves to the Implementation View when technology-specific API commands are added to the Contract to implement the Tasks, e.g., we have decided to implement our Contract using Web Services. At each stage the relevant authority or domain expert is expected to input into the Contract from analyst to developer.

In this way, a Contract provides the context and purpose for a set of interactions between two systems in a way that fully specifies all the requirements for implementation. Importantly, this specification is independent of the integration technology, so can be applied to all technologies.

A Contract is targeted at a specific business purpose, but in principle any given interface will support a number of different business purposes. The SP is free to select which Contracts it will implement based on the specific purpose of the integration.

In addition, because the Business and System view Contract is technology neutral, the Contract provides the perfect mechanism to harmonize different interfaces standards. This is important as new standards appear for emerging technologies. It also accelerates the process of standardizing emerging technologies, by providing pre-built business frameworks.

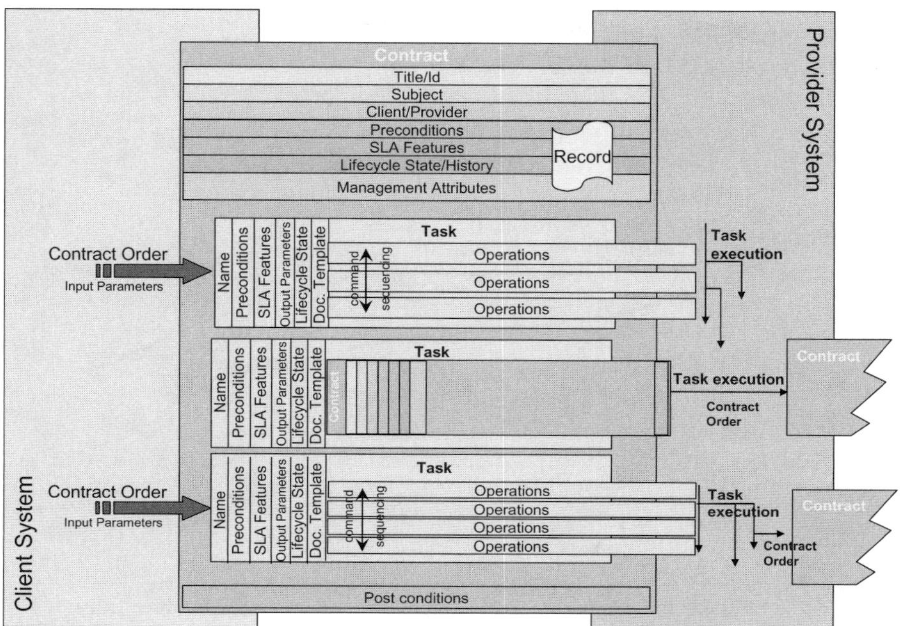

Figure 6-25. Representation of Business Service Contract and components. GB942, NGOSS Contracts—Concepts and Principles (http://www.tmforum.org) ref 15

Looking Figure 6-25, it will be noted that the Contract also supports features for multiple Service Level Agreements (SLAs), and Conditions (pre- and post-condition). The SLAs have initially been mined from one of the TM Forum teams specializing in that area and while pre- and post-conditions exist in older work (Programming by Contract) the Business Service interpretation of these rules is not as rigorous as previously defined in that methodology. Essentially, the Contract is an agreement between a client and a provider system. In return for submitting the request for work of the correct form (pre-condition), a provider guarantees to provide a response of the correct form.

A Contract can be a long-running activity; this means it must maintain state information about the current status of a Task. This may seem at odds with other definitions of SOA, which talk of SOA being stateless, however, while a Contract is definitely stateful, each individual Task or part of a Task itself be stateless. A further feature of the Contract is that it records its lifecycle history as it runs; this may be queried later by another system to determine why a Contract may have failed or to analyze SLA performance.

In order to instantiate a Contract, a client first needs to send the provider useful configuration data in a Contract order (which contain individual Task orders). This could, for example, be a Web Service document specifying run time parameters, requested SLAs etc. It is the Contract order, which flows across the wire, not the actual Contract.

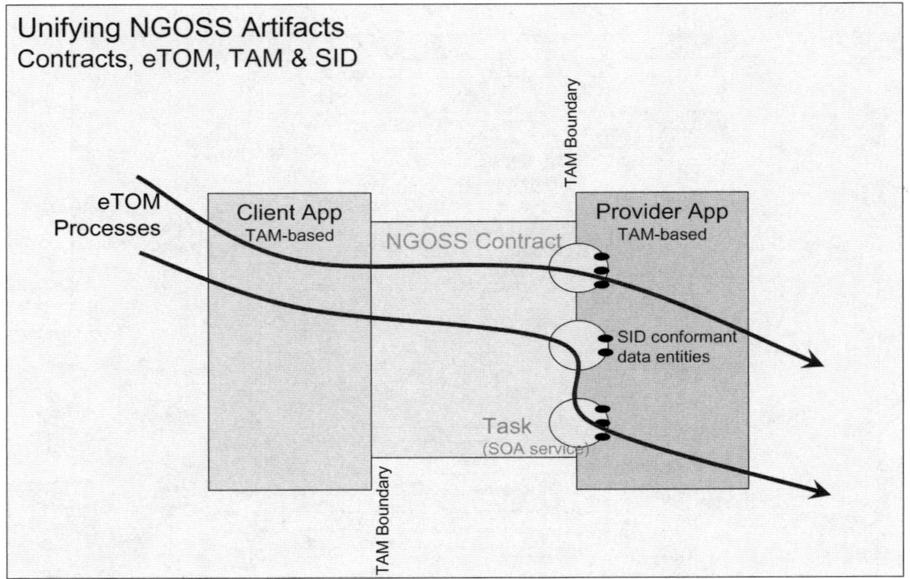

Figure 6-26. Process Flow between TAM components accomplished by three tasks. GB942, NGOSS Contracts—Concepts and Principles (http://www.tmforum.org) ref 16

Contracts are managed and reusable, much more so than individual APIs can be, because they specify common business requirements over the interface rather than individual interface commands. This means that interfaces are specified "top-down" according to business requirements, rather than "bottom-up" by vendors offering a broad range of APIs, many of which are not used.

However, where standardized interface technologies are used (a Contract could be constructed around existing such APIs, each corresponding to a Task), the Contract becomes a fully standardized interface specification, enabling plug-and-play integration.

Figure 6-26 shows two applications as identified by the Telecoms Application Map (TAM, TM Forum standard that maps OSS application platforms to specific eTOM layers and utilization of data). The provider in this case supports three Contracts. In order to implement two business processes, a Contract exists between client and provider. The client sending the provider an order for three Tasks to be implemented by three SOA services delivers this process successfully. The parameterized data sent by the client is based on standard data. This data is mapped into the Contract either directly from standardized data models on client on provider, or more likely by mapping technology (self-written or created by commercial tool).

Figure 6-27 shows how these processes, data models, and their implementation exist at differing levels and are defined by respective standards.

Contracts may be linked in the form of a value chain, as in Figure 6-28, showing an interaction between four platforms exposing multiple Contracts. An aspect that is important in the long-term design of such architectures is that of life-

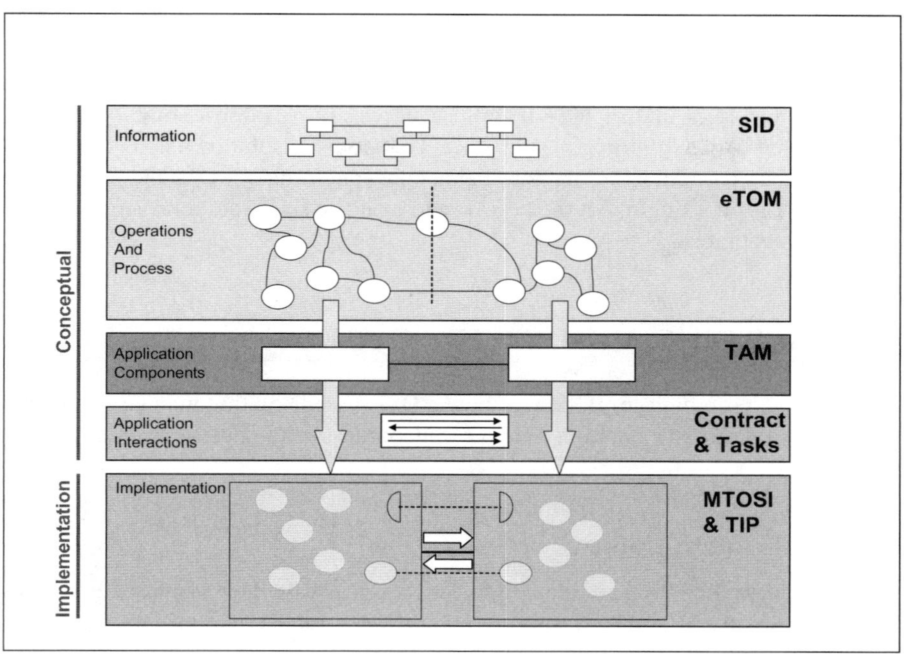

Figure 6-27. Relationship between different standards from data to implementation. GB942, NGOSS Contracts—Concepts and Principles (http://www.tmforum.org/) ref 17

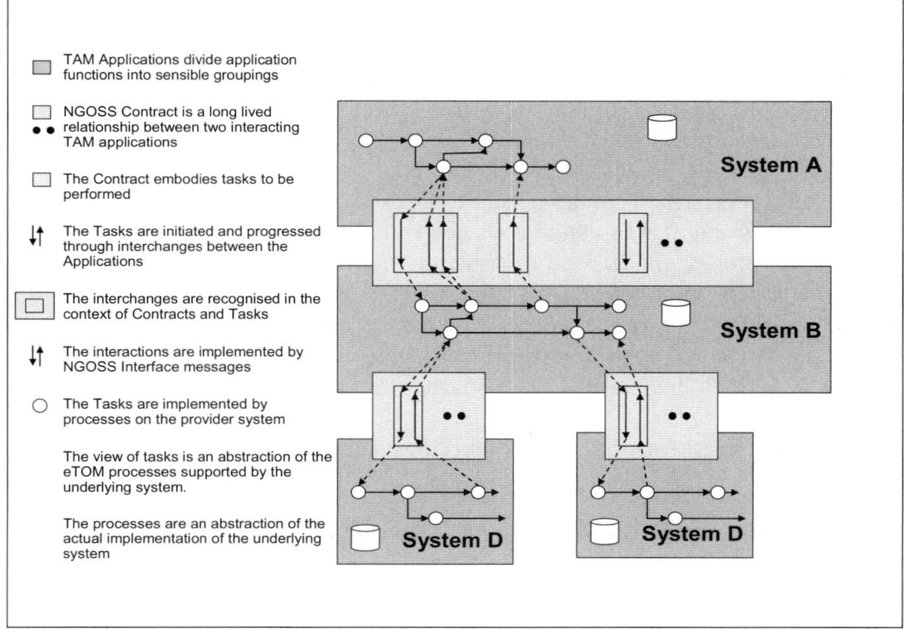

TAM Applications divide application functions into sensible groupings
NGOSS Contract is a long lived relationship between two interacting TAM applications
The Contract embodies tasks to be performed
The Tasks are initiated and progressed through interchanges between the Applications
The interchanges are recognised in the context of Contracts and Tasks
The interactions are implemented by NGOSS Interface messages
The Tasks are implemented by processes on the provider system
The view of tasks is an abstraction of the eTOM processes supported by the underlying system.
The processes are an abstraction of the actual implementation of the underlying system
System A
System B
System D
System D

Figure 6-28. Multiple platforms and Contracts as part of a value chain. GB942, NGOSS Contracts—Concepts and Principles (http://www.tmforum.org/) ref 18

cycle management. What happens if the bi-directional flow of the Contract between system B and D is interrupted? It will be necessary to inform some controlling authority that such a failure has taken place. One possible solution here is to develop a standard based on SDF, IPSphere utilizing concepts from WS-DM and OSGi. This is still work for the future, but the best way to evaluate such solutions is by building test implementations.

6.24 BUILDING A DSL-BASED SOLUTION

The following is a high-level view of how a DSL-based solution may be attempted. Of course, a normative solution requires collaboration and consensus among many stakeholders.

6.24.1 Problem Context

Create an automated standard that aligns business process, enterprise data, generates runtime code, increase agility, improves quality, and reduces overall cost.

6.24.2 Proposed Initial Feature Content

Such a solution is likely to contain references to an enterprise UML model captured in UML 2.1. It will have a reference to a process repository—either document-based or database. It may involve the use of a database repository containing existing Web services deployed in the OSS. There may be requirements for the support of Service Level Agreements, Policy, and Pre and Post Conditions. Generation of PDF documentation and Java/XSD runtimes may be required.

Rather than define such things such as Policy and Service Level Agreements from first principles, it pays to see whether there are any existing specifications for such features. Fortunately, in the case of Business Services, there do exist specifications for Policy and SLAs so these are a good place to start in the modeling process. Likewise for process models and information models.

The following, figure 6-29 is the kind or architecture we are aiming for. Eclipse sits at the heart of the process. Since it has native support for UML through it's open-source UML projects it seems reasonable to leverage this rather than create a whole new UML implementation (also saves time).

6.24.2.1 Desired Inputs The desired inputs are:

- A representation of a UML model
- A representation of a process model extracted from a tool or database
- A representation of an application map detailing OSS components, their location, capabilities, and requirements for data.

In the TM Forum the process and application map repositories are commercial tools with proprietary formats. In a service provider environment these may be relational

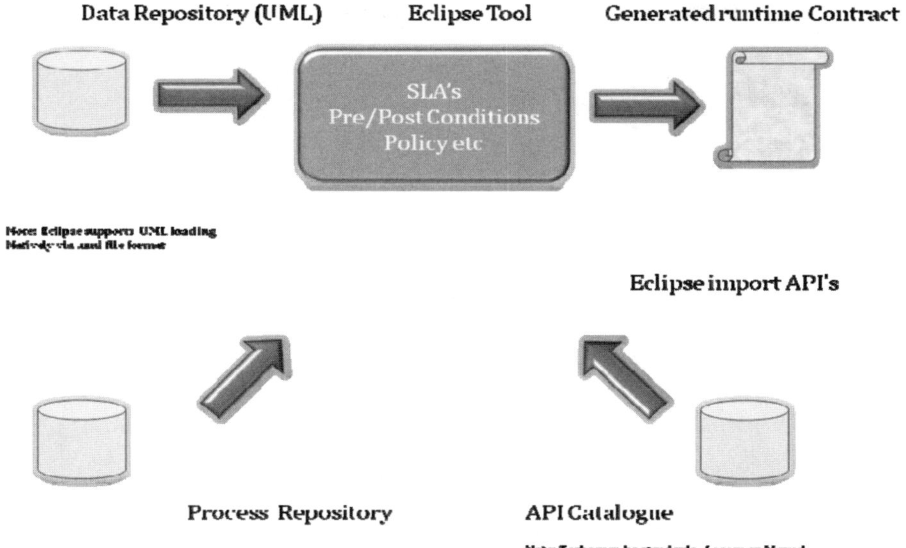

Figure 6-29. Relationship between Eclipse and repositories

databases so the solution may depend on the implementation context.

6.24.2.2 Desired Outputs The desired outputs are:

- An electronic blueprint (Contract) encompassing all business, solution design, and implementation view data and meta-data
- A runtime representation of that data in Java and XSD format
- A text-based document that contains the Contract design features (Word, PDF, or HTML)
- A stored persistable record of the Contract for re-uses
- A reference implementation of a "possible" technology specific solution

6.24.3 Open-source Tool Environments

Having decided to embark on a MDSD-based solution, we need a suitable tool. Fortunately there are both commercial tools and free open-source tools such as the open-source Eclipse. This has the benefit that a development community can share in the development unhindered by the requirement to use a specific tool (lowering barriers to adoption is an important aspect to standards work). Eclipse supports all the required tools a tool-smith could require, such as EMF DSL creation, GMF editor creation, OCL, QVT, document generation via BIRT and code generation via M2T (model to text project incorporating Xp and—Java is just text until compiled!).

Starting from the above high-level design we decided to make our solution modular so that new features could be added (e.g., Service Delivery Framework,

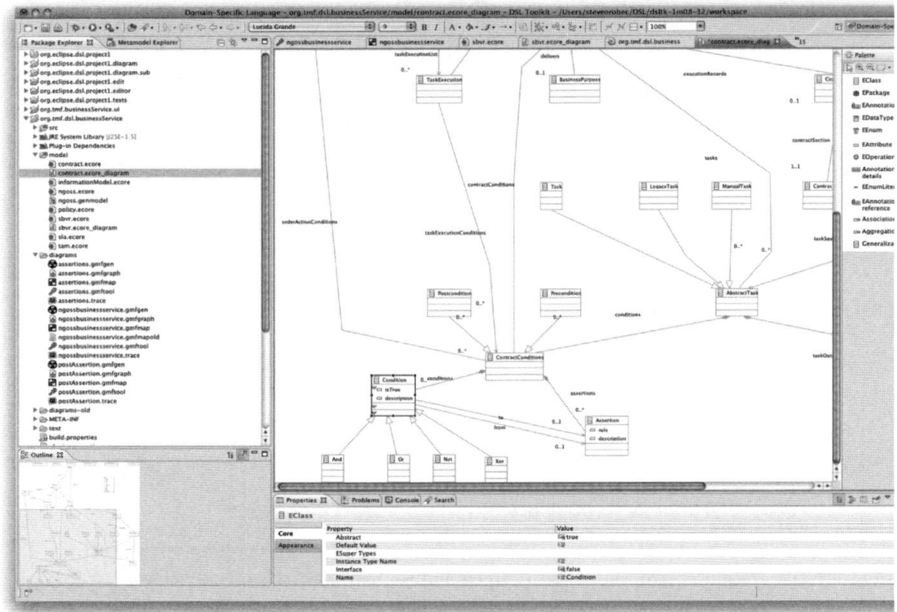

Figure 6-30. Screenshot of Contract DSL in Ecore

Service Quality Management as mentioned previously). Hence we created several DSLs to capture our problem space; one DSL captures the essence of the Business Service Contract, another captures Policy, another captures SLAs etc. These child DSLs are linked by a common parent that contains special links to each model (containments). We want to give our solution the ability to understand UML so we have used a feature of Eclipse to reference the Eclipse UML plug-ins.

Having done this we may design our DSLs. Figure 6-30 is a representation of the Contract Business and System view artifacts as of May 2009 (it may evolve over time—that is the nature of standards). The tool smith just drags the appropriate Ecore entity onto the canvas (e.g., EClass from Ecore diagram fig Y) and names it, e.g., Contract, and adds appropriate meta-attributes. The whole process is similar to UML modeling but for two things, the models are generally smaller and secondly we are dealing with an extra layer of abstraction so a Contract may be a Billing Contract or a Service Activation Contract.

Eclipse has an in-built code generator plug-in that allows the Figure 6.30's pictorial representation to be transformed into editor code (actually all these editors use XMI to store the entered data and render it according to certain rules). Hence the code generator is just transforming text in the form of XMI into another form of text (Java for the editor code) according to a set of rules in pre-built Eclipse templates written in special meta-languages such as Xpand or Jet. This is how we will ultimately generate Java from our editor; hence we know the process does work, since our editor will be 100% machine generated in a similar way (there are techniques for further customization).

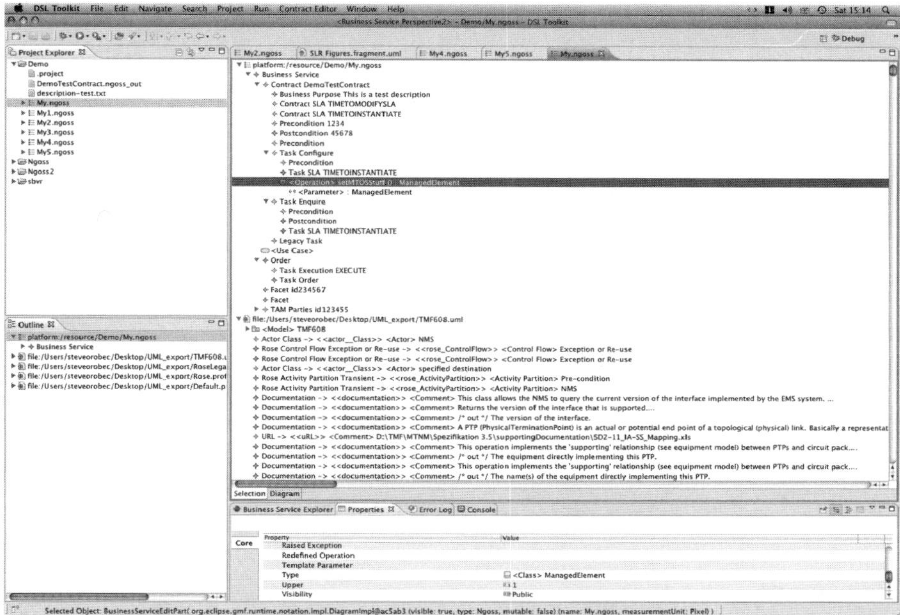

Figure 6-31. Machine-generated tree editor that a user has partially filled in data. Note that the tree has a structure defined by the DSL

We can force Eclipse to create two types of editor, a graphical tree editor, see figure 6-31, or a rich GUI editor rendered by the GMF plugin. Natively, the generated code produces a set of simple white boxes but the key aspect is that the relationship between these boxes is semantically constrained by our DSL. Much of what has been described so far makes use of an architecture principle called "separation of concerns"; here the visual depiction of the editor is de-coupled from the actual data behind it. This gives the opportunity to tune an editor to a specific set of users. For instance, you may encourage analysts with little modeling skills to use it by representing aspects of the model by pictures rather than UML type boxes, and hence no new special skills are required. Since the tool is created from a DSL, which constrains the rules by which artifacts can be modeled, it is impossible for the user to model a Contract incorrectly.

Figure 6-32 depicts a beta version of the normative Contract tooling. It shows a graphical display of the same data. Note the left-hand palette now contains items, which were defined as nodes in the DSL. We are now working at the model level as opposed to the meta-model level, and attributes are plain attributes rather than meta-attributes. This is the same level of abstraction that UML resides at but, rather than defining a UML class model, we are defining a Business Service Contract model in Business Service language.

One can take the DSL building to arbitrarily sophisticated levels and add modules to define implementation detail and business logic (perhaps by leveraging existing or new Eclipse plugins, for example). If so inclined, a tool smith could link

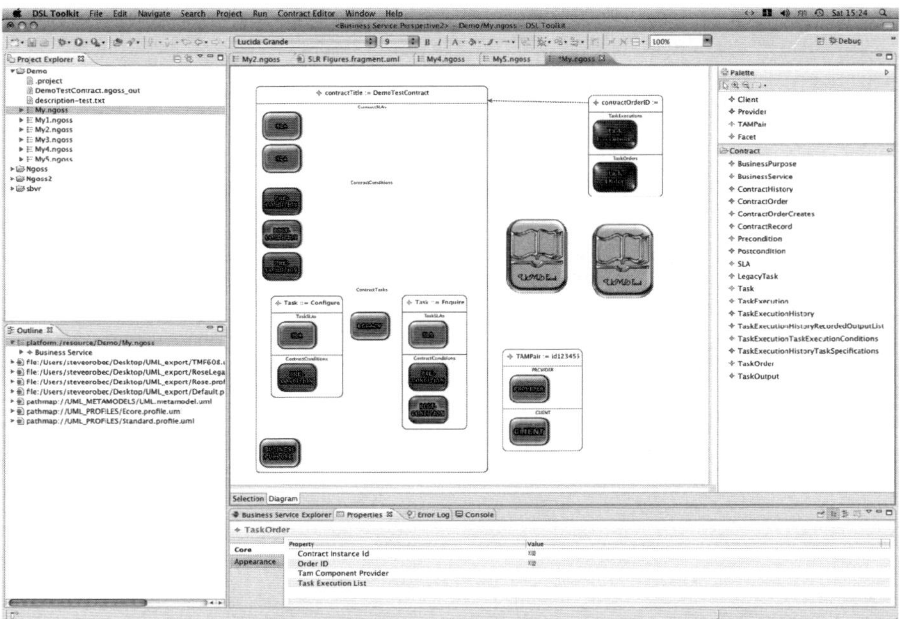

Figure 6-32. Simple graphical representation of similar data to Figure 6-31

the above tool with an open source BPMN plug-in or a new OMG-based SBVR (Semantics of Business Vocabulary and Business Rules) plug-in. It pays to do regular research for new features, as well as old features that may once have been a technology that was ahead of its time!

It is possible to link other DSLs to leverage their functionality, e.g., UML. This is achieved by linking Contract nodes to a referenced model node, in this case, UML::Class (the definition of Class in the UML kernel). Having taken care of this, Eclipse takes care of the implementation for the developer.

Once the user or users have entered the necessary data via the property window attributes, they can use the code generation capability of Eclipse to render the stored data into a Java and XSD runtime artifacts. The tool smith writing specific Xpand templates to create a Business Service achieves this. The easy way to do this is to actually create a Java prototype realizing that much of the content of the source code will be boilerplate scattered with occasional data items.

Having created a reproducible but data independent solution, a user (solution designer or developer) may now create hundreds of Contract variants in the space of hours simply by changing the entered data and generating code. (It still needs plumbing in to back-end systems but the level of automation can be increased if required.) The data that a user may enter may be further constrained by use of enumerations in the DSL, which present the user with a drop-down pick list, or additionally with OCL. Running tool smith–written audits and metrics prior to code generation may also check data quality. The result of this is a fully documented, extremely agile solution of consistent quality and style(s).

The reader may ask at this stage, How did you build the initial prototype for the runtime? This is a good question. At present there are no normative implementations for a Business Service. There have been several prototypes and hand-crafted solutions, either implemented as catalyst projects at trade shows or as prototypes in service provider environments. The key here is to realize that iteration is the key. By taking the concepts of established and working prototypes and catalysts and building a testable reference implementation we may learn what does work and what doesn't (similar to the approach taken by REST advocates). For example, a suggestion is that the Contract "scaffolding" that manages Contract Task choreography may use Java Space or Jini (a Java zero configuration service—a mature technology before its time perhaps). Some aspects of an OSGi standard may implement lifecycle management. Eclipse plugin lifecycle management is based on OSGi and there is work to deliver similar features in an application server. Data mapping of standard SID data in an API could be achieved by using commercial products that run in the Eclipse platform, so without a doubt there are technologies that do support our goals.

While this may seem slightly contrived, is the alternative of a proprietary solution delivered on the basis of an untested premise superior?

6.25 FINAL THOUGHT

The world is moving at an ever-faster pace and customers, managers, and shareholders demand speed of delivery. If OSS standards are going to play a part in the development of a true 21st century OSS they also need to be built quicker and faster. The old 2–3 year gestation period is no longer going to be tolerated. The only way of achieving quality standard output is via automation. Fortunately, the barrier to adoption of a model-driven approach has reduced from almost mysticism to a valuable skill in the space of just 2 years or so, thanks to a growing amount of documentation and examples. The Business Services model introduced here is just an initial example of this methodology, but you the reader could do worse than become an early adopter and potential expert with such skills.

6.26 BIBLIOGRAPHY

1. Figure 6.3 Zachman Framework
http://commons.wikimedia.org
2. Figure 6.4 eTOM levels 0 and 1
GB921, The Business Process Framework (eTOM) http://www.tmforum.org/
3. Figure 6.5 eTOM Process levels
GB921, The Business Process Framework (eTOM) http://www.tmforum.org/
4. Figure 6.6 Example eTOM level 3 process GB921, The Business Process Framework (eTOM) http://www.tmforum.org/
5. Figure 6.7 Example eTOM level 3 Process decomposed to level 4 GB921, The Business Process Framework (eTOM) http://www.tmforum.org/
6. Figure 6.8 Example additional meta data enrichment in the transitioning to different view—Business view enriched with new System view attributes GB922 Information Framework (SID) Solution Suite Release 8.0 (SID Primer) http://www.tmforum.org/

7. Figure 6.15 High level MTOSI integration
   TMF517 Business Agreement for MTOSI Release 1.1
   http://www.tmforum.org/
8. Figure 6.18 Gradual migration from a legacy OSS of mixed applications to a capability based collection of platforms BTs Matrix Architecture W G Glass
   http://www.btplc.com/Innovation/Journal/BTTJ/current/HTMLArticles/Volume26/08Matrix/Default.aspx
9. Figure 6.19 3GPP IMS model http://commons.wikimedia.org
10. Figure 6.20 TISPAN High level model
    http://portal.etsi.org/
11. Figure 6.21 Example of 3GPP relationships from TS32.172 v8.0
    Telecommunication management; Subscription Management (SuM) Network Resource Model (NRM) Integration Reference Point (IRP): Information Service (IS) http://www.3gpp.org/ftp/Specs/html-info/32-series.htm
12. Figure 6.22 NOSI concept—note similarity to SDF model
    Telecommunications and Internet converged Services and Protocols for Advanced Networking (TISPAN); Network and Service Management; Subscription Management; Part 3: Functional Architecture TS 188 002-3
    http://portal.etsi.org/
13. Figure 6.23 OMG MOF-UML and explanation of meta-levels
    http://commons.wikimedia.org/
14. Figure 6.24. Ecore meta-meta-model representation. Simplified representation of Ecore Class Hierarchy
    http://www.devx.com/Java/Article/29093/1954
15. Figure 6.25 Representation of Business Service Contract and components GB942, NGOSS Contracts—Concepts and Principles
    http://www.tmforum.org/
16. Figure 6.26 Process Flow between TAM components accomplished by 3 Tasks GB942, NGOSS Contracts—Concepts and Principles
    http://www.tmforum.org
17. Figure 6.27 Relationship between different standards from data to implementation GB942, NGOSS Contracts—Concepts and Principles
    http://www.tmforum.org/
18. Figure 6.28 multiple platforms and Contracts as part of a value chain GB942, NGOSS Contracts—Concepts and Principles http://www.tmforum.org/

CHAPTER 7

MANAGEMENT OF WIRELESS AD HOC AND SENSOR NETWORKS

Mehmet Ulema

7.1 INTRODUCTION

Wireless ad hoc networks and wireless sensor networks are relatively new types of wireless networks with drastically different architectures and properties than traditional wireless networks, such as cellular and wireless local area networks.

In a wireless ad hoc network, nodes communicate with each other without the help of any pre-existing structure. The "network" is autonomously formed among many nodes (laptops, notebooks, etc.) with varying functionalities, processing capacities, and power levels. A wireless sensor network typically consists of a large quantity of low-cost, low-power radio devices dedicated to specific functions, such as collecting environmental data and sending it to processing nodes. Like wireless ad hoc networks, wireless sensor networks are usually autonomously formed and may include large number of nodes (sensors, etc.) with varying resource and power levels as well. Although many ad hoc network techniques are applicable, wireless sensor networks differ in several areas, including its very high deployment density and, most notably, in the number of nodes, which may be in the thousands.

Wireless ad hoc networks and wireless sensor networks are well positioned to be among the primary enablers for ubiquitous computing and communications. They are well-suited to perform critical functions during natural disasters where pre-existing infrastructure may be destroyed. Another area of greater interest is in military applications, which require dynamic autonomous architectures formed on-the-go.

Given their relatively more complex and highly dynamic environments, the wireless ad hoc networks and wireless sensor networks present enormous challenges, demanding new network management approaches that must be different than traditional ones. For example, a wireless ad hoc network can re-configure its topology by itself, dynamically, depending on the environment. This may require a management approach that also dynamically adapts itself. In a wireless sensor network, nodes are assumed to be discardable. This requires that the fault management for wireless sensor networks has a totally different approach than for a traditional

Next Generation Telecommunications Networks, Services, and Management, Edited by Thomas Plevyak and Veli Sahin

cellular network. In addition to filling out the "matrix of network management functional areas vs. network management layers" with new microsolutions, there needs to be new approaches and new management architectures.

The chapter addresses the network management aspects of wireless sensor networks and wireless ad-hoc networks. It presents a survey of what has been done in the area of managing these types of networks. The remainder of the chapter is organized as follows:

Section 7.2—Provides a brief overview of the wireless ad hoc networks, wireless sensor networks, technologies, and architectures. It also introduces a framework that the chapter uses to discuss network management specific aspects of these networks.

Section 7.3—Provides a discussion about the functional and physical management architectures for wireless ad hoc networks and wireless sensor networks.

Section 7.4—Discusses the logical management architectures for wireless ad hoc and sensor networks. This includes management layers and management functional areas under the configuration, fault, performance, security, and accounting management for wireless ad hoc networks and wireless sensor networks.

Section 7.5—Discusses information architecture for managing wireless ad-hoc networks and wireless sensor networks. This includes manager-agent models, interfaces, and protocols as well as the information modeling on these interfaces.

Section 7.6—Provides a summary and brief conclusion.

7.2 OVERVIEW

This section first provides a brief discussion of the wireless ad hoc networks and wireless sensor networks. Also provided in this section is a management framework that the succeeding sections of the chapter use to discuss network management-related aspects of these networks.

Wireless networks and systems come in many flavors. In general, the following provides a rough classification: wireless cellular networks, wireless local area networks, wireless fixed networks, satellite networks, infrared networks, etc. Wireless cellular networks are typically designed to provide wide area coverage. Their architecture is highly hierarchical, although the latest trend is towards a peer-to-peer flat structure. The wireless phones can be mobile and were originally designed mainly for voice transmission, but later designs of cellular networks handle data transmission as well. Generally, the air interface and the protocols (GSM, CDMA, WCDMA, etc.) on the phone separate a cellular technology from another. A wireless Local Area Network (LAN) has a rather simple architecture that includes one or more access points attached to a wired Ethernet. The nodes with wireless access capabilities can have limited mobility and are typically attached to an access point. Some of the examples of this category include IEEE 802.11 type wireless LANs and Bluetooth. Wireless fixed networks (a.k.a., Wireless Local Loops) typically replace the wire between customer premises and carrier central office. They rely on the existing telephone network infrastructures. Some networks of this type are designed to provide high-speed data and video services in a distributed way with a separate architecture. Fixed WiMAX is a good example of this type of network. There are a number of other types of wireless networks that do not fit well into one of the above categories. For example, wireless networks utilizing satellite communications have different architectures and offer services that are suitable for large area coverage.

Wireless ad hoc networks and wireless sensor networks are drastically different than the other wireless networks mentioned briefly above. The following subsections discuss wireless ad hoc networks and wireless sensor networks in more detail.

7.2.1 Wireless Ad Hoc Networks

In addition to the fact that the nodes are mobile, a major defining feature of a wireless ad hoc network is the way in which the nodes establish a "network" autonomously without the help of any pre-existing structure. According to the Internet Engineering Task Force (IETF), Mobile Ad-hoc Networks (MANET) [1] is an autonomous system of mobile routers (and associated hosts) connected by wireless links—the union of which form an arbitrary graph. The routers are free to move randomly and organize themselves arbitrarily; thus, the network's wireless topology may change rapidly and unpredictably. A mobile ad hoc network may operate in a standalone fashion, or may be connected to the larger Internet [2].

As mentioned above, a wireless ad hoc network is autonomously formed with a large number of heterogeneous nodes, without the aid of any pre-existing communication infrastructure. The topology may be highly dynamic due to energy limitation (nodes come and go), constrained bandwidth and quality, as well as node mobility. These limitations can result in frequent partitioning of the network. The type of communications among nodes is characterized as "multi-hop," where a node is capable of receiving and forwarding data destined for other nodes.

In general, nodes can be mobile or static. The complexity of the nodes in a wireless ad hoc network can range from sensors to hand-held devices and fully functional laptops to routers. Since the nodes usually run on battery power, they must be energy-conscious, as the functions offered by a node depend on its available energy level and capability. Another interesting characteristic of wireless ad-hoc networks is that a group of nodes can dynamically collaborate among themselves to carry out specific tasks. In other words, tasks are distributed over and carried out by groups of collaborating nodes.

There are many application areas for wireless ad hoc networks. The rapid establishment of military communications during the deployment of forces in unknown and hostile terrain requires the use of wireless ad hoc networks. Rescue missions that require communication in areas without adequate wireless coverage, rapid establishment of communication infrastructure during law enforcement operations, and setting up communication in exhibitions or conferences are some other highly suitable applications for wireless ad hoc networks. Also, for communication during national crisis, where the existing communication infrastructure is non-operational due to a natural disaster or a global war, wireless ad hoc networks can be very useful.

An excellent source of up-to-date publications on this topic can be found at the National Institute of Standards and Technology (NIST) web site [24].

7.2.2 Wireless Sensor Networks

A wireless sensor network consists of large numbers of sensors, which are tiny, low-cost, low-power radio devices dedicated to performing certain functions such as collecting various environmental data and sending them to infrastructure processing nodes. Therefore, wireless sensor networks "embed numerous distributed devices to monitor and interact with physical world: work-spaces, hospitals, homes, vehicles, and the environment" [9]. In other words, wireless sensor networks connect the physical world to computer networks by deploying hundreds or thousands of sensor nodes. In essence, a wireless sensor network is an intelligent information infrastructure that can detect, store, process, and integrate situational and environmental information (Figure 7-1).

A tiny node of a wireless sensor network contains a number of tiny sensors, a tiny memory, and a tiny central processor that processes information gathered from the sensors. This is made possible by the advances in integrated circuit technology, manufacturing of more powerful yet inexpensive sensors, radios, and processors, and mass production of sophisticated systems. Varying from a few millimeters to 1 or 2 meters, sensor nodes combine Micro Electro Mechanical Systems (MEMS) technology, new sensor materials, low-power signal processing, computation, and low-cost wireless networking in a compact system [6], [7]. Figure 7-2 shows the major components of a typical sensor node. Note that sensor nodes may vary considerably not only in size, but also in capabilities such as processing power, energy level, radio bandwidth, and mobility.

There are many areas in which wireless sensor networks may be used. Applications for wireless sensor networks can be in both civilian and military fields [8]. Civilian applications include environment and habitat monitoring, healthcare, home automation, and intelligent transport systems. Among many application areas, these networks will perform significant functions during natural disasters where pre-

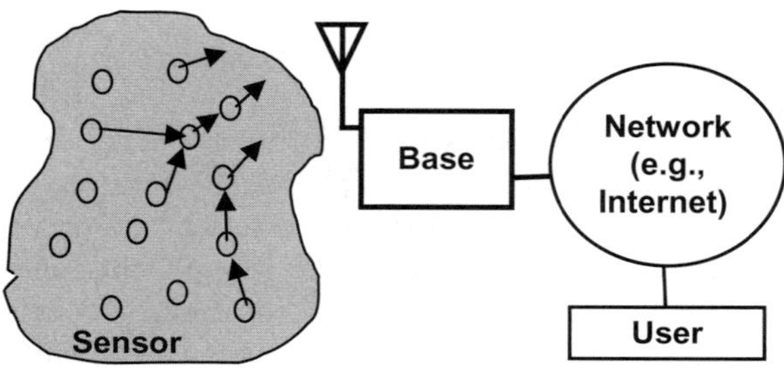

Figure 7-1. A wireless sensor network

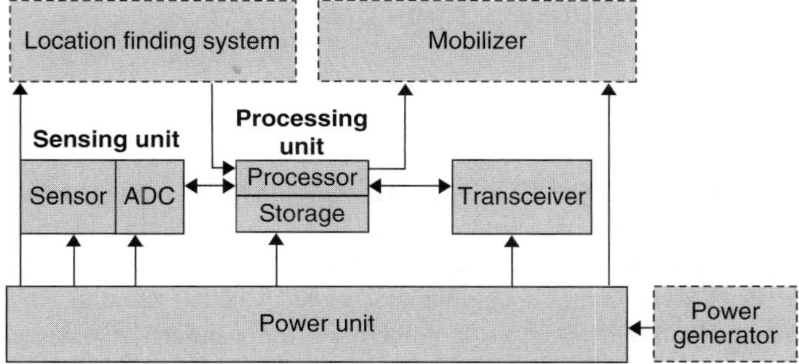

Figure 7-2. Major components of a sensor node [10]

existing infrastructure may be destroyed. Another significant area in which wireless sensor networks are expected to be useful is in so-called smart spaces and smart office applications, which may have significant impact on the enterprise networking.

7.2.3 Wireless Ad Hoc Networks *vs.* Sensor Networks

Wireless ad hoc networks and wireless sensor networks share many characteristics and therefore many papers in the literature use them interchangeably. However, there are some subtle differences, most notably in the number of nodes, which may be in thousands, and in their deployment density [6]. It can be argued that although most wireless sensor networks are ad hoc networks, one can put together a wireless sensor network based on *a priori* infrastructures. By the same token, one can have an ad hoc network that does non-sensory functions. The differences are discussed in the following categories [6]:

- *Quantity:* The number of nodes in a wireless sensor network can be several orders of magnitude higher than the nodes in a wireless ad hoc network. Also, nodes in a wireless sensor network may be deployed more densely than those of a wireless ad hoc network.
- *Topology:* Due to severe limitations of the resources in sensor nodes and worse environmental conditions, the topology of a wireless sensor network may change much more frequently than that of a wireless ad-hoc network. Also, sensor nodes may not have global identification (ID) because of the large amount of overhead and large number of sensors.
- *Communications:* Sensor nodes in a wireless sensor network typically use the broadcast communication paradigm whereas most wireless ad hoc networks are based on point-to-point communications.
- *Nodes:* Sensor nodes, especially those in a wireless sensor network, are much more prone to failures. Additionally, sensor nodes in a wireless sensor network are much more limited in power, computational capacities, and memory. They also tend to be deployed in harsh environments that greatly restrict communications and increase the risk of malfunction, lost, and security.

7.2.4 Network Management Aspects and Framework

Traditionally, aspects of network management are discussed within an outline defined by the Telecommunications Management Network (TMN) framework [15]. This includes functional, logical, information, and physical architectures complete with layers and management areas. The TMN layers include business management, service management, network management, element management, and network element components. Also, the management areas include configuration management, fault management, performance management, security management, and accounting management. The TMN framework, principles, and standards are based on the overall goal of efficient management of telecommunication and traditional data networks and services. Regardless of wire-line or wireless, networks and services are managed based on the principles that reliability and availability are maximized and delays are minimized.

As mentioned earlier, wireless ad hoc networks and wireless sensor networks present drastically different challenges compared to the traditional networks and services. The limited capabilities of the nodes and the sensors in a wireless sensor network have already been discussed. It was also emphasized that the management approaches must take these limitations into the account. The hardware limitations like the processor and memory with varying capabilities may play a significant role in determining the element level management capabilities. The power limitations due, most likely, to the battery with a rather limited power and lifetime may become important in many management aspects, including fault management and performance management. The transmission media-related factors are also important since transmission over the air in unpredictable environments, with possibly intentional or unintentional interference, impacts the link quality as well as security. Finally, since the nodes are assumed to be discardable, the cost of a node must be significantly low.

Specific characteristics of the wireless ad hoc networks and wireless sensor networks discussed in the previous sections translate into a number of significant factors that must be considered in devising new approaches to the management of these networks. For example, the topology could be an important factor in managing these networks. Nodes may be deployed by dropping them from an airplane. These nodes are expected to dynamically self-configure themselves, depending on the environment, to form the topology of the network. Here, scalability is important since the number of nodes in a network can change dramatically because many of the nodes can be lost, can malfunction, or can simply be out of power. Another important factor has to do with the insecure, perhaps hostile, environment in which sensor nodes may operate. Since the potential for eavesdropping and node tampering is rather high, the communications for data transfer as well as transfer of management information must be as secure as possible. Also, as node failures are now an expected outcome, management must include fault-tolerant aspects as well.

Furthermore, because of their specific characteristics such as dynamicity, heterogeneity of devices, bandwidth, and energy constraints, a wireless ad hoc network management architecture presents different requirements from those for fixed networks. The self-organized aspect of ad hoc networks leads to management architectures that are capable of adapting towards node heterogeneity and mobility; the unstable nature of most ad hoc networks requires management architectures capable of providing robustness to the network and services [6]; Also, the constraints concerning bandwidth and energy consumption impose management architectures that provide optimizations in various levels of the operation of the network. For example, routing is an area where management plays an important role in providing strategies that save communications costs. Even the traffic generated by the management operations is a matter of concern. Many of these concerns lead us to the following basic principle: The main goal of the management of wireless ad hoc and sensor networks should be to maximize the efficiency in utilization of the network resources and to maintain the services that these networks offer.

Many features built into the nodes, including hardware, software, and protocols, also support various network management functions. For example, at the network layer, an efficient routing protocol with alternate routing capabilities may provide fault management and performance management functions.

Unfortunately, the research in this area has been slow, therefore, there are only few publications: *Anmp: Ad-hoc Network Management Protocol* [4], *Guerilla, an Adaptive Management Architecture for Ad-hoc Networks* [5], *Management of Mobile Ad-hoc Networks: Information model and probe-based architecture* [23], and *MANNA: Management Architecture for wireless Sensor Networks* [10]. Although the first two publications are designed with ad hoc networks in mind, they can be used as references for wireless sensor network information as well [25].

This chapter discusses some of the significant and directly related solutions published in managing wireless ad hoc networks and wireless sensor networks. In doing so, the chapter tries to follow the TMN framework to a certain extent by first examining the approaches related to network management architectures.

7.3 FUNCTIONAL AND PHYSICAL ARCHITECTURES

The *functional architecture* describes the distribution of functionality within the management architecture. It identifies function blocks (e.g., Operations Systems Function (OSF) block) and reference points between functional blocks (e.g., q-reference point). This is an abstract model developed by standards organizations [12] [13] [14]. There have been no functional architecture-related discussions in the literature for wireless ad hoc networks and wireless sensor networks. This is perhaps due to the fact that the function blocks defined in the TMN context are generic enough and cover wireless sensor networks and wireless ad hoc networks.

The *physical architecture* describes physical implementation of function blocks and reference points identified in the functional architecture. The physical architecture includes physical entities, interfaces, and protocols. The entities may contain one or more function blocks. For example, an OS (a physical entity) includes the OSF block and Q physical interface includes q reference point. This is basically an exercise of mapping functional blocks onto physical entities; therefore, there are many possible scenarios. Since there are no functional architecture-related studies for wireless ad-hoc networks and wireless sensor networks, the lack of work related to the physical architecture is understandable. However, the issue of agents, managers, and their appropriate locations has been addressed in several studies [4] [5]. These will be discussed under the topic of manager-agent communication models below.

7.4 LOGICAL ARCHITECTURES

To address the complexity involved in managing networks, a *logical architecture* decomposes "management" horizontally into layers and vertically into functional areas as mentioned earlier, briefly. Business management, service management, network management, element management, and element layers are the components of the layered architecture; and configuration management, fault management, performance management, security management, and accounting management are the components of the functional areas. (Note that the functional areas in the logical architecture are different from the function blocks covered in the functional architecture.) Figure 7-3 shows management layers and functional areas in a two-dimensional matrix.

In general, configuration management is responsible for designing, installing, testing, initializing, and operating network and service resources. In the context of wireless ad hoc networks and wireless sensor networks, configuration management includes procedures, protocols, and activities related to the planning, deployment, topology discovery, and provisioning of the resources. Both wireless ad hoc networks and wireless sensor networks are considered as self-organizing networks where no prior infrastructure plan is required. However, whereas wireless

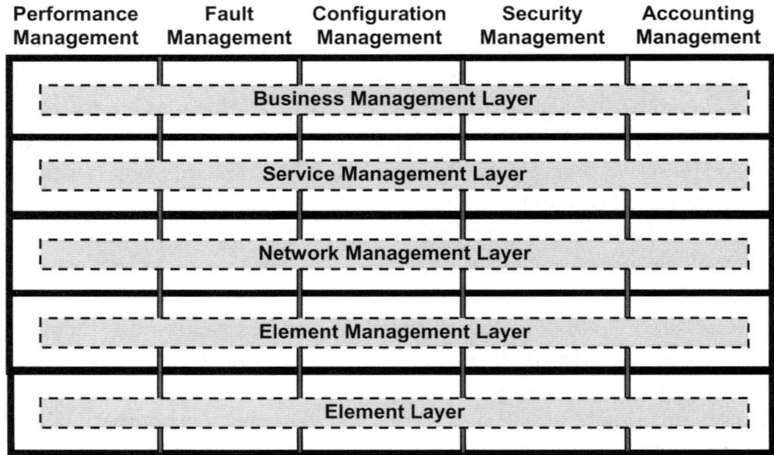

Figure 7-3. Management layers and functional areas

ad hoc networks require no specific plan for the deployment of the nodes, the deployment of the nodes in a wireless sensor networks may require careful planning, by paying closer attention to the node density (i.e., number of nodes per unit area or unit volume) and deployment methods (e.g., thrown in via an airplane or rocket, placed manually). Both wireless ad hoc networks and wireless sensor networks are self-organizing and self-optimizing and the topology of the network is expected to be highly dynamic. Therefore, the topology, the current state of the network connectivity/reachability, and the state of the nodes need to be monitored continuously. An up-to-date network topology could be used for a number of management functions such as redeployment of nodes, setting operating parameters (e.g., routing tables, timeout values, position estimation), network maintenance, and predict future states [11]. There are a number of published works on the topic of topology discovery. Typically, these algorithms rely on nodes' neighborhood information, which can be obtained rather easily due to the broadcast nature of communications. See [11], [18], [19], and [26] for a detailed discussion of this topic.

Fault management deals with network anomalies and failures in real time. Fault management in wireless ad hoc networks and especially in wireless sensor networks must include drastically different functions and features from the national networks. Due to the rather limited resources, especially in energy, harsh/hostile environments, and poor/interfered communications, faults are assumed to happen and the disappearances of the nodes are expected. Therefore it is important that fault management approaches employ self-diagnostics and self-repair functionalities as well as self-organizing in the case of disappearing nodes [22].

Performance management is responsible for providing reliable and high-quality network performance and maintaining the quality of service for users. Performance management in wireless ad hoc and sensor networks, in addition to traditional performance parameters such as throughput and delay, must also be cognizant of the energy consumption. Furthermore, depending on the application, coverage area, locality of nodes, quality and timing of data produced, network settle time, network join time, network depart time, and network recovery time need to be taken into account.

Security management, as the name implies, is responsible for providing protection against threats that may compromise network resources, services, and data. Security management is another unique area in managing wireless ad hoc and sensor networks due to the vulnerabilities of the nodes in the face of a possible harsh and hostile environment [21]. Self-protection and self-recovery from various threats and compromises should be essential elements of any approach here. The security-related procedures need to recognize the facts about limited resources, energy, and bandwidth.

Finally, accounting management is responsible for managing and administering user-related matters, especially billing. Accounting management wireless ad hoc and sensor networks should include functions related to the use of resources and corresponding reports.

As indicated earlier, energy is a critically important factor in the lifetime of the nodes in wireless ad hoc and sensor networks. Therefore, an energy map

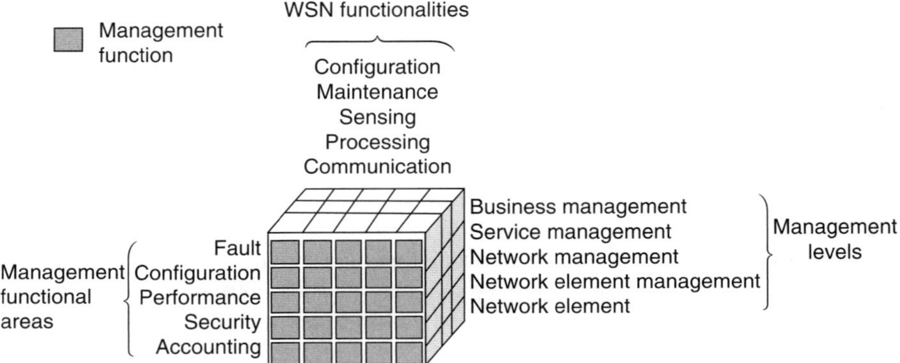

Figure 7-4. MANA: Management functionality abstractions [10]

constructed by using a model of the dissipated energy in each node would be another useful tool for managing these types of networks. See [27], [28], and [29] for a detailed discussion on the algorithms used in constructing energy maps. Provisioning resources, a primary function of the configuration management area, is used in redeploying additional nodes either to replace malfunctioning ones or to simple add more nodes in a specific area. This operation will also require configuring, automatically or manually, a number of node and related parameters.

Briefly, the business management layer includes functions related to business aspects such as profitability, the service management layer includes functions (such as billing) related to the services that the network offers, the network management layer includes functions related to the resources that make up the network, and the element management layer includes functions related to a group of, perhaps, similar network elements. Finally, the element layer includes those management functions (within the element) related to a specific network element.

Ruiz et al., in [10], present a management architecture, called "MANNA," for wireless sensor networks. In addition to the physical and information architectures, MANNA also proposes an extended three-dimensional logical architecture. It considers three management dimensions: functional areas, management levels, and wireless sensor network functionalities, as shown in Figure 7-4.

Wireless sensor network functionalities are grouped by configuration, maintenance, sensing, processing, and communication areas where appropriate; network, not management-related functions, reside. For example, the actual coverage area of the sensors and their corresponding map is part of the "sensing" network functionality.

7.5 INFORMATION ARCHITECTURES

The TMN *information architecture* is based on the "manager-agent" model [16] as depicted in Figure 7-5. An "agent" system in a managed entity communicates with

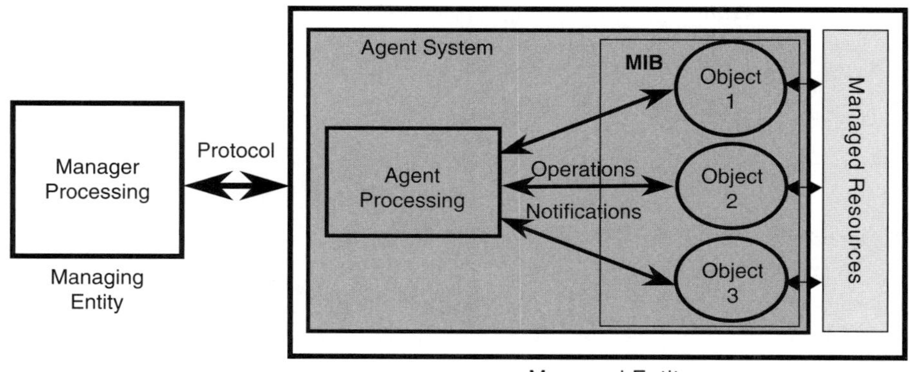

Figure 7-5. Manager-agent based generic communication model

a typically external "management entity" on behalf of the resources of a network element by utilizing "objects" organized into a structure referred to as Management Information Base (MIB). Information architecture, therefore, includes an object-oriented methodology for defining the management behavior of managed devices and representing management information, a protocol for exchanging information, and a scheme for naming and addressing.

There are a number of recent works related to the information architectures for both wireless ad hoc networks and wireless sensor networks. These studies are reviewed under manager-agent models, in the following section.

7.5.1 Manager-Agent Communication Models

The following is a more general discussion on various architectures from the previous section and the more detailed aspects are presented here. First, various proposals in literature related to the manager-agent communications models are discussed.

Chen et al., in [4], propose a three-level hierarchical architecture where the manager, the highest level, communicates with agents, lowest levels, through "cluster heads," or intermediaries, as shown in Figure 7-6. These cluster heads have the capability to collect, aggregate, and filter data before forwarding it to the manager.

Cluster heads are formed for management purposes only and are different from those formed for routing purposes [20]. Clusters are used to simplify network management and provide more efficient and fault-tolerant message transmission. The structure is dynamic in the sense that the number and composition of nodes may change over time. This means that nodes acting as cluster heads may also change.

Shen et al., in [5], propose a similar hierarchical architecture, called Guerilla, based on the "Supervisor/agency" model. The Supervisor includes the manager functions and the "agency" functions are carried out by clusters. Here, the cluster heads are called "nomadic managers" and are dynamically selected. Nomadic managers utilize mobile code techniques and "collaborate" with each other and a supervisor as shown in Figure 7-7.

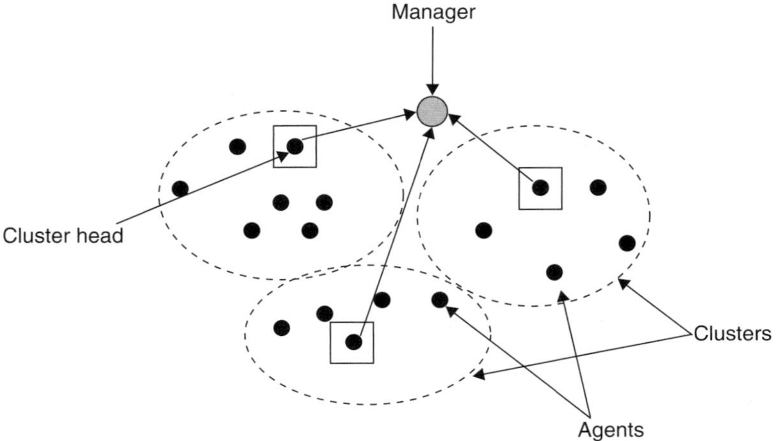

Figure 7-6. Anmp manager-agent communications architecture [4]

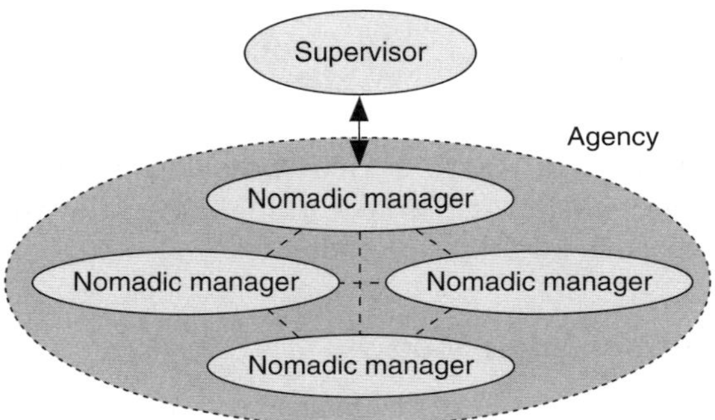

Figure 7-7. Supervisor/agency model [5]

Nomadic managers collaborate autonomously to manage the nodes with the help of "agent" and "probe" capabilities residing in them. Guerilla architecture therefore identifies three types of nodes, as illustrated in Figure 7-8. Guerilla architecture relies on so-called probes to carry out some network management tasks to minimize management traffic. A probe is a "lightweight management script," performing either a monitoring task or some specific management tasks through interactions with the local SNMP agent. These probes can process the raw data locally, then filter and aggregate it before sending it to the manager.

A nomadic manager is a software program that can be hosted by a node (i.e., network element) with appropriate resources. It should be noted that nodes in a wireless ad hoc network or a wireless sensor network may have different capabilities and resources and, furthermore, only those nodes with greater capabilities will be allowed to host the nomadic managers. Nodes are clustered into groups with at least one nomadic manager in each group. As in [4], clusters are dynamically formed; this means that:

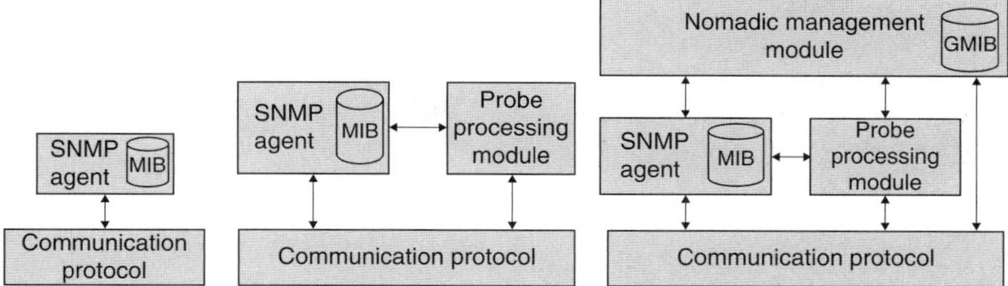

Figure 7-8. Node types in Guerilla architecture: SNMP-capable node, Probe-capable node, and, Nomadic Manager-capable node [5]

- Nodes may leave a group and join another.
- Nomadic managers may move from one node to another depending on the changes to the attributes such as topology, energy level, etc.
- A nomadic manager may clone itself to another full node to share the management load (within the same cluster).
- A nomadic manager may spawn itself into a newly created partition.
- Two nomadic managers may merge when the corresponding clusters merge.

The two schemes defined above seem to be more suitable for wireless ad hoc networks, although they may be applicable to wireless sensor networks with different and more capable nodes. MANNA architecture defined by Ruiz et al. in [10] is specifically designed for wireless sensor networks. However, MANNA does not provide a specific manager-agent communications model, rather, it allows a number of different models depending on the type and architecture of the wireless sensor networks. If a wireless sensor network is hierarchical where nodes are grouped and organized into certain relationships, MANNA suggests that the manager function can be placed either externally or at the cluster heads in a distributed fashion and the agents can be placed either at the cluster heads or in the nodes capable of handling this function. Figure 7-9 shows an alternative architecture where the agents and distributed managers are placed in the network.

If the wireless sensor network has a flat architecture where no hierarchical structure exists among the sensor nodes having the same type of capabilities and a sink node provides access to the network, the manager can be located externally, at the sink node, or internally and the agent can be located at the sink node or internally in the network. Figure 7-10 shows an example where both managers and nodes are located in the network in a distributed fashion.

Although it allows greater flexibility as far as the placement of the managers and agents, Reference [10] does not provide any suggestions, for example, as to which node can assume the responsibility of the manager or agent, and also what procedures and algorithms are used to make these assignments. Also, it should be noted that MANNA assumes that some nodes in a wireless sensor network may not have the capability to host either an agent or a manager.

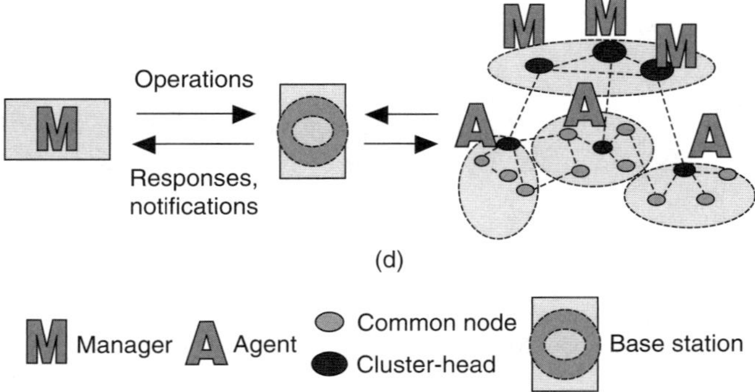

Figure 7-9. Location of manager and agent in a hierarchical wireless sensor network [10]

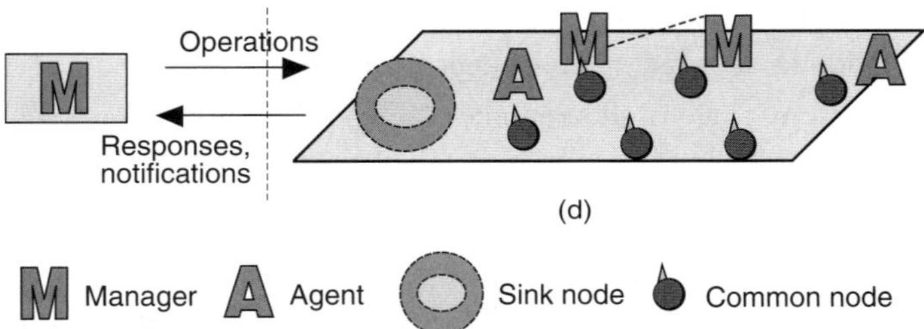

Figure 7-10. Location of manager and agent in a flat wireless sensor network [10]

In the case of hierarchical manager-agent communication architecture, the network is organized into clusters (for network management purposes only) and one or more nodes in the clusters are selected dynamically as the cluster heads to assume intermediate agent (*a.k.a.* nomadic manager) responsibilities. Both [4] and [5] propose solutions to form more efficient clusters and assignment of cluster heads. Before discussing these solutions, the factors involving an ideal cluster need to be discussed.

- *Size of the cluster:* To be selected to minimize the network management related traffic among manager/supervisor, nomadic managers/cluster heads, probe-capable nodes, and other nodes.
- *Cluster maintenance:* Minimize the re-computation possibly due to node mobility and at the same time speed up the process to incorporate the newly arrived nodes into clusters so that they can be included in the management process [4].

Anmp proposes two algorithms for clustering [4]:

- *Graph-based clustering:* A graphic view of the net is formed by showing the nodes and undirected links between two nodes if they are within an acceptable transmission range. Anmp uses node ID and placement of the node in selecting

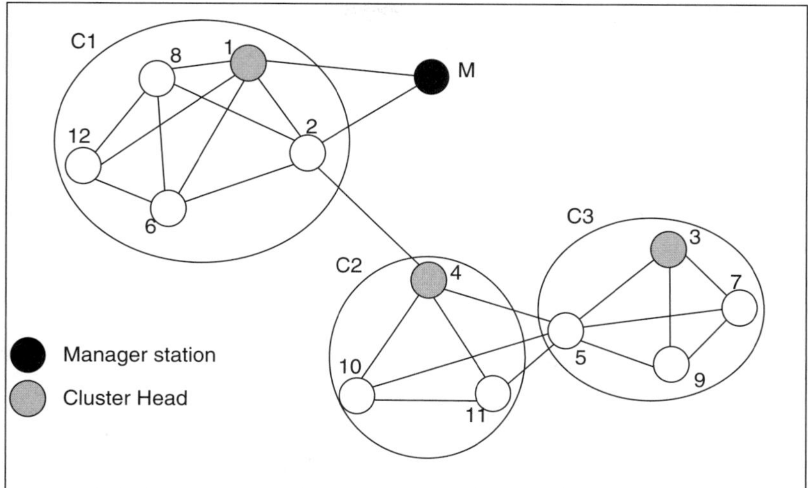

Figure 7-11. An Example of Graph based Clustering Algorithm [4]

cluster heads. A node with a minimum ID forms a cluster and becomes the cluster head (see Figure 7-11). Each node in a cluster is no more than one or two hops away from the cluster head. The algorithm includes in detail how the cluster is maintained in the case where nodes move into or out of the clusters with the help of various lists (e.g., neighbors with one hop away) that each node maintains.

- *Geographical clustering based on spatial density of nodes using latitudes and longitude information provided by Global Positioning System (GPS):* In the first stage of the algorithm, nodes are grouped into rectangular shaped clusters based on the spatial density. Around the initial box covering the whole area, nodes are placed and bar graphs are formed to indicate the frequency of nodes, and then the box is simply divided into smaller boxes (clusters) along the lines of "valleys" as in Figure 7-12. A node in the center of the box is selected as the cluster head. This stage is executed periodically since the node movement changes the spatial density. The second stage takes place between periodic executions and handles node movement within each cluster.

Extensive performance analyses comparing these two algorithms were provided in [4]. The authors found that the message cost is little better in Graphical Based Clustering, even though it leaves high number of nodes unmanaged.

Guerilla architecture [5] uses a protocol called Adaptive Dynamic Backbone (ADB) to form clusters with various sizes. This protocol elects the nomadic managers as well. (The architecture also allows the supervisor to designate an initial set of nomadic managers.). As a result, ADB results in many small clusters, if the nodes are quite mobile, to minimize the maintenance since there will be more frequent changes. In a less mobile environment, the cluster sizes will be larger since there will be few changes expected.

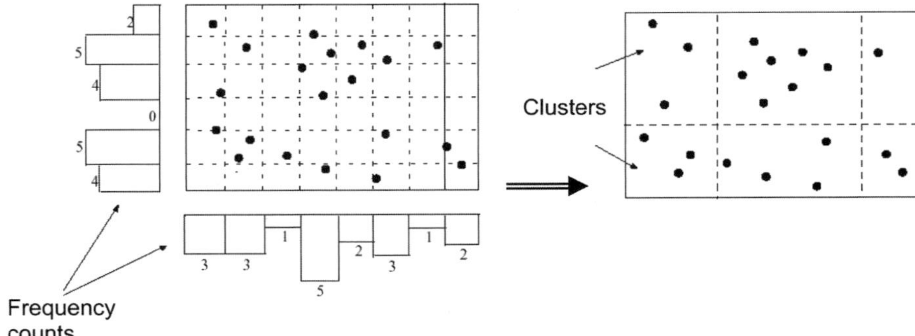

Figure 7-12. An example of periodic clustering algorithm [4]

As discussed previously, nomadic managers are responsible for the maintenance of the clusters. This involves reconfiguration of clusters based on the information obtained from its nodes through active probes. To select an appropriate action a, each nomadic manager evaluates

$$\arg\max_{a \in A} \frac{U(T(s,a)) - U(s)}{C(a)}$$

In this process, s is the perceived view of the cluster, $U(s)$ is the utility function representing the policy assigned by the supervisor, $T(s, a)$ is the transition function associated with a, and $C(a)$ is the cost function if the action a is taken by the manager. The action set provided may include actions such as spawning a new manager at another node, or adding links between two nodes. Guerilla architecture assumes that the utility takes into account only the metrics for the following:

- *Resource consumption*—includes po wer usage for packet transmission and management processing load (CPU Load).
- *Quality of management*—includes observability (average amount of time taken for a particular event generated at a node to be observed by the nearest manager) and isolation probability (probability of a node being isolated from management).

Note that the increase in the first set of metrics will have a negative impact while the second set will have a positive impact on the utility.

This procedure is repeated periodically and action, if necessary, is taken only if the utility gain exceeds a certain threshold value to minimize temporal changes. Figure 7-13 shows a scenario given in [5] to illustrate the action taken by the manager. The action in this case is the linking of nodes B and C, and C and F. MANNA does not provide any procedure to form clusters and assign cluster heads.

Both [4] and [5] are developed for wireless ad hoc networks. Also, both are applicable to wireless sensor networks with hierarchical architectures where the capabilities and resources of each node vary. For a wireless sensor network with a flat architecture, it is understood that the capabilities and resources of the nodes are so limited to allow additional management responsibilities. Therefore, developing

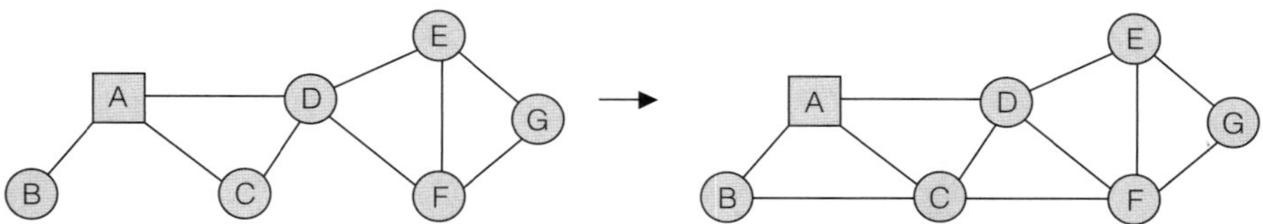

Figure 7-13. A sample scenario showing the result of action taken by manager [5]

appropriate specific management architecture for this type of wireless sensor network is an open research area.

7.5.2 Management Interfaces and Protocols

A manager-agent communications model must also identify how these entities communicate with each other. What are the physical characteristics of the interfaces that will be used and what is the protocol suite running on these interfaces? There is a limited amount of discussion on this area in the literature on the wireless ad hoc and sensor networks. It is generally assumed that Simple Network Management Protocol is a good choice; however, others are also mentioned in some work. There is no mention of the physical characteristics and bandwidth of the interfaces to be used for management purposes. There is also a lack of any discussion of how the instrumentation (agent to node resources) is to be done.

In [4], a protocol for managing mobile wireless ad hoc networks, called Anmp, is defined. Anmp is fully compatible with SNMPv3 and includes enhanced security features that can be fine-tuned. SNMP is also assumed in [5] without any further detail. MANNA physical architecture [10] suggests the use of a lightweight protocol stack, but no details are given. However, MANNA defines a number of new object classes by following an object-oriented methodology as discussed in the next section. Although it is not explicitly mentioned in [10], this suggests that MANNA is favoring Common Management Information Protocol (CMIP) for manager-agent communication.

7.5.3 Structure of Management Information and Models

As discussed above, the information architecture includes an object-oriented methodology for defining the management behavior of managed devices and representing management information. An unambiguous structure of this information and how it is collected is also part of the information architecture. References [4], [5], and [10] provide relevant models based on the SNMP MIB.

In Anmp, data collection is done via SNMP MIBs [4]. Every node is responsible for storing its information in the MIB. To represent those features specific to the wireless ad hoc networks, Anmp adds a new MIB, called Anmp MIB, to the MIB-II defined in SNMP standards. Figure 7-14 shows Anmp MIB and its four groups.

The *powerUsage* group is used to record the node's power-related information such as the battery power and power consumption, as shown in Figure 7-15.

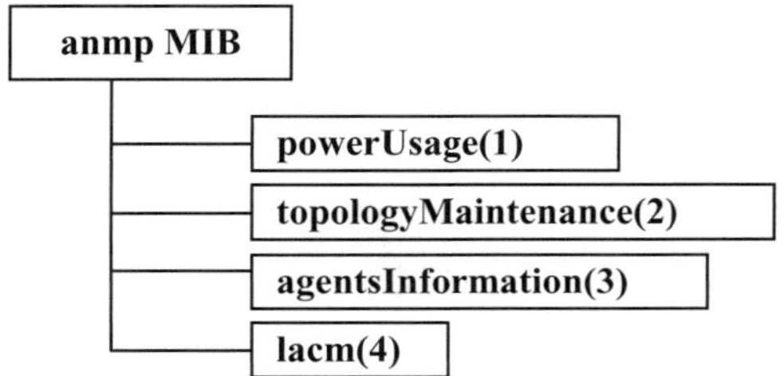

Figure 7-14. Anmp MIB groups

```
powerUsage(1)
    powerExternalPowerStatus(1)
    powerRemainingPower(2)
    powerRadioInterfacePowerTable(3)
        powerInterfacePowerEntry(1)
            powerInterfaceIndex(1)
            powerTransmitPower(2)
            powerReceivePower(3)
            powerIdlePower(4)
            powerSleepPower(5)
    powerSystemPowerUsageTable(4)
        powerSystemPowerEntry(1)
            powerSystemPowerIndex(1)
            powerSystemPowerMode(2)
            powerPowerConsumed(3)
    powerCurrentSystemPowerIndex(5)
    powerBatteryDrainTable(6)
        powerBatteryDrainEntry(1)
            powerBatteryDrainEntryIndex(1)
            powerBatteryDrainModelText(2)
            powerBatteryDrainModelFunction(3)
    powerBatteryDrainIndex(7)
    powerBatteryType(8)
    powerBatteryAge(9)
    powerBatterySerialNumber(10)
    powerLastRecharged(11)
```

Figure 7-15. The objects defined under the *powerUsage* group [4]

The *topologyMaintenance* group is used for the topology maintenance-specific information such as neighbors, clusters, and protocols as shown in Figure 7-16. This group also houses two subgroups (*graphicalClustering* and *geographicalClustering*) containing information related to the corresponding algorithm used to form clusters.

The *agentsInformation* group is used mainly by the cluster heads and managers to keep the information collected from the agents. As shown in Figure 7-17, the *agentsInformation* group contains several subgroups such as alarms and events. Most of these subgroups are used simply as logs to store the information collected from the agents. RMON-MIB is used as a model to construct these subgroups [4].

Finally, the *lacm* group is used to store the security specific information regarding the security classification and security clearance of a node as shown in

```
topologyMaintenance(2)
    neighborTable(1)
        neighborTable rEntry(1)
            neighborIndex(1)
            neighborIpAddress(2)
    protocolTable(2)
        protocolTableEntry(1)
            protocolTableIndex(1)
            protocolID(2)
            protocolIndex(3)
            protocolDesc(4)
    geographicalClustering(3)
    graphicalClustering(4)
```

Figure 7-16. The objects under the *topologyMaintenance* group [4]

```
agentsInformation(3)
    informationUpdateInterval(1)
    percentageAgentManaged(2)
    percentageAgent UpdateInfo(3)
    mobileGatewayFlag(4)
    mobileGatewayStatus(5)
    statisticsInfo(6)
    hardwareInfo(7)
    systemInfo(8)
    powerInfo(9)
```

Figure 7-17. The objects defined under the *agentInformation* group [4]

Figure 7-18. Anmp provides extensive security specific features, which are discussed under the security management category.

A shown in Figure 7-8, Guerrilla architecture [5] assumes the SNMP information structure and that all nodes keep an SNMP MIB (SMIB). Furthermore, a node having the Nomadic Manager Module (NMM) includes a MIB called Guerilla MIB (GMIB). This is a data structure equivalent to an SNMP MIB and contains all management-related data such as an aggregation of management information collected from neighbor nodes via probes. Another part of GMIB is that the aggregated information is stored in nodes capable of handling probes. No further details are provided in [5] as to the specific objects in GMIB.

The MANNA's information architecture [10] is based on Open Systems Interconnection (OSI). OSI's object-oriented information modeling and defines a number of managed object classes mostly derived from the standard OSI objects. MANNA defines two object class types:

- *Managed object classes*—These object classes are used to represent network-specific resources. Table 7-1 shows a list of these classes and some examples of new attributes added by MANNA.
- *Support object classes*—These object classes are used to provide information to management functions. Some example object classes that are derived mostly from the standard classes are: *log, stateChangeRecord, ttibuteChangeValueRecord, alarmRecord, eventForwardingDiscriminator, managementOperationSchedule.*

```
lacm(4)
    objectSecurityLevelTable(1)
        objectSecurityLevelEntry(1)
            objected(1)
            objectSecurityLevelIndex(2)
            objectSecurityLevel(3)
            objectSecurityProject1(4)
            objectSecurityProject2(5)
            objectSecurityProject3(6)
            objectSecurityProject4(7)
            objectSecurityProject5(8)
            objectSecurityProject6(9)
    agentSecurityRankTable(2)
        agentSecurityRankEntry(1)
            agentSecurityName(1)
            agentSecurityRankIndex(2)
            agentSecurityRank(3)
            agentSecurityProject1(4)
            agentSecurityProject2(5)
            agentSecurityProject3(6)
            agentSecurityProject4(7)
            agentSecurityProject5(8)
            agentSecurityProject6(9)
```

Figure 7-18. The objects defined under the *lacm* group [4]

TABLE 7-1. List of MANNA's object classes and new attributes [10]

Object Class	Represents	Examples of New Attributes
Network	Group of managed objected connected in some fashion to communicated among themselves and outside entities	• network identifier • composition type (homogeneous, heterogeneous) • organization type (flat, hierarchical) • organization period • mobility (stationary, stationary nodes and mobile phenomenon, mobile node, mobile phenomenon) • data delivery (continuous, event driven, on demand, programmed) • type of access point (sink node, base station) • localization type (relative, absolute)
Managed Element	Sensor nodes, acting nodes or other wireless sensor network entities providing sensing, processing, and communication services	• localization (relative, absolute) • element type (common node, sink node, gateway, cluster head) • minimum energy limit • mobility (direction, orientation, acceleration)
Equipment	Physical aspects of the sensor node	*(specialized in the following object classes: battery, processor, etc.)*
Battery		• battery type • capacity • remaining energy level • energy density • max current

TABLE 7-1. Continued

Object Class	Represents	Examples of New Attributes
Processor		• clock • state of use • available memory • endurance • AD channel • operating voltage • IO pins
Sensor		• sensor type • current consumption • voltage range • minmax range • accuracy • temperature dependence • version • state current
Transceiver		• type • modulation type • carrier frequency • operating voltage • current consumption • throughput • receiver sensitivity • transmitter power
System	A set of hardware and software that is capable of processing and transferring information	• operational system type • version • code length • complexity • synchronization type (mutual exclusion, synchronization of processes)
Environment	Environment where the wireless sensor network is operating	• environmental type (internal, external, and unknown) • noise ratio • atmospheric pressure • temperature • radiation • electromagnetic field • humidity • luminosity
Phenomenon	Phenomenon behavior in the environment where the wireless sensor network is operating	• phenomenon type • occurrence frequency • media type
Connection	Actual connections in the wireless sensor network	• communication type (simplex, half duplex, full duplex)

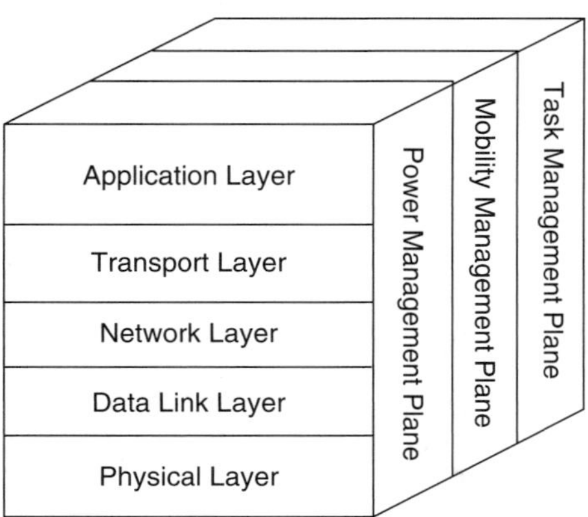

Figure 7-19. Protocol architecture for wireless sensor network [6]

7.5.4 Others

Although it is not specifically for manager-agent communication architecture, the protocol stack described in [6] shows three management planes: power, mobility, and task management, as depicted in Figure 7-19. Although these management planes play critical roles in providing efficient utilization of resources and therefore efficient operation of the nodes and network as a whole, they are not directly involved in the manager-agent communications and, therefore, they are not included in this chapter. Furthermore, there are numerous other proposals in the literature regarding power and energy savvy protocols operating at various layers.

Akyildiz et al., in [6], discuss the need for three different types of application layer protocols: Sensor Management Protocol (SMP), Task Assignment and Data Advertisement Protocol (TADAP), and Sensor Query and Data Dissemination Protocol (SQDDP). These protocols are discussed briefly and identified as open research areas in [6].

Shen et al., in [17], proposes a protocol/language called Sensor Query and Tasking Language (SQTL) for the application programs to interact with the resources in a node. This is part of their middleware architecture called Sensor Information Networking Architecture (SINA), which is designed to facilitate the information gathering and communication with the nodes.

7.6 SUMMARY AND CONCLUSIONS

Wireless ad hoc networks and wireless sensor networks are relatively new network architectures that are drastically different from traditional wireless and wired networks. Design, deployment, and operations of these new networks require new management technologies and approaches. Unfortunately, there has been relatively little research and publications in the area of managing the wireless ad hoc and wireless sensor networks. A few significant published works are discussed in this chapter.

There are a large number of significant potential application areas (humanitarian as well as military) for these types of networks. As more and more of these applications begin using wireless ad hoc and sensor networks, the need for managing these networks and services they offer will become mission critical. Therefore, there will be a strong demand for network management solutions that are optimized for wireless ad hoc and sensor networks. New network management solutions will be enabled by the advances in policy-based management, autonomic management (self-organizing, self-optimizing, self-protecting, and self-healing with self-diagnostic capabilities), and intelligent mobile agents with minimum human interactions.

7.7 REFERENCES

1. Internet Engineering Task Force (IETF). "Mobile Ad-hoc Networks (manet)." http://www.ietf.org/dyn/wg/charter/manet-charter.html (accessed October 1, 2009).
2. Corson S, Macker J. 1999. Mobile Ad hoc Networking (MANET): Routing Protocol Performance Issues and Evaluation Considerations. RFC 2501. Fremont, CA: Internet Engineering Task Force (IETF).
3. Haas ZJ, et al. 2002. Wireless ad hoc networks. In *Encyclopedia of Telecommunications*, J Proakis, ed. Hoboken, NJ: John Wiley & Sons.
4. Chen W, Jain N, Singh S. 1999. Anmp: Ad Hoc Network Management Protocol. *IEEE Journal on Selected Areas in Communications* 17(8):1506–1531.
5. Shen C-C, Srisathapornphat C, Jaikaeo C. 2003. An adaptive management architecture for ad hoc networks. *IEEE Communications Magazine* 41(2):108–115.
6. Akyildiz IF, Su W, Sankarasubramaniam Y, et al. 2002. A Survey on Sensor Networks, *IEEE Communications Magazine* 41(8):102–114.
7. Akyildiz IF, Su W, Sankarasubramaniam Y, et al. 2002. Wireless sensor networks: a survey. *Computer Networks* 38(4):393–422.
8. Eicke J. 2002. *Networked Sensors for the Objective Force*. Adelphi, MD: U.S. Army Research Laboratory.
9. Estrin D. 2000. Some Distributed Coordination Schemes for Wireless Sensor Networks. Stanford Network Seminar Series, November 16, 2000.
10. Ruiz LB, Nogueira JM, Loureiro AAF. 2003. MANNA: a management architecture for wireless sensor networks. *Communications Magazine, IEEE* 41(2):116–125.
11. Deb B, Bhatnagar S, Nath B. 2001. A Topology Discovery Algorithm for Sensor Networks with Applications to Network Management. Technical Report DCS-TR-441, Department of Computer Science, Rutgers University.
12. ITU-T Recommendation M.3010: Principles for a Telecommunications management network, 2000. Geneva, Switzerland: International Telecommunication Union Telecommunication Standardization Bureau.
13. ITU-T Recommendation M.3100: Generic Network Information Model, 2005. Geneva, Switzerland: International Telecommunication Union Telecommunication Standardization Bureau.
14. ITU-T Recommendation M.3400: TMN Management Functions, 2000. Geneva, Switzerland: International Telecommunication Union Telecommunication Standardization Bureau.
15. Wang H. 1999. *Telecommunications Network Management*. Hightstown, NJ: McGraw-Hill.
16. Raman L. 1999. *Fundamentals of Telecommunications Network Management*. Piscataway, NJ: IEEE Press.
17. Shen CC, Srisathapornphat C, Jaikaeo C. 2001. Sensor information network architecture and applications. *IEEE Personal Communications* 8(4):52–59.
18. Bao L, GarciaLunaAceves JJ. 2003. Topology Management in Ad Hoc Networks. MobiHoc '03, June 1–3, 2003, Annapolis, MD.

19. Li L, Bahl V, Wang YM, et al. 2001. Distributed Topology Control for Power Efficient Operation in Multihop Wireless Ad Hoc Networks. Proceedings of the IEEE Conference on Computer Communications (INFOCOM 2001).
20. Lin CR, Gerla M. 1997. Adaptive clustering for mobile wireless networks. *IEEE Journal on Selected Areas in Communication* 15(7):1265–1275.
21. Chiou GH, Chen WT. 1989. Secure broadcasting using the secure lock. *IEEE Transactions on Software Engineering* 15(8):929–934.
22. Staddon J, Balfanz D, Durfee G. 2002. Efficient Tracing of Failed Nodes in Sensor Networks. Mobicon '02 Wireless Sensor Network, September 2002, Atlanta, GA.
23. Badonnel R, State R, Festor O. 2005. Management of mobile ad hoc networks: information model and probe-based architecture. *International Journal of Network Management* 15(5):335–347.
24. Wireless Communication Technologies Group. "Wireless Ad Hoc Networks Bibliography." http://w3.antd.nist.gov/wctg/manet/manet_bibliog.html (accessed October 1, 2009).
25. Ulema M, Nogueira JM, Kozbe B. 2006. Management of Wireless Ad Hoc Networks and Wireless Sensor Networks. *Journal of Network and Systems Management* 14(3):327–336.
26. Nezhad AA, Miri A, Makrakis D. 2008. Efficient topology discovery for multihop wireless sensor networks. *Proceedings of the Communication Networks and Services Research Conference*. Washington, DC: IEEE Computer Society, pp 358–365.
27. Mini RAF, Val Machado M, Loureiro AAF, et al. 2005. Prediction-based energy map for wireless sensor networks. *Ad Hoc Networks Journal* 3(2):235–253.
28. Ci S, Guizani M. 2006. Energy map: mining wireless sensor network data. *Proceedings of the IEEE International Conference on Communications*, June 2006, pp 3525–3529.
29. Gungor VC. 2008. Efficient available energy monitoring in wireless sensor networks. *International Journal of Sensor Networks* 3(1):25–32.

CHAPTER 8

STRATEGIC STANDARDS DEVELOPMENT AND NEXT GENERATION MANAGEMENT STANDARDS

Michael Fargano

8.1 INTRODUCTION

Standards provide for a competitive economic environment where R&D can flourish, best-in-class technologies can thrive in the marketplace, and innovative solutions are enabled. However, at its core, standards are about legitimacy via the consensus process.

The most successful standards are often taken for granted because wide deployment and usage results in true ubiquity. We can recognize some familiar standards in our everyday life, e.g., the RJ-45 jack for wireline Ethernet and the RJ-11 jack for wireline phone service, etc. Management standards are rarely visible to the general public, but that is changing with Next Generation Network (NGN), as more and more complexity is managed in areas that are visible to customers, e.g., wireless networks and home networks.

Each standards area has it own challenges and network management has its share. Management standards development exists at the nexus of two of the most dynamic technology areas in modern times, i.e., communications technology and information technology (a.k.a, Information and Communications Technology or ICT). This dynamic existence coupled with service provider business process linkage is the key challenge facing NGN management standards.

This chapter will cover the strategic standards development process and strategic aspects of NGN management standards. A landscape of these topics will be provided rather than tutorials on specific management standards or a laundry list of standards.

Next Generation Telecommunications Networks, Services, and Management, Edited by Thomas Plevyak and Veli Sahin

8.1.1 General Drivers for Standards

There are several general drivers for standards. For a given standards project, some are emphasized more than others. The most compelling driver that could be in play comes from law or regulation. This could take the form of standards that support regulations, e.g., for citizen-to-authority emergency communications (911 in USA), or could take the form of a legal safe-harbor standard, e.g., for Lawful Intercept (LI) or Lawfully Authorized Electronic Surveillance (LAES). In the safe-harbor case, laws and/or regulations allow implementation of the standards to provide protection from penalty or liability. Many times, especially in the USA, the standards development process is used as a means for the industry to get ahead of regulation by developing voluntary standards that either negate the need for regulation, help to guide the regulation development process, or limit the depth of the regulations. The most common driver for standards development in the NGN/ICT environment is for interoperability in support of multi-supplier/manufacturer/vendor environments, multi-service provider/operator environments, and multi-content provider environments.

The general drivers above apply to management standards.

8.1.2 Management Standards History

This section is intended to provide a brief history for the reader to get a sense of how we got to where we are regarding management standards. Management standards development started to take off in the USA after the AT&T divestiture in 1984. The two key players were Bellcore (a consortium owned by the divested Regional Bell Operating Companies [RBOCs]) and Committee T1 (an American National Standards Institute [ANSI]-accredited body within the Alliance for Telecommunications Industry Solutions [ATIS]). Bellcore was not a Standards Development Organization (SDO) or open forum per se. However, requirements and specifications produced by Bellcore supported the same interoperability goals of a SDO/Forum. The work of Bellcore focused on evolving Operations Support System (OSS) functional and interface requirements with greater consideration for the existing RBOC networks and operations environment. Committee T1 (with T1M1, i.e., the Operations Administration Maintenance and Provisioning [OAM&P] Technical Subcommittee) was able to initiate greater evolutionary steps regarding OSS functional and interface requirements. Bellcore and T1M1 interacted and the work of one played into the other. The key divergence occurred when T1M1 moved down the Common Management Information Service Element (CMISE) path for OSS interfaces and Bellcore continued primarily with Transaction Language 1 (TL1) based on Man-Machine Language (MML). T1M1 was also the key driver for Telecommunications Management Network (TMN) architecture. T1M1 was a key contributor to International Telegraph and Telephone Consultative Committee (CCITT) to develop TMN and CMISE. At that time CMISE was seen in the telecom industry as the preferred target Open Systems Interconnection (OSI) application layer service for management information transfer among and between OSSs and Network Elements (NEs). At the same time Internet Engineering Task Force (IETF) saw CMISE as too computer processing intensive and resource intensive for the needs of lightweight enterprise/data applications, thus, IETF moved down the Simple Network Management Protocol (SNMP) path. In the early 1990s the Network Management Forum (NMF) came on the scene to help promote OSS-OSS interoperability with a TMN emphasis. NMF was a global forum that developed interoperability requirements and specifications, not standards per se. As it turned out, CMISE was too costly for the telecom industry and as more efficient object-oriented platforms/protocols were developed for a general Information Technology (IT) environment, such as Common Object Request Broker Architecture (CORBA), work on CMISE started to be phased out (but the management information models were salvaged and reused). CORBA came on the scene in the late 1990s. As we moved in to the early 2000s eXtensible Markup Language (XML) and Web Services became the preference for OSS interfaces.

The morphing of SDO/Forums:

- Committee T1 was retired with its Technical Subcommittees moving directly under ATIS. ATIS became a Standards Development Organization (SDO). This was part of a broad restructuring of ATIS.
- T1M1 became the Telecommunications Management & Operations Committee (TMOC) under ATIS.
- CCITT became ITU-T (Telecommunication Standardization Sector of the International Telecommunication Union [ITU]).
- Bellcore ceased to be a consortium owned by the RBOCs and became Telcordia.
- NMF became the TeleManagement Forum (TMF).

8.2 GENERAL STANDARDS DEVELOPMENT PROCESS

This section covers the following key aspects of the general standards development process:

- Key Attributes of the Standards Development Process
- General SDO/Forum Types and Interactions
- General Standards Development and Coordination Framework

Key Attributes of the Standards Development Process provide principles that apply to all SDOs and some Industry Forums. The General SDO/Forum Types and Interactions provide a high level model to categorize SDO/Forums and show their interactions. The General Standards Development and Coordination Framework provides a standards project execution and management framework that can be utilized for standards projects that are either executed in a single SDO/Forum or projects that require coordination across SDO/Forums.

The following are general terms used in this chapter:

- *Standards Development Organization (SDO):* An organization that develops standards (i.e., requirements, protocols, interface specifications, etc.). The attributes and organizing principles of the organization are of high caliber (per the attributes in Section 8.2.1 in this chapter). An SDO is officially sanctioned, accredited or accepted for the development of standards within a jurisdiction (i.e., international, national and/or regional).

- *Forum (or Industry Forum):* An organization that creates interoperability specifications, other related documents and/or provides added value based on standards. The attributes and organizing principles of the organization can be of high caliber (per the attributes in Section 8.2.1 in this chapter). The organization is not officially accredited, sanctioned or accepted as an SDO, but can be recognized as an authority on selected matters.
- *SDO/Forum:* A general term that refers to an organization that is either an SDO or Industry Forum.

8.2.1 Key Attributes of Standards Development Process

The key attributes of all well-run standards development processes include: due process and consensus process, openness, transparency, balance of interests, lack of dominance, patent policy, and an appeals process as follows:

- Due process and consensus process includes consensus based opening and closing of standards projects (e.g., via voting) as well as appropriate management of the consensus-driven standards development process between the opening and closing of projects (e.g., with the use of Parliamentary Procedures).
- Openness refers to allowing all materially interested parties to participate in the standards development process. Participation fees could apply.
- Transparency refers to providing visibility into the standards development process, i.e., appropriate meeting notes/minutes are available, proper meeting notifications are provided, project status and summaries are available, etc.
- Balance of Interest refers to ensuring that there is not inherent dominance by a given interest category among the membership. Typical interest categories are producers (e.g., equipment providers), users (e.g., service providers) and general (e.g., other stakeholder such as consulting firms or government agencies).
- Lack of Dominance refers not having any entity (i.e., person, company, group, etc.) dominate the process.
- Patent Policy refers to having a well defined and well communicated policy regarding patents that are essential for standards implementation/compliance. This can include early disclosure notification and assurances regarding patents, such as royalty-free or reasonable and non-discriminatory terms.
- Appeals Process refers to having a well-defined process for appeals regarding the standards development process. This does not allow for the appeal of technical consensus results or outcomes. Just the process and/or procedural matters can be appealed.

In the USA, these attributes are core principles for the American National Standards process as outlined by ANSI. However, the adoption of such principles is not restricted to ANSI-accredited organizations. The US Office of Management and Budget (OMB) Circular A-119 (under the subject of *Federal Participation in the Development and Use of Voluntary Consensus Standards and in Conformity Assessment Activities*) [1] includes guidance regarding US Federal Agency participation in standards development and procurement involving standards created by "voluntary consensus standards bodies." The key attributes outlined in this section are core to defining "voluntary consensus standards bodies" per OMB Circular A-119.

8.2.2 General SDO/Forum Types and Interactions

The following are the general types of SDO/Forum:

- *International Standards Development Organization (SDO)*—An SDO that develops international recommendations and standards. The key examples are ITU, ISO, and IEC.
- *End-to-End Deployment and Interoperability Forum*—A forum that works on the entire suite of topics involved in full deployment of a given technology. From a telecom industry perspective, some examples are the Broadband Forum, Metro Ethernet Forum, and Optical Interoperability Forum.
- *National/Regional SDO*—An SDO that develops national or regional standards. From a telecom industry perspective, members of the Global Standards Collaboration (GSC) are key organizations in this category. Some examples are ATIS (USA), ETSI (Europe), TTA (Korea), and TTC (Japan).
- *Focused Interoperability Forum*—A forum that provides detailed interoperability standards, specifications and/or added value to existing standards or industry specifications by focusing on detailed interoperability matters but not necessarily end-to-end interoperability or broad deployment matters. From a telecom industry perspective, some examples are IEEE and IETF.[†]
- *Work Initiation Forum*—A forum that initiates new standards work, e.g., by providing new or incremental requirements for a new or existing technology area, proof-of-concept specification in a new technology area, or strategic direction in the form of a framework and work plan. An example of the proof-of-concept case is the IPDR.org forum which was absorbed into TMF.

See the Figure 8-1 for the primary added value information flows between the general types.

As SDO/Forums are mapped into this model it becomes clear that these *SDO/Forum Types can apply as primary roles or can apply as subtending committee roles in a SDO/Forum.* This can be seen in large SDOs with many subtending committees and forums. For example, ATIS hosts "focus groups" or "incubators" that can take on the role of a Work Initiation Forum, even though ATIS's primary role in this model would best fit that of a national SDO (in the USA).

8.2.3 General Standards Development and Coordination Framework

Standards development frameworks in SDO/Forums across the world have much in common regarding the top-down nature of the work, usually starting with require-

[†] Note that from a general electrical/electronic perspective, IEEE is considered an SDO. And even though IEEE and IETF are considered "Forums" from a telecom perspective in this model, in practice some of their standards have greater force than that of many telecom SDO generated standards.

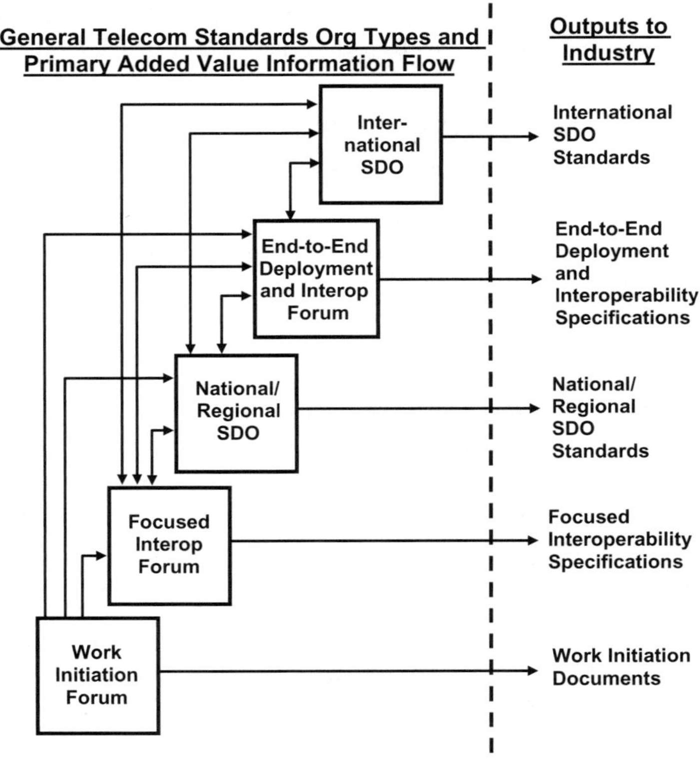

Figure 8-1. General SDO/forum types and interactions

ments, and then drilling down with analysis and design of the standard. Not only are these frameworks based on sound systems engineering development process principles, the framework provides for an incremental consensus building process for various standards stakeholders. The information provided in this section has been in use implicitly and explicitly in ATIS (e.g., T1M1/TMOC, Ordering and Billing Forum [OBF], Emergency Services Interconnection Forum [ESIF], and IPTV Interoperability Forum [IIF]) for several years and variations are used in other SDOs. The Unified Modeling Language (UML) can be used as the overall methodology, where applicable, such as for interface standards development.

The framework provided in this section can be used within a single SDO/ Forum for standards development or can be used where close cross-organization collaborations are required.

The key to successful end-to-end standards development in any domain is well executed cross-organization (e.g., committee) coordination. In some cases close collaborations are required. The key to successful cross-organization collaborations is to have the collaborations built on a sound standards development and coordination framework that provides for well defined cross-organization interactions and handoff points to facilitate structured cross-organization cooperation.

At the high level, the Standards Development and Coordination Framework is a simple four step process:

1. *Requirements* for systems, networks and interoperability points (e.g., network interoperability points) to drive Analysis and Design. Output contains the following for the application/release (or problem or domain) at issue:
   - Use cases
   - Business and technical requirements

   A structured requirements approach is preferred where requirements are identified (e.g., numbered) in some logical manner (e.g., hierarchy). The technical requirements can be relatively high level and the Analysis step could define the detailed technical requirements.
2. *Analysis* of systems, networks, and interoperability points based on Requirements to drive Design. The output contains the technical analysis of the application/release (or problem/domain) at issue. Analysis can include details such as:
   - Detailed technical requirements
   - Architecture (e.g., network architecture)
   - System interaction/message diagrams
   - Procedure outline/steps
   - Flow charts
   - Data definitions, data/information models
   - Process models
   - State diagrams
   - Object models

   All analysis descriptions and modeling should strive for commonality, reuse and extensibility—within reason. Specifically the analysis descriptions and modeling should strive to be technology/protocol independent to drive the extensible technology/protocol dependent design (next step) and to allow for analysis output reuse with multiple technologies/protocols and to allow for technology/protocol evolution. A structured linkage to the requirements that drove the analysis should be implicitly or explicitly provided.
3. *Design* of systems, networks, and interoperability points based on Requirements and Analysis where the design specification is appropriate to be published as a normative interface standard (e.g., American National Standard in the USA). The output contains the technical design for the application/release (or problem/domain) at issue. The design can include details such as:
   - Detailed architecture (e.g., detailed network architecture)
   - System or subsystem architecture
   - System or subsystem interaction/message diagrams
   - Detailed procedure steps
   - Schema (XML)

- Protocol
- Transport technology

The design specification is technology/protocol dependent and should be positioned for extensibility. A structured linkage to the parts of the analysis that drove the design should be implicitly or explicitly provided.

4. *Implementation Guidelines* in support of system implementation, interoperability and interconnection. The implementation guidelines include technology/protocol dependent details to facilitate efficient and effective implementations, but may not be appropriate to be included in the normative interface standard (from the design stage).

8.2.3.1 Project Execution and Cross-Organization Interactions and Handoff Points With the Standards Development and Coordination Framework in place, multiple organizations (e.g., committees) can provide the integrated outputs in a multi-volume (and cross-referenced) set. Several factors should drive the organization handoff point decision, i.e., charters, missions/scopes, expertise, etc. As an example, with just two organizations collaborating, a reasonable handoff point is from the Requirements step to the Analysis step (if the business requirements experts are in one committee and the technology experts are in another committee). Having a cross-organization team (or sub-team) that can move among the various organizations with the work will facilitate smooth handoffs and work transitions.

Notes on Generic Gantt Chart (Figure 8-2):

- All projects/tasks should be Deliverable or Achievement focused, e.g., "Delivery of Application X Requirements." These Deliverables or Achievements should be end products of the organization (e.g., Published Standards Documents).
- Sub-tasks (i.e., specific intermediate tasks done by an organization to work toward a project/task) in cross-organizational programs should not be managed at the overall cross-organizational program management level. Sub-tasks and their management should be defined and managed by the responsible organizations.

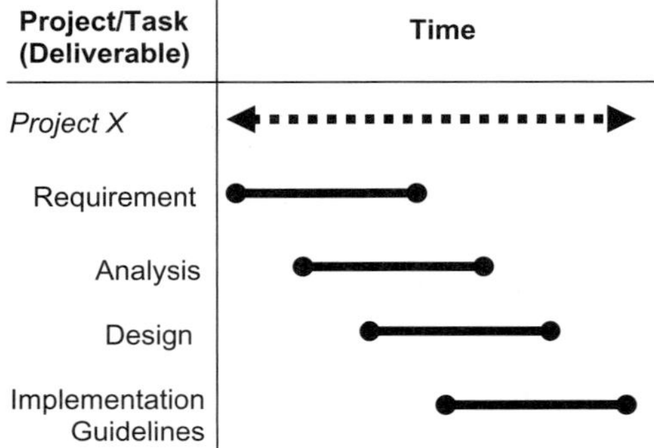

Figure 8-2. Generic Gantt Chart for Project/Program Management

- The overlap of projects/tasks is intended to make note that the some aspects of the next phase of the work can start in advance of completion of the previous phase.
- The duration of the overlap of projects/tasks and duration of the projects/tasks will vary depending on the scope of the work, base standards available, contributions available, resources available, etc.

8.3 MANAGEMENT SDO/FORUM CATEGORIES

General aspects of the standards development process, from the previous section, apply to management SDO/Forums, i.e., Key Attributes of Standards Development Process, General SDO/Forum Types and Interactions and General Standards Development and Coordination Framework. In addition, there is a further categorization of SDO/Forums, from a management standards perspective, that can help to understand the domain. It is as follows:

8.3.1 General Network/Service SDO/Forum

An SDO/Forum that has a broad based scope from a network/services perspective and includes management standards development. The primary organizations in this category are derived from the National/Regional SDO and International SDO general types (per Section 8.2.2 in this chapter). The following are key example SDO/Forums in this category:

- International Telecommunication Union—Telecommunication Standardization Sector (ITU-T)
- Alliance for Telecommunications Industry Solutions (ATIS)
- European Telecommunications Standards Institute (ETSI)

8.3.2 Specific Network/Service SDO/Forum

An SDO/Forum that has a relatively narrow scope from network/services perspective and includes management standards development. The primary organizations in this category are derived from the Focused Interoperability Forum or the End-to-End Deployment and Interoperability general types (per Section 8.2.2 in this chapter). The following are key example SDO/Forums in this category:

- Broadband Forum
- Internet Engineering Task Force (IETF)
- Metro Ethernet Forum (MEF)
- Optical Internetworking Forum (OIF)

8.3.3 Information Technology SDO/Forum

An SDO/Forum that focus on IT areas that management standards development can utilize. The primary organizations in this category are derived from the Focused

Interoperability Forum general type (per Section 8.2.2 in this chapter). The following are key example SDO/Forums in this category:

- Organization for the Advancement of Structured Information Standards (OASIS)
- World Wide Web Consortium (W3C)

8.3.4 Management-Standards Focused SDO/Forum

An SDO/Forum that focuses on management standards development. The primary organizations in this category fall into the Focused Interoperability Forum or the End-to-End Interoperability Forum general type (per Section 8.2.2 in this chapter). The following is a key example SDO/Forum in this category:

- TeleManagement Forum (TMF)

The interactions shown in Figure 8.1 (General SDO/Forum Types and Interactions) provide the overriding interaction model that applies to all SDO/Forum types. Thus this interaction model applies to management SDO/Forums per their mapping to general SDO/Forum types.

8.4 PRINCIPLES, FRAMEWORKS, AND ARCHITECTURE IN MANAGEMENT STANDARDS

This section describes the key principles and concepts in management standards development and the key frameworks and architecture.

8.4.1 Principles and Concepts in Management Standards Development

Aspects of the general standards development process, from Section 8.2 in this chapter, apply as underlying principles and fundamental concepts in management standards development. In addition, there are key principles and concepts that apply to management standards as follows:

- *Scope of Management* (network operations and management vs. real-time and/or transient network services management): Generally speaking, the concept of "management" that is addressed in this book is related to networks and services operations and management and the automation thereof.
- *Separation of Concerns:* This is a basic principle of computer science that deals with minimizing overlaps of functionality. It manifested itself in paradigms/methodologies such as modularity, layered design, encapsulation, information hiding, object-oriented modeling, and Service Oriented Architecture (SOA).
- *Structured Approach to Standards:* Section 8.2.3 in this chapter (General Standards Development and Coordination Framework) describes the general standards development approach. Management SDO/Forums are early adopters of structured standards development tools such a Unified Modeling Language (UML).

- *Functional and Interface Frameworks:* Since the advent of TMN, management SDO/Forums has relied on solid functional and interface frameworks.
- *Protocol Independent Information Modeling:* This key concept is related to all of the above items and is a principle that was developed both for structure as well as a hedge for reuse when the information model could survive independent of the particular protocol for an interface. It is common to see this principle widely used today. An early use of this principle (in the early 1990s) can be seen in Bellcore's Advanced Intelligent Network (AIN) operations/management information model (see operations/management sections and related information model appendix of early issues of Bellcore/Telcordia GR-1298, *AIN Switching System Generic Requirements* [2]).[†]

8.4.2 Frameworks and Architecture

The suite of TMN standards and Enhanced Telecom Operations Map (eTOM) standards provides very useful functional and process architectures, respectively, for the management of communications networks and services.

Whereas the TMN functional architecture was developed primarily from a bottom-up perspective (i.e., with a primary bias toward the network management), eTOM provides a business process framework that was developed primarily from a top-down perspective (i.e., with a primary bias toward customer management and business management). The eTOM was developed by TMF and was adopted in the ITU-T M.3050.x [3]. Enhanced Telecom eTOM series of ITU-T Recommendations. The eTOM provides the framework for business processes that guides the development and integration of Operations Support Systems (OSS) and Business Support Systems (BSS). See Chapter 6 in this book and ITU-T M.3050.x. for more detail regarding eTOM.

Key aspects of the TMN architecture (per ITU-T Recommendation M.3010 [4]) are:

- *Logical Management Layers:* Business Management Layer (BML), Service Management Layer (SML), Network Management Layer (NML), Element Management Layer (NML), and Network Element Layer (NEL).
- *Management Functional Areas:* Fault, Configuration, Accounting, Performance, and Security Management (FCAPS).
- Interoperability Reference Points and Interfaces: Key reference points are "q or Q" and "x or X." where q/Q are interface reference points internal to a service provider's business (e.g., between an OSS and an NE) and x/X are interface reference points external to a service provider's business (e.g., to a customer or another service provider). Note that lowercase designations refer to logical reference points; uppercase designations refer to system interfaces.

FCAPS are characterized as follows:

- Fault Management is characterized primarily by the detection, correlation, isolation and correction of faults in the network and its environment.

[†] Note that the author of this chapter was the engineer/creator of the protocol independent information model in the early versions of GR-1298.

- Configuration Management is characterized primarily by planning, installation, provisioning and control of the network.
- Accounting Management is characterized primarily by usage measurements collections and billing of network services.
- Performance Management is characterized primarily by monitoring the behavior and the effectiveness of the network.
- Security Management is characterized primarily by security administration of the network.

Logical Management Layers are characterized as follows:

- Business Management Layer is characterized primarily by policy and goal setting.
- The Service Management Layer is characterized primarily customer facing interactions such as service requests and billing.
- The Network Management Layer is characterized primarily by visibility and broad controls of the entire network.
- The Element Management Layer is characterized primarily by visibility and narrow controls of a subset of the network (usually based on some commonality of Network Elements [NEs]).
- The Network Element Layer is characterized primarily by visibility and narrow controls of a NE itself (including its components).

In Figure 8-3 the TMN functional architecture/matrix of Logical Management Layers and Management Functional Areas can be seen. At each intersection of a Logical Management Layer and Management Functional Area the cell would contain function sets. Figure 8-4 shows some example functions sets.

For more detail on TMN, see ITU-T Recommendations: M.3010, *Principles for a Telecommunications Management Network* and M.3400 [5], *TMN Management Functions*.

Figure 8-5 shows a simplified reference model, a view of the Q and X interfaces and how the core OSSs, Element Management Systems (EMSs), and Network infrastructure relate. The key TMN interface reference points are shown (Q, X).

	Fault Mgt.	**Config. Mgt.**	**Acct. Mgt.**	**Perf. Mgt.**	**Sec. Mgt.**
BML					
SML					
NML					
EML					
NEL					

Figure 8-3. TMN functional architecture

	Fault Mgt.	Config. Mgt.	Acct. Mgt.	Perf. Mgt.	Sec. Mgt.
BML	Trouble Report Policy	Service Feature Definition	Settlements Policy	Network Performance Goal Setting	Security Policy
SML	Trouble Reporting	Customer Service Planning	Bill Assembly and Sending Bill	QoS Performance Assessment	Admin. of External Access Control
NML	Network Fault Event Analysis	Network Resource Selection and Assignment	Usage Short-term Storage	Data Aggregation and Trending	Admin. of Internal Access Control
EML	Alarm Reporting	NE(s) Configuration	Usage Accumula-tion and Assembly	NE(s) Threshold Crossing Alert Processing	Admin. of Keys for NEs
NEL	Failure Event Detection and Reporting	Access to Service Features in NEs Self-inventory	Usage Generation	Detection, Counting, Storage and Reporting	Admin. of Keys by an NE

Figure 8-4. TMN functional architecture with example function sets

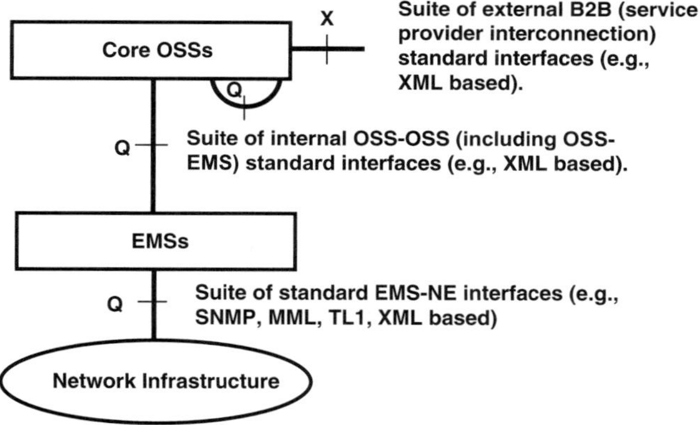

Figure 8-5. Simplified OSS/EMS/network interoperability reference model

Note that the Q interface shows the internal service provider (intra-TMN) nature of the interface whereas the X interface shows the external service provider to service provider (inter-TMN interconnection) nature of the interface. The previously mentioned SDO/Forums in this chapter publish standards that cover these interfaces either generically (e.g., ITU-T or TMF frameworks) or for specific network technologies/services (e.g., Broadband Forum for Digital Subscriber Line [DSL]). EMS is called out specifically because of the reality of its use as a system that mediates between the OSSs and the network (this includes network supplier mediation).

8.5 STRATEGIC FRAMEWORK FOR MANAGEMENT STANDARDS DEVELOPMENT

This section provides the strategic questions for standards engagement determination, strategic progression of the standards work and strategic human side of standards development.

8.5.1 Strategic Questions for Standards Engagement Determination

For a specific standards topic, area or project the following strategic questions should be asked:

What is the interest level in a specific standards topic, area, or project? The answer can range from very low to very high. It depends first on the general category of the network management area with respect to the entity (i.e., person, business unit, corporation, etc.) doing the analysis. The general categories that the standards could fall under are sustaining technology (or operations process or function) or emerging technology (or operations process or function). An example of a very high interest in a sustaining technology would be if the technology involves existing (current, operational) large scale infrastructure or deployment with extended life. Standards impacts here can be very costly due to forced upgrades. An example of a very high interest in an emerging technology would be if the technology involves definite (or highly likely) large scale deployment plans. Standards impacts here can be very costly due to changes to future deployment plans. Examples of very low interest would be cases where there are no existing or planned deployments and the likelihood of a future deployment is very low for the standards in question.

What is the approach to the "front-game" for standards development? The answer can range from "just monitoring the standards work" to being the project champion or leader or primary contributor. The front-game refers to what occurs in the SDO/Forum itself.

What is the approach to the "back-game" for standards development? The answer can range from "no back-game needed" to "significant alignment needs to be developed" (e.g., alignment within an interest group such as service providers). The back-game refers to what occurs behind the scenes (i.e., sidebar discussion, joint contributions, etc.), i.e., outside of the SDO/Forum. (*Caution must be exercised here to avoid running afoul of anti-trust laws; it is strongly recommended that standards representatives contact their standards organization coordinator and legal counsel regarding ant-trust compliance.*)

What general type of forum and/or specific forum should be engaged? The answer is highly dependent on (personal and business) relationships in the likely forums and the likelihood of success in those forums. The answer can include a roadmap for standards development that can move specific work from a Work Initiation Forum to an International SDO with all forum types in between. This decision should include consideration of the Key Attributes of Standards Development Process.

8.5.2 Strategic Progression of Standards Work

Once the above questions are answered, the General Standards Development and Coordination Framework can be utilized to advance the work, starting with Requirements. It's always best to start with some type of requirements and/or framework to initiate a consensus process. This is true even if there is a massive body of work available (behind the scenes) to contribute. Starting small and growing the work in SDO/Forum results in ownership by the SDO/Forum. In addition, appropriate adjustment can be made incrementally as the work progresses.

8.5.3 Strategic Human Side of Standards Development

The foundation of standards development (like any group endeavor) is good human relations. The goal of standards development is a consensus solution; this is built on group and individual attributes such as respect, relationships, self-interests, persuasion, and interest alignment. If the human side of standards development is neglected, projects can be delayed or abandoned due to lack of support. The golden rule applies: "do unto others as you would have them do unto you." There is inherent good in the rule but there are also self-interest reasons for the rule in a standards development setting. For example, the rule can be augmented as follows: "especially when others will have ample opportunity to do unto you."

There are lots of books on the market to help in human relations, especially with difficult people and situations. However, it is important to know that even having the best human relationships will not make a standard representative work counter to their interests. The key in the human game is to have good relationships that have a reasonable level of trust such that creative compromises and middle ground solutions can be identified to move consensus development forward.

Some tips:

- Refer to the Chair with the utmost respect; especially in formal settings, e.g., refer to the Chair as Mr. Chair or Madam Chair or Chairman, per the bylaws or operating procedures.
- Refer to the participant/members with the utmost respect; especially in formal settings, e.g., refer to the participant/members by their formal name, if appropriate (especially in a contentious debate).
- Network well with participant/members outside of the formal meetings; this goes a long way toward relationship building.

8.6 SAMPLING OF NGN MANAGEMENT STANDARDS AREAS AND SDO/FORUMS

Table 8-1 maps key SDO/Forums with their General Standards Organization Type and Management Standards Category; examples of NGN Management Standards Areas are also provided.

TABLE 8-1. Mapping of key SDO/Forums to their General Standards Organization Type and Management Standards Category

SDO/Forum	General Standards Org Type	Management Standards Category	Example NGN Management Standards Areas
3GPP (3rd Generation Partnership Project, www.3gpp.org)	End-to-End Deployment and Interoperability Forum	Specific Network/Service SDO/Forums	Wireless Network Management, e.g., 3G, Long Term Evolution (LTE), IP (Internet Protocol) Multimedia Subsystem (IMS)
ATIS (Alliance for Telecommunications Industry Solutions, www.atis.org)	National SDO (USA)	General Network/Service SDO/Forum	Multiple Management Areas, e.g., Security, Accounting Management, Interconnection (e.g., Trouble Administration, Service Ordering); and NGN specific areas (e.g., network management for IPTV)
Broadband Forum (www.broadbandhyphen;forum.org)	End-to-End Deployment and Interoperability Forum	Specific Network/Service SDO/Forums	DSL (Digital Subscriber Loop) and related Customer Premises Equipment (CPE) Management Areas
CCSA (China Communications Standards Association, www.ccsa.org.cn/english/)	National SDO (China)	General Network/Service SDO/Forum	Multiple Management Areas for NGN, e.g., FCAPS for Wireless, Transport, and Access Networks
ETSI (European Telecommunications Standards Institute, www.etsi.org)	Regional SDO (Europe)	General Network/Service SDO/Forum	Multiple Management Areas, e.g., Security, Accounting Management; and NGN specific network management
IETF (Internet Engineering Task Force, www.ietf.org)	Focused Interoperability Forum	Specific Network/Service SDO/Forums	IP Centric Management Areas
ITU (International Telecommunication Union, www.itu.int)	International SDO	General Network/Service SDO/Forum	Multiple Management Areas, all FCAPS and NGN specific areas (e.g., network management for Optical Transport Network)

MEF (Metro Ethernet Forum, www.metroethernetforum.org)	End-to-End Deployment and Interoperability Forum	Specific Network/Service SDO/Forums	Metro Ethernet Management Areas, e.g., Configuration Management (Provisioning)
IP/MPLS Forum (www.ipmplsforum.org); note that the work of this forum is planned to migrate to the Broadband Forum	End-to-End Deployment and Interoperability Forum	Specific Network/Service SDO/Forums	IP/MPLS (Multi Protocol Label Switching) Management Areas, e.g., Configuration Management (Provisioning)
OASIS (Organization for the Advancement of Structured Information Standards, www.oasishyphen;open.org)	Focused Interoperability Forum	Information Technology SDO/Forums	XML Schemas (used for OSS Interfaces, e.g., for NGOSS)
OIF (Optical Internetworking Forum, www.oiforum.com)	End-to-End Deployment and Interoperability Forum	Specific Network/Service SDO/Forums	Optical Management Areas, e.g., Configuration Management (Provisioning)
TM Forum (TeleManagement Forum, www.tmforum.org)	Focused Interoperability Forum	Management Standards Focused SDO/Forums	OSS Process and Architecture, e.g., eTOM and NGOSS
TTA (Telecommunication Technology Association, www.tta.or.kr/English)	National SDO (Korea)	General Network/Service SDO/Forum	Multiple Management Areas for NGN, e.g., FCAPS for Wireless Networks and IP Centric Networks
TTC (Telecommunication Technology, Committee, www.ttc.or.jp/e)	National SDO (Japan)	General Network/Service SDO/Forum	Multiple Management Areas, e.g., Performance Management; and NGN specific network management
W3C (World Wide Web Consortium, www.w3.org)	Focused Interoperability Forum	Information Technology SDO/Forums	Web Services (used for OSS Interfaces, e.g., NGOSS)

8.7 SUMMARY AND CONCLUSIONS

8.7.1 Chapter Summary

At its core, standards development is about legitimacy via the consensus development process. The most successful standards are taken for granted because of such wide deployment and usage that there is true ubiquity.

A key challenge for management standards development is that it exists at the nexus of two of the most dynamic technology areas in modern times, i.e., communications technology and information technology, a.k.a, Information and Communications Technology (ICT).

Standards Drivers: The most compelling driver that could be in play comes from law or regulation. However, the most common driver for standards development in the NGN/ICT environment is for interoperability in support of multi-supplier/manufacturer/vendor environments, multi-service provider/operator environments, and multi-content provider environments.

8.7.2 General Standards Development Process

The following are key aspects of the general standards development process: Key Attributes of Standards Development Process, General SDO/Forum Types and Interactions, and General Standards Development and Coordination Framework. The Key Attributes of Standards Development Process provides principles that apply to all well-run SDO/Forums. The General SDO/Forum Types and Interactions provide a high-level model to categorize SDO/Forums and show their interactions. The General Standards Development and Coordination Framework provides a standards project execution and management framework that can be utilized for projects that are either executed in a single SDO/Forum or projects that require coordination across SDO/Forums.

8.7.3 Management SDO/Forum Categories

The general standards development process aspects above apply to management SDO/Forums. In addition there is a further categorization of SDO/Forums from management standards perspective that can help to understand the domain, as follows:

- General Network/Service SDO/Forum
- Specific Network/Service SDO/Forum
- Information Technology SDO/Forum
- Management Standards SDO/Forum

8.7.4 Principles, Frameworks, and Architecture in Management Standards

8.7.4.1 Principles The general standards development process aspects from Section 8.2 in this chapter apply as underlying principles and fundamental concepts in management standards, i.e., Key Attributes of Standards Development Process, General SDO/Forum Types and Interactions, and General Standards Development and Coordination Framework apply. In addition there are key principles and concepts that apply to management standards as follows:

- Scope of Management (network operations vs. network services)
- Separation of Concerns
- Structured Approach to Standards
- Functional and Interface Frameworks
- Protocol Independent Information Modeling

8.7.4.2 Frameworks and Architecture The suite of Telecommunications Management Network (TMN) standards and Enhanced Telecom Operations Map (eTOM) standards provide very useful functional and process architectures, respectively, for the management of telecommunications networks and services. Whereas the TMN functional architecture was developed primarily from a bottom-up perspective (i.e., with a primary bias toward the network management), the eTOM provides a business process framework that was developed primarily from a top-down perspective (i.e., with a primary bias toward customer management and business management). The eTOM was developed by the TeleManagement Forum and was adopted in the M.3050.x. Enhanced Telecom eTOM series of ITU-T Recommendations. The eTOM provides the framework for business processes that guides the development and integration of Business and Operations Support Systems (BSS and OSS respectively). Figure 8.5 shows a simplified reference model and of the view Q and X interfaces and how the core OSSs, EMSs, and Network infrastructure relates. The key TMN interface reference points are shown (Q, X).

8.7.5 Strategic Framework for Management Standards Development

Strategic Questions for Standards Engagement Determination: For a specific standards topic, area or project the follow strategic questions should be asked:

1. What is the interest level in a specific standards topic, area, or project?
2. What is the approach to the “front-game” for standards development?
3. What is the approach to the “back-game” for standards development?
4. What general type of forum and/or specific forum should be engaged?

8.7.5.1 Strategic Progression of the Standards Work Once the above questions are answered, then the General Standards Development and Coordination Framework can be utilized to advance the work, starting with Requirements.

8.7.5.2 Strategic Human Side of Standards Development The foundation of standards development (like any group endeavor) is good human relations. The goal of standards development is a consensus solution; this is built on group and individual attributes such as respect, relationships, self interests, persuasion, and interest alignment. If the human side of standards development is neglected, projects can be delayed or abandoned due to lack of support. The key in the human game is to have good relationships that have a reasonable level of trust such that creative compromises and middle ground solutions can be discovered to move consensus development forward.

8.7.6 Key Lessons Learned for Strategic NGN Management Standards Development

- *Choose a strong foundation from frameworks, structured standards development, and sound principles:* Frameworks, structured standards development, and sound principles have been successful and will continue to lead to success.
- *Choose a path that includes reusable schemas, information models, and other analysis and design:* Some standards don't make it in the market however if the standard is structured well, parts of the standard can be reused. Protocol independent information modeling is a good example of this.
- *Choose a path of simple solutions that can scale well and leverage IT standards:* IT standards (including web standards) tend toward simple interoperability frameworks that can be utilized by many industries (i.e., telecom, finance, medical, etc.); this has great economies of scale. It is best to leverage IT standards because of the economies of scale even if compromises must be made.

8.7.7 Challenges and Trends

- *Nexus of IT and Telecom:* Given that management standards are at the nexus of IT and telecom, this has the challenge of riding the dual innovation and change curves. The challenge will be to continue to bring in the best of IT solutions for telecom operations/management automation and efficiency in an ever changing telecom technology environment.
- *Standards Development Resources Spread Thin:* Almost every ICT/NGN SDO/Forum has a management standards development group. This has a tendency of spreading the human resources (standards representatives) thin. This trend will continue. This adds risks such as quality risk and broad consensus risk to the standards development process.

8.8 REFERENCES

1. U.S. Office of Management and Budget. "Circular No. A-119. Federal Participation in the Development and Use of Voluntary Consensus Standards and in Conformity Assessment Activities." http://www.whitehouse.gov/omb/rewrite/circulars/a119/a119.html (Accessed October 1, 2009).
2. Telecordia. "AIN Switching System Generic Requirements." http://www.telecordia.com/store/ (Accessed October 1, 2009).

3. 13. ITU-T Recommendation M.3050: Enhanced Telecom Operations Map (eTOM)—Introduction, 2007. Geneva, Switzerland: International Telecommunication Union Telecommunication Standardization Bureau.
4. ITU-T Recommendation M.3010: Principles for a Telecommunications Management Network, 2000. Geneva, Switzerland: International Telecommunication Union Telecommunication Standardization Bureau.
5. ITU-T Recommendation M.3400: TMN Management Functions, 2000. Geneva, Switzerland: International Telecommunication Union Telecommunication Standardization Bureau.

CHAPTER 9

FORECAST OF TELECOMMUNICATIONS NETWORKS AND SERVICES AND THEIR MANAGEMENT (WELL) INTO THE 21ST CENTURY

Roberto Saracco

This final chapter focuses on new opportunities provided by the unrelenting technology evolution to further develop the telecommunications business well into the next decade. It takes into account the evolution of storage, processing, sensors, displays, statistical data analyses, and autonomic systems and discusses how such an evolution is going to reshape markets and business models into a new era where business ecosystems supplement value chains. It includes a forecast on two new areas for telecommunications, Internet with Things and Digital Shadow. The former, Internet with Things (it is indeed a "with," not an "of") stems from the opportunities provided by a number of identification technologies to enable a seamless continuum between the physical and the digital world. The latter, Digital Shadow, is a discussion of the role that may be played by telecommunications in the management of all the digital information created by individuals to the benefit of the individual-self and the community. This discussion is of particular relevance at a time when the BWPOTS (Broadband, Wireless and Plain Old Telecommunications Services) revenues are declining and new revenue streams have to be found. It is also very much relevant to practitioners and graduate students who need to get the broader picture. Although most of them are and will be working with yesterday or state-of-the-art systems, they need to understand the overall evolution since today's decisions and actions are creating the legacy systems of tomorrow. The chapter closes on a few examples of "living in the future" with communications as the enabling fabric.

Next Generation Telecommunications Networks, Services, and Management, Edited by Thomas Plevyak and Veli Sahin

9.1 HAVE WE REACHED THE END OF THE ROAD?

The evolution we have witnessed in these last 50 years, in electronics, optics, smart materials, biotech, and in all those fields using these technologies, has been relentless. Although today there is no sign of having reached a plateau we suspect that a physical limit to progress lies somewhere. The fact that in many fields, like electronics, this ceiling seemed to have been approaching and engineers have found ways to circumvent it does not change the fact that a physical limitation exists. In economics we have seen what happens when we reach a ceiling, such as when we run out of liquidity: the downward spiral of stock markets in the second half of 2008 was a clear statement of the havoc happening when progress is suddenly stopped. The technology evolution has progressed with such regularity that it no longer surprises us. We have got used to it. Actually, we have built a world that is relying on it. If technical evolution were to stop next year we would need to reinvent the way we are doing business and that would cause tremendous problems. Looking at the physical barriers, like the speed of light, the quantum of energy, the smallest dimension that exists, we can determine where the ultimate limit lies [1]. The good news is that such a limit is very, very far from where we are today. At the present pace of evolution we won't be reaching it for the next few centuries. This does not mean, however, that such limits will ever be reached. Actually, I feel that we will discover unsolvable issues much before getting to those physical barriers. The investment required for chip production plants is growing exponentially and pay back requires huge revenues. As we will see, this is pushing towards huge volumes, with individual products costing less and less in order to be sustainable by the widest possible market. The economics are already slowing down the creation of new plants but new production processes may circumvent what we see as an upper boundary today.

The progress of technology has continuously increased the amount of energy being consumed. It is estimated that the power consumed by residential households in Europe to access broadband networks in 2015 will reach 50 TWh. To give a sense of the number, 10 years ago there was no power consumption in accessing the network, all that was needed was provided by the network itself. Networks in the period 1980–2000 have doubled the power consumption and in 2008 the overall power consumption of networks in Western Europe has reached 20 TWh. This means that we are expecting that a consumption that was non-existent 10 years ago will more than double power consumption of all networks in Europe.

Energy is becoming a bottleneck to evolution. In 2008, China consumed as much energy as the USA. But the per-capita consumption is a fraction of the USA. We are expecting between 8 and 10 billion people to populate the Earth in the next decade. In terms of energy, this is not a 25–80% increase but is likely to be an 800% increase since that population will consume on average what is now consumed on average by a USA citizen thanks to a widespread better life. Quite simply, we do not have that much energy available to meet these assumptions. This means that either global wealth, in terms of energy consumption, will not be, on the average, what a USA citizen had in 2008 or that we will have found ways to dramatically decrease our power consumption and increase energy production (both are indeed required to come anywhere close to meeting those requirements).

The energy issue is going to influence overall evolution in the next decade, in terms of availability and cost. The shift toward a "greener" world, although absolutely important, is going to increase the impact of energy on evolution. The bright side is that the energy "crunch" will force investment into alternative energy sources and into decreasing consumption. This, rather than slowing evolution is likely to shift the direction of the evolution, accelerating the deployment of optical networks that are much more energy savvy than copper networks, pushing towards radio coverage made by smaller cells, since the energy required to cover a given surface decreases (approximately) with the square of the number of cells being used (on average joule per bit transport over radio is 15 times greater than on fibre).

Another important aspect of evolution, core to this book, is its management and the management of its outcome. On both sides we have initiatives, doubts and a sense, sometimes, of an unbreakable wall facing us and stopping future progress.

The Internet is being forecast periodically to be on the brink of collapse because "it is overused and it is not managed." Airwaves are filled to capacity and the explosion of mobile Internet will bring it to a halt. Investment is in the billions of dollars range, partly in coordinated initiatives, some public and cross-national, others private. But there is a sense that it is not steering evolution effectively (someone is saying it is just wasted money) since evolution "just happens" as result of a world without boundaries. If once the "fire was in the valley" [2], now the fire is basically everywhere and it is difficult to tell which will sweep the market and which will just die out. Applications are sprouting when each GB of information is created and made available, apparently with no regulation mechanisms.

Now, for the cognoscenti, the Internet "is managed" and it works well because it lays on a networked infrastructure that is carefully tweaked day in and day out by telecom operators all over the world. Social communication (on and outside of the web) is effectively regulating the adoption of applications and access to information. When people say that there is no management in place for the Web they are just telling half the story. The classic management of networks, of their resources and access, cover only a part of the overall management. The rest takes place using different paradigms. These are not new. They are new to engineers, not to ecosystems.

Living beings have very little that is centrally controlled and managed. The same goes for communities of living beings, from bee hives to jungle. Their management is the behavior of the ecosystem not something separate from it that somehow is regulating it. Lions eating gazelles are examples of resource management in the savannas' ecosystem. Actually there is very little in the behavior of an ecosystem that can be scrapped for not being "management." This may be seen as contrary to the design of things by an engineer. For a long time it has been said that one should design at the same time an object and its management. In the future the object and its ecosystem will be undistinguishable from management. We have already reached this "conceptual" approach in design with the novel paradigm of interactive design. The object is designed around its usage. This paradigm can be extended to the area of management within the ecosystem framework. A product is going to be managed by the ecosystem as a whole. CRM is no longer a task that the object producer has to undertake. That object, if successful in the marketplace, will be supplemented by several CRM services provided by third parties, harvesting the business opportunity being created by that evolving ecosystem. We will see a few examples of this new approach to management at the business layer.

For the time being, and for the horizon that is reasonable to consider today, not beyond 2050, we can be confident that technology evolution will continue, at a different pace in different sectors as has been the case in these last 50 years, but overall at a similar rate to what we have been experiencing in these last decades. Some believe we will see near-exponential growth of knowledge across many technologies, e.g., telecom, information, biology, transportation, etc., that effectively will act as a multiplier, leading to increasing evolution speed. Technology will become reality in the marketplace because it is managed, actually, it is self-managed. Many remember the forecast in the 1920s proclaiming the increase of telephone density would cross the point where all people would need to be employed as switchboard operators. The forecast proved to be accurate but in an unexpected way. The introduction of automatic switches forced all people to act as switchboard operators including the area code before the called number to route the call. Technology will be even more present in everyday life and in business than what we have today but in a way it will disappear from perception. Complexity will be swept under the carpet of everyday experience.

Now, saying that the pace will be substantially unchanged should not deceive us in believing that it is going to be "business as usual" for three reasons.

First, as "Moore's Law" claims, in the next 18 months we are going to have evolution that will double today's performances This means that in the next 18 months we will go a distance that has taken us the last 40 years to walk. That same distance will be covered in just 18 months. Now this is quite a change!

Second, performance increase has a linear effect on the ecosystem until a certain threshold is reached. Beyond that point, it is no longer seen as a performance increase, rather as a change of rules. Think about electronic watches. There was a time, in the 1970s, when owning an electronic watch was very, very expensive. As price went down, more and more people could afford an electronic watch. At a certain point the watch cost dropped, basically, to zero (it passed the threshold of cost perception) and there was no further market for electronic watches. The industry had to reposition itself into the fashion industry. Long gone are the times of ads claiming better and better precision for a particular watch. The marketing value of a Swiss Certified Chronograph dropped to zero. A similar thing will happen with the deployment of broadband networks and its enabling optical infrastructures. Once you reach a bandwidth of 1 Gbps, and possibly before that time, it will become impossible to market increased bandwidth at a premium. Bandwidth value will drop to zero. Marketers will need to find new slogans. Notice how the threshold is linking technology with market value and these two are changing the business model, the rules of the game. A disruption takes place. As this happens, consolidated industries need to reinvent themselves; new ones find leverage to displace the incumbents. In the discussion on technology evolution we will be coming to this point over and over. The question to consider, therefore, is not if a technology is reaching its evolution limit, rather if that technology is leading to a disruption threshold.

Third, the increase of players, i.e., the fuzzy boundaries between producers and consumers and the fruition of systems and services that are a cluster of components created and delivered by different parties, will require a different approach to management.

9.2 "GLOCAL" INNOVATION

Innovation used to be easier to predict when few companies and countries were leading the way. The evolution of infrastructures was so easy to predict that the International Telecommunications Union (see Chapter 8) published, every year, a table with the status of Telecommunications Infrastructures and Telecommunications Service penetration. Maps showed the progress made and others showed what penetration will be reached in ten to twenty years. For example, it would have taken 19 years for a country to move from 1 to 10% telecommunications penetration but only 12 years to move from 10 to 20%, 8 from 20% to 30%, and so on. The advent of wireless communications has changed the world. Countries like India that used to have less than 3% telecommunication density (and that is still the quota of the fixed lines) moved in 10 years to 20% density and are likely to reach 50% penetration within the next decade. China moved in 10 years to over 30% (via wireless). However, more than in the infrastructure domain, where globalization has an impact is in the decrease of price that in turn enables a progressively broader market, thus affecting, indirectly, the local situation. Service domain has a distribution cost that is basically "zero." Once an application is developed, it can be made available over the network with no distribution cost hampering its marketing.

The network is also playing another trick. An application that can make sense in the US market can be developed and "marketed" from India. Innovation is no longer confined to a few wealthy companies or countries because of huge investment barriers. The real barriers have moved from money availability to education availability. It is the increase in production of engineers in India and China that is placing these two countries at the forefront of innovation, not their huge (potential) market. The market of innovation is no longer local but global. This global-localization of innovation is going to continue. It is not by chance that the new President of the United States has set the goal of increasing scientific education and the number of engineers as the way to remain at the forefront of progress, both scientific and economic.

Optical and wireless networks will further shrink the world. Distance is already irrelevant for information flow. It is becoming irrelevant for delivery cost of a growing number of products and services and surely it is becoming irrelevant for innovation. Out-sourcing will be increasingly practiced by big industries. Smaller companies will take innovation from anywhere and will make a business out of their ability to localize a global world. The latter is going to stay; the former will be viable only until a labor cost differential will "plague" the world. Eventually, it will disappear. Politics, regulations, and cultures will be the determining factors in the evolution. From a technological point of view the Earth will be no bigger than a village.

The Web 2.0 paradigm will evolve from being a network of services to applications made available by a plethora of (small) enterprises. It will also be made available, thanks to the huge investment of a few (big) enterprises, as Web 3.0 where the interaction will take place among services and applications to serve users' context. Questions like "*when is the next train*?" will become answerable because somewhere in the Web there is an understanding of the context. It will be obvious from this understanding that I am looking for the train to Milan. It is not a small step.

Again, it is a matter of "glocalization." We are moving from the syntax, from infrastructures providing physical connectivity, to semantics, to the appreciation of who I am. This includes the understanding of who I was (the set of experiences shaping my context) and the forecasting of who I will be (the motivations and drives to act). This might seem Orwellian, scary, and definitely not the way to go. However, the evolution, to happen over a long period of time, has to be beneficial; otherwise it is not going to become entrenched. Because of this, we can expect that the balance between what is technologically possible and what the market is buying will depend on us, as individuals and as communities. "Contextualization" can give rise to many issues, from privacy to ownership, from democracy to the establishment of new communities, continuously reshaping themselves.

Contextualization is not likely to result from an "intelligent, Orwellian" network, but rather from an increased intelligence of my terminal, and that is under my control. I will make decisions, most of the time unconsciously, on what to share of my context and the network will be there to enable it.

My terminal, and the "my" is the crucial part, will act as an autonomous system, absorbing information from the environment, both local and, thanks to the network, global, and will let me communicate with my context, my information, my experiences, my environment and, of course, my friends and acquaintances, in the same seamless way that, today, I walk into a room and act according to my aims, expectations, and what is there.

Therefore, "telepresence," one of the holy grails of communications, will also be "glocal." I will communicate locally and remotely as if both remote and local are present at the same time. There is still another aspect of glocal that may play a significant role in the future—the advent of bottom-up infrastructures. Infrastructures require a huge investment and therefore only a few players have had the financial strength to deploy them. Similarly, their operation and up-keep require a lot of resources and again place them beyond the private citizen or small entrepreneur. Technology evolution is going to change that, at least to a certain extent. Progress in technologies on terminals will transform them into nodes of a local area network that can connect to the main infrastructure on one side but also to other nearby terminals, thus creating an alternative infrastructure, provided a sufficient number of terminals are available. If we think about cars, cell phones in an urban environment, smart buildings, etc., we have a picture of millions of networked elements that can create alternative infrastructures, thus changing the rules of the game.

All of this will be possible because of technology evolution. Although the list of technologies to consider would be very lengthy we might be contented in looking at a few of them, more in terms of the evolution of functionalities that such a technology evolution makes available than in terms of the evolution of single technologies. Let's do it.

9.3 DIGITAL STORAGE

Digital storage capacity has increased by leaps and bounds. The original solutions to digital storage have basically disappeared (magnetic cores, drums, tapes, etc.), giving way to new technologies, magnetic disks, solid state memory, optical disks, and polymer memories.

In 2008, hard drives, that is, devices using magnetic disks for storage, reached 2 TB in the consumer market and 37.5 TB disks are announced to appear in 2010. 100 TB will become common place by the end of the next decade. The new leap in magnetic storage density is achieved through HAMR (Heat Assisted Magnetic Recording).

Solid state memories have advanced significantly. Compact flash cards are now cheap and ubiquitous. They were invented in 1994 and have moved from 4 MB to 64 GB in 2008. 128 GBs are already in sight and have become available in 2009. Solid State Disks (SSD) based on flash technology appeared in 2007. Announcements at the end of 2008, of new etching processes, able to reach the 22–15 nm level (down to the current 60–40 nm) clearly show that more progress in capacity is ahead. This increase in capacity is placing flash memory on a collision course with magnetic disks in certain application areas, e.g., Mp3 players and portable computers. They consume only 5% of the energy required by a magnetic disk and are shock resistant up to 2000 Gs (corresponding to a 10 foot drop).

Bit transfer rate has already increased significantly and there is a plan to move the interface to the Serial Advanced Technology Attachment (SATA) standard, the one already used by magnetic disks. This raises the transfer speed to 3 Gbps. The current Parallel Advanced Technology Attachment (PATA) interface tops out at 1 Gbps.

Optical disks and more generally optical memories are a common sight, CD and DVD are plenty. Blue Rays extended the capacity of a single disk to 50 GB and new technology promises to reach 500 GB within five years. Holographic storage promises several TBs of capacity in a space of a thimble. The limits of this technology are in the much slower recording rate but it may be of interest in certain kind of applications like massive storage units.

Polymer memories have seen an increased effort by several companies to bring them to market. Commercial availability is likely in 2010. These memories are made by printing circuit components on plastic and are a precursor of full-fledged printed electronics. Their big advantage over other types of memories is in their extremely low cost and potentially huge capacity. In an area the size of a credit card one could store several TB of data. Most recent development has led to creation of polymer nano cells, 3 nm across, each able to store a bit. A square inch would be able to store up to 10 Tbit, 10 times more than the highest storage density available in the spring of 2009.

Data are going to be stored both at the edges and in the network. A 1 TB cell phone may be available in 2012, media centres in the home will host the entire life production of a family in their multiTB storage, EB (a billion billion of bytes) will become commonplace as data warehouses in data-based companies and whoever is going to come on the scene in the next decade. Institutions and governments will harvest on a day-by-day basis the digital shadow of their constituencies to offer better welfare.

Raw data generated by sensors, will be transformed into economic value through statistical data analyses approaches. Storage is becoming one of the most important enabler for business in the next decade.

What is the consequence of this continuous increase of storage capacity? Clearly, we can store more and more data and information. However the real point is that this huge capacity is changing paradigms and rules of the game, affecting the value of the network and impacting on its architecture. Since data is everywhere the flow of data will no longer be restricted from the network towards the edges. The other direction will be just as important. In addition, we are going to see the emergence of local data exchange, edges-to-edges and terminal-to-terminal. The first evolution makes the uplink capacity as important as the downlink (leading to the decommission of ADSL). The second emphasizes the importance of transaction-oriented traffic. "Updates" achieve greater importance and possibly are perceived as the real value that some providers may deliver. Raw data (and this also applies to information that causes us to drown in information) make sense only if they can be converted into perceptible chunks of information, relevant to the here and now for a given user (person and machine).

As we will discuss in the following, storage may disappear from sight, being replaced by small "valuets," a mixture of applications and sensors/displays able to represent a meaning valuable to a user. We are starting to see this appear as tiny applications on the iPhone. They mask the data, the information, the transactions required and even the specific applications being used. The new way of storage cards, embedding communications and applications, is a further hint on where the future will take us.

Having huge amounts of data immediately creates the problem of its management. The offer by a number of service providers to host private data on their service factories has been taken up by many residential customers in terms of "back up." It has not resulted in a change of habits, since most, if not all users have their local HD to store the information. Enterprises are making use of storage services to a larger extent than private users although we are still far from seeing a centralized shared storage replacing the locally owned one. This situation might change with the advent of cloud computing, since, at that point, storage will be just one out of many services offered by the cloud.

Today, the problem with centralized-remote storage is still the answer to the question "what if I need the data and there is no connectivity available? We can see people taking storage for granted, i.e., whatever terminal they are using, as long as it is possible to characterize it as "my terminal," will have the desired data available (access fades away). By the end of the next decade, once connectivity disappears from perception, we might presume that 1 Gbps fixed connectivity plus 10 Mbps wireless connectivity plus flat rate access will do the trick.

9.4 PROCESSING

Processing evolution is no longer on the sole axis of increased performances. Other factors, like reduced energy consumption and ease of packaging, are becoming more and more important.

As in the past, the continuous increase in processing performance has expanded the market. Now, decreasing energy consumption and cheaper ways to package the chip in a variety of objects are dramatically opening up new markets. In 2005, it was declared that chips are targeting a 100 times reduction of energy per GFLOP (Billion of Floating Point Operations) by 2010, and as of 2008, they are right on target. A decrease in power consumption enables the packaging of more processing power in hand-held devices, like cell phones. The issue is not the resulting reduced drain on the battery, rather the reduced heath dissipation. A 500 watt cell phone will burn your hand long before it runs out of battery.

Secondly, and possibly with far reaching consequences, very low consuming devices may be powered using alternative sources of power, such as conversion of sugar circulating in the blood into energy for tiny medical devices. These will deliver drugs and monitor certain parameters in the body. Other examples are conversion of surface vibration into energy for sensors placed in the tarmac of roads to measure traffic and conversion of wireless radio waves into energy using evanescent waves to power sensors in a closed environment. As the cost of producing sensors decreases, the economics shifts to their operation with powering becoming a crucial factor. Because of this progress, in decreasing power consumption we can be confident that the next decade will see an explosion of sensors and along with that an explosion of data.

In 2005, only a tiny fraction of microprocessors (99.5%) ended up in something that can be called a "computer" and in 2008 it has been even more so. By far, most microprocessors ended up in devices like microwave ovens, remote control, cars, and electronic locks just to mention a few categories. In the next decade most objects will embed a microprocessor and most of them will have the capability to be connected in a network (to the Network). This will change dramatically the way we will perceive objects and the way we will be using them.

Part of this change will be enabled by the rising star of "printed electronics." This manufacturing process, based on a derivative from ink jet print tech, is very cheap, i.e., two to three orders of magnitude cheaper than silicon etching. Additionally, printed electronics is cheaper to design (again three orders of magnitude cheaper than etching silicon) and can embed both the processing/storage and the antenna for radio communication and if needed a touch-based interface avoiding the cost of packaging. In principle, it will be possible to write on goods as easily as we are sticking labels on them today. This evolution is in the direction of what can be called "micro-processing." We will continue to see evolution also in the opposite direction, that of "super crunchers." In this direction, we are seeing a continuous increment of processing speed achieved through massive parallel computing with hundreds of thousands of chips within a single machine exceeding the PFLOPS today and the EFLOPS in the next decade (billions of billions of floating point operations per second). In addition we are also going to see more diffused usage of the cloud computing paradigm both in the business environment and in the business-to-consumer environment. The consumer is unlikely to appreciate what is really going on behind the scene: the fact that some of the services he or she is using are actually the result of massive processing achieved through a cloud computing infrastructure.

Looking at a longer time frame, we can speculate that cell phones and wireless devices in general, may form a sort of cloud computing for resolving interference issues, thus effectively multiplying spectrum usage efficiency. The major hurdles on this path, that have already been demonstrated as technically feasible from an algorithmic point of view, is the energy required by the computation and communications among the devices that practically makes it impossible today and for the coming years.

9.5 SENSORS

Sensors are evolving rapidly; they are getting cheaper and more flexible. They embed the communications part, thus are ready to form local networks. Sensors open up a Pandora's box of services. Think about the thousands of applications that are coming up on the iTouch and iPhone, exploiting the accelerometer sensor. Drug companies are studying new ways to detect proteins and other substances. What in the past required long and expensive tests, executed by big and very very expensive machines, can now be done cheaper, quicker and easier by one or several sensors in combination. Actually, a new area of management is going to spring up from the existence of hundreds and thousands of sensors whose raw data can be analysed in a statistical way to derive information (see also discussion on sensors in Chapter 7).

Some of these sensors are being targeted for embedding in cell phones, like the one able to analyze breath as the person talks and assess, over time, the presence of markers for lung cancer. SD (Secure Digital)-like cards containing tens, and soon hundreds, of substances will be plugged into the cell phone enabling the detection of a variety of illnesses well before clinical signs appear. Now, this is not just an application, although an interesting and valuable one. It is a driver to miniaturize sensors, to make them more flexible and responsive to the environment and thus able to pick up telling signs. More than that, hundreds of sensors will be constantly producing data that will become a gold mine to derive meaning. Communications is the enabling factor since these data need to be seen as a whole, to derive meaning. We'll see this in a moment when considering statistical data analysis.

Other researchers are investigating e-textiles, special fibers that can be woven in a shirt, in pants, basically in any clothing to sense a variety of conditions and presence of special substances like sugar and proteins and thus provide the data to detect several pathologies.

Printed electronics will contribute to the slashing of cost to produce and deploy sensors in any object. Pick up something and that something knows it and gets ready to interact. Sensors are also providing what it takes to transform a collection of objects into an environment. Context awareness will make significant advances because of sensors presence everywhere. Intel has announced, at the end of 2008, a research program, Wireless Identification and Sensing Platform (WISP). They expect WISP will be available in the next decade and will be able to provide identification of any object, including our body—a sort of miniaturized Radio Frequency Identification (RFID) forming a continuous interconnected fabric. Present RFID technology, over time, will transform itself into active components with sensing capabilities, as price of sensors goes down. This probably won't happen before the end of the next decade. In the meantime, more and more objects will embed sensors and part of these will act as identification, thus avoiding, for many purposes, the need for an RFID.

The transformation of an object into an entity that can communicate and can become aware of its environment leads to a change in the business space of a manufacturer. In fact, this opens the possibility to remain in touch with the user of the product and, as a matter of fact, this is transforming the object into a service. In parallel, this enables new business models and requires a transformation of the producer's organization. Most producers will not be prepared for this change but it will be difficult to resist this evolution since the competition will be ready to exploit the marketing advantages provided by these new "context aware" objects. Some producers will decide to open up their product communications and on board flexibility to third parties to let them further increase the features and hence the perceived value of the product. This openness, in turn, will give rise to a variety of architectures, making network platforms and service platforms true service factory and delivery points.

Research work on sensors will create ripples in today's established dogmas, like the ubiquity of IP. Energy efficiency considerations are driving sensors networks to use non-IP communications and there will be many more sensor networks using ad hoc protocols than local and backbone networks using IP. Identity and authentication will need to cover objects and this might bring to the forefront new approaches to assess identity. The SIM card is very effective as identification and authentication goes but it has not satisfied the banking system and it may not be the future of identification. In fact, cell phones equipped with sensors detecting biometric parameters may provide even better authentication mechanisms and would make it possible to separate the terminal from the user (and that would appease the banking system).

The need for having a self-standing set of sensors within an environment and the need to cut energy consumption on each sensor, is pushing researchers to work out ever better autonomous systems theory and application. This is going to have a profound effect on network ownership and management architecture since autonomous systems destroy the principle that one needs to have central control to deliver end-to-end quality and hence the very foundations of today's telecom operators.

9.6 DISPLAYS

Display technology has brought us the wide flat screens everybody loves. It has also populated, with a screen, a growing number of devices, from digital cameras to cell phones. Digital frames have invaded our homes where probably we have some hundred electronic screens if we just care to count them around us. Some dreams have not come to pass, yet. Like the holographic screen that was supposed to take center place in our living room according to futurists in the 1960s. There are many basic technologies available that are bound to progress even further, particularly in the direction of lower cost to the end user. The bettering of production processes is the single most important factor in this progress. Lower cost makes it possible to have screens popping up everywhere and this is in synch with our perception of the world that is based on visual communications. The telephone has been a compromise, a very successful one indeed. So successful that it created a new communication paradigm. So strong that now most people prefer talking rather than video communicating (the latter is considered in general much more intrusive, it brings you very close to the other party).

There are, however, other directions of progress that are important because of the perception impact they have. The resolution of our eye is approximately equivalent to 8 Mpixel. Our brain composes the signals received from the eyes in a bigger window whose resolution is roughly equivalent to 12 Mpixel. Present HD television screens have a 2 Mpixel resolution (achieved using 6 Mdots, a triplet of red, green and blue makes up one resolution pixel). Hence, although we are in awe of the quality of images, our brain is not fooled. We are looking at a screen, not at reality. We are watching a show. We are not "at the show." The Japanese have the goal of arriving at a 32 Mpixel screen (and the required production chain) by the end of the next decade. A few 4K screens are available on the Japanese market, reaching the 8 Mpixel threshold. If we look straight at one of these screens we can't tell the difference with reality.

We already have, in consumer electronics, 8Mpixel resolution. Many digital cameras have that kind of resolution and even higher resolution is available. However, most of the time, we are never looking straight at something. Even without noticing, our eyes scan the environment and it is this scanning that allows the brain to create a larger image and to get the feeling of "being there." To replicate this sensation, we need to have our eyes scan only the screen real estate. That is, we need to be sufficiently close to the screen and the screen dimensions need to be such as to create an angle with our eyesight exceeding 160 degrees (when we are looking straight the angle captured by the eyes is slightly less than 130 degrees). The increasing dimension of entertainment screens and their increase in resolution will lead us into make believe situations in the next decade. The bandwidth required to transmit that amount of information exceeds 100 Mbps, therefore only optical fibre connection will do.

Although we will have LTE and LTE+ (the more performing successor), there will always be a better one on the horizon that will be able to handle those kinds of speeds. It does not make economic sense to chew up all available spectrum for these types of services. Within the home, the situation is different. The optical fiber may well terminate into a gateway that will beam information wirelessly at speeds up to 1 Gbps.

Smart materials will become more and more available to display images and clips. We already have special varnish that can change its colors to create images.

Electronic ink, on smaller scale, can display black and white text and by the end of this decade, it will be able to display color images. We can expect significant progress in this area that will lead, by the end of the next decade, to ubiquitous display capabilities on most kinds of objects. This is going to change the look and feel of products and, as pointed out in the case of sensors, it is bound to change the relation between producer and user. More than that, as indicated for sensors, these capabilities, coupled with open systems and open service creation platforms, will enable third parties to provide services on any object.

Displays are the ideal interface for human beings because we are visually orientated. The coupling of touch sensors or other kinds of "intention" detectors opens up the way to new services. The underlying assumption is that objects will be connected to the network, either directly or, more often, through a local ambient network. This connection in many instances will be based on radio waves, although a strong competitor might be the power lines within a given ambient. The fibered telecommunications infrastructure is likely to stop at the entrance of the ambient on the assumption that the less wires you have around your home the better.

It is interesting to note the work of Ishi [3] that ties together images and their manipulation and applies this to the management of networks and services. The "manipulation dimensions, particularly when associated to haptic interfaces providing the sense of touch and force feedback, helps significantly in the management simplifying the interaction.

So far, very little has been done by the management groups working in service management and network management to exploit these new interactions media. In the next decade these may take the upper hand and change significantly the way we look at management.

9.7 STATISTICAL DATA ANALYSES

The quantity and variety of data that are becoming available through networks is growing exponentially. In the next decade, cell phones will be equipped with sensors and this will create an avalanche of data. Five billion cell phones generating data on their position, temperature, movement (direction and speed), and special sensors will provide a huge base of information. Telecom operators will have to come to an agreement on what to do with these data and how to make them available in a neutral way that preserves privacy. These data can be used to monitor traffic in an urban area, to detect the emergence of some epidemics, to plan and monitor the effect of public transportation, to get precise maps on environment pollution, to understand social networks and the interest generated by an advertisement or event. They may be used in an emergency or for urban planning. They can be used by shop managers to dynamically change their shop window. Clearly, data derived from cell phones are only a fraction of the total data that will be harvested by sensors and that can potentially be made available for analyses. As we have seen, many objects will embed sensors and most of them will be connected to the network, thus the data they gather could be made available to third parties. Homes, department stores, schools, hospitals, parks, cars, etc., basically any ambient, can generate data and make them available. The possibilities are endless.

There is a growing understanding on how to analyze massive data banks. This will further develop in the next decade to include the analyses of distributed data banks leading to the generation of metadata. Public access to metadata will stimulate the development of many services and new industries will be created. The step from raw data to metadata is a crucial one. Metadata should be able to capture what is of interest in a set of data, masking, in a secure way, any detail that can be used to trace the owner/generator of any data used to create the metadata. The absolute guarantee of a decoupling between data and metadata is a fundamental pre-requisite to enable the publication of metadata.

It is likely that most of these metadata will remain invisible to the public and will be transformed into useful information through services. We are already seeing this trend in several applications available on the iPhone. Here, as an example, the user is asking for a weather forecast in a certain area. The related information is presented without the user having to know from where the information is actually derived. It might even be that the service (the touch of a button) is actually integrating several weather forecasts based on a success track. These data will be distributed in several databases and their harvesting and exploitation will generate traffic with varied characteristics. On the harvesting side, it will likely be in forms of billions of tiny transactions. On the usage side, it will probably be in the form of bulk data transfer to feed statistical data analysers and in the form of images, video clips to the end users.

Wireless and optical networks will be the supporting infrastructure. Cloud computing may offer the computational capabilities required for data analysis in several cases. Enterprises will sprout with business ideas to leverage potential data value. They will not require massive investment in computation structures if these will be made available on-demand through the network. Note how a significant portion of these enterprise-generated services may actually take the form of applets residing on users' terminals, like cell phones, that, once activated, may generate traffic and computation in the network.

Summarizing, the availability of massive quantities of data is changing the approach to data analysis from algorithmic processing to statistical data processing. The latter implies the access to distributed data banks and significant computation capabilities that will be better satisfied by cloud computing, i.e. decreasing the cost of computation. Communications fabric is the key enabler both at the level of raw data harvest and, at the level of processing into metadata, processing of metadata versus specific application and customer demand and distribution of results.

Networks and services management will use both a variety of data existing at the edge of the networks and in the terminal/environment and will make part of network data available to the environment and to third party services creators. This is a great cultural challenge for operators who tend to shield themselves, and their networks and services, behind the curtain of security and need to preserve data integrity. They are currently as far as can be from the wave of Web 2.0, even thought they claim to be moving in that direction.

9.8 AUTONOMIC SYSTEMS

The growth of independent network providers, particularly at the edges of the "old time network" gives rise to a new paradigm for management and exploitation. In addition, what were "dumb terminals" connecting at the edges of the network, are, more and more, sophisticated devices whose level of intelligence is comparable to, and sometimes exceed, that of the network. These devices no longer terminate communications and services arriving from the network but have their own local network of relationships with other devices, sometimes acting as intelligent gateways that use the network as a tunnel for pure connectivity.

Intelligence has moved to the edges of the network and with that comes the need for a different paradigm to understand, create, and operate services, the paradigm of autonomic systems. The basic assumption of telecom engineers has always been to plan, design, engineer, control, control and control. All measurements going on in the network are to provide information for control. The more complex the network, the more sophisticated the control. In the 1980s, with the dissemination of computers, the dream of having a fully centralized control became possible and telecom operators built their own control centers. As more equipment and technology found their way into the network, engineers and researchers developed standards to ensure the centralized control of diversity. Those in the field remember CMIP and SNMP, two different paradigms for controlling network elements and both based on the fact that a single comprehensive view of the network was required to ensure fault control, maintenance and fair usage of resources.

Advent of the Internet, with the creation of a network that was the result of thousands of interconnected networks having different owners, was looked at with suspicion by most telecom engineers. The single fact that the quality of the Internet was based on best effort disqualified Internet in their eyes. As traffic kept growing, we heard voices of imminent collapse. The Internet is not centrally managed, it will collapse. As a matter of fact, although local parts of the Internet have, and are experiencing, outages and disservice, traffic keeps growing and no collapse has ever occurred. On the contrary, we have seen in a number of occasions, like September 11, 2001, that catastrophic events creating havoc in the telecommunications infrastructure left the Internet unscathed. On that particular day, the only way to communicate with people in New York was through messaging via Internet.

The Internet is not a different network in the sense that it uses different wires. The wires are in common with the telecommunications network and one of the reasons why Internet works so well is because it is using one of the most reliable infrastructures on the planet, telecommunications infrastructure. The reason the Internet is so resistant to local faults is its basic absence of a hierarchy of control. Control is local and distributed. Something may go wrong at point A but it is not going to affect communications transiting through A since there are so many equal alternatives in the network bypassing A. The Internet is not, yet, an autonomic system in a full sense. Within its boundaries various components act, with regard to the routing of information, as autonomic systems.

A more apt example of an autonomic system is Roomba, the vacuum cleaner available on the market since the beginning of this century. It is a robot that has its own goal (this goal can be finely tuned by the owner), that is, vacuuming the spaces around it as far as it can reach. This space is cluttered with obstacles, table legs, furniture and people (it is claimed in the ads that Roomba can avoid stepping on the cat although it looks more likely that the cat will take appropriate action to step away). Over time, it learns how the space is structured and works out the best vacuuming strategy. If the space changes (new furniture, reconfiguration of the present furniture) it will change its strategy accordingly. The home environment is an ideal lab for experimenting with autonomic systems. Seen as a whole, it can become an autonomic system. All devices dialogue with each other and benefit from one another. As new devices are inserted in the home, they make themselves known in the environment and become part of it. As the external environment changes (new connections may become available, active ones may be malfunctioning) the home reacts and reconfigures itself.

The evolution of wireless in the direction of more and more wireless local areas will exploit the technology of autonomic systems. Devices, by far, will have a wireless cloud around them. At any moment, a device will check any overlapping with other clouds and will try to establish connection at a semantic level, that is, it identifies itself and exposes its characteristics, objectives and needs and will expect to receive similar information from other devices sharing the cloud. This exchange of information will be followed, where appropriate, by handshaking, to bind the devices in forming a single system. The set of all devices will indeed behave as a single system. Because of the continuous sensing of the environment, the system will evolve in terms of participants to the system and in terms of behaviour.

9.9 NEW NETWORKING PARADIGMS

The advent of autonomic systems, the multiplication of networks, the presence of huge storage capacity at the edges of the old network (more specifically in the terminals, cell phones, media centre) and the growing intelligence outside the network, will change significantly the networking paradigms. The efforts in the past 30 years have focused on exploiting the progressive penetration of computers in the network to make the network more intelligent. A simple economic drive motivated this evolution, i.e., the network is a central resource whose cost can be split among the users. It makes more sense to invest in the network to provide better services at low cost to low cost low intelligent edges. The Intelligent Network finds economic justification in that fact. The first dramatic shift happened with cell phones, with the mobile network. If you were to develop a network from scratch and you decide to use a fixed line network to provide services, you would have to pay almost 100% of the investment. On the other hand, if you were to deliver the same services using the mobile network approach, the overall cost would be split 30% in the network and 70% in the terminals (and this latter part is likely to be sustained by customers). This reflects the shift of processing, storage, and intelligence from the network to the edges, to the terminals using it.

A possible, and likely vision, for the network of the future is a bunch of very high-capacity pipes, several Tbps each, having a meshed structure to ensure high reliability and to decrease the need for maintenance, in particular, for responsive maintenance (the one that is most costly and affects most of the service quality). This network terminates with local wireless drops. These drops will present a geographical hierarchy in the sense that we will see very small radio coverage through local wireless networks, dynamically creating wireless coverage through devices, very small cells (femto and picocells), cells in the order of tens of metres (WiFi), larger cells belonging to a planned coverage like LTE, 3G and the remnants of GSM or the likes, even larger cells covering rural areas (such as WiMax when used to fill the Digital Divide) and even larger coverage provided by satellites. In this vision, the crucial aspect is ensuring seamless connectivity and services across a variety of ownership domains (each drop in principle may be owned by a different party) and vertical roaming in addition to horizontal roaming (across different hierarchy layers rather than along cells in the same layer). Authentication and identity management are crucial. This kind of evolution requires more and more transparency of the network to services.

The overall communications environment will consist of millions of data and service hubs connected by very effective (fast and cheap) links. How could it be "millions" of data and services hubs? Trends are toward shrinking the number of data centres. The technology (storage and processing) makes it theoretically possible to have just one data centre for the whole world. Reliability requires that it be replicated several times in different locations, but still we can be talking of several units!

The fact is that the future will see the emergence of data pulverization in terms of storage. Basically every cell phone can be seen as a data hub, any media center in any home becomes a data hub. When all these data hubs are added together, millions of data hubs is actually a very low figure. How can one dare to place, on the same level, a TB of storage in a cell phone, a 10TB in a media centre and several EB in network (service) data centers? The fact is that from an economic point of view, if we do the multiplication, the total storage capacity present in terminals far exceeds the one present in the network-service data centre (TB * Gterminals = 1000EB). The economics of value is also on the side of the terminals. The data we have in our cell phone will be worth much more (to us) than the ones in any other place. People will consider local data as "The Data" and the ones in the network as very important back-up. Synchronization of data will take care of reliability but at the same time asynchronous (push) synchronization from the network and service DBs to the terminals will make, perceptually invisible, those centralized DBs.

The same is happening for services. Services are produced everywhere, making use of other services, of data, of connectivity, and are perceived "locally" by users. They are bought or may be gotten for free, possibly because there is some indirect business model in place to generate revenues for the service creator and to cover its operational cost. Services may be "discovered" on the open Web or may be found in specific aggregator places. The aggregator usually puts some sort of mark up on the service but at the same time provides some sort of assurance to the end user (see Apple Store). We'll come back to this in a moment.

Once we have a network that conceptually consists of interconnected data-service hubs, one of which is in our hand, possibly another in our home, what are the communications paradigms used?

Point-to-point communications, i.e., calling a specific number, is going to be replaced by a person-to-person or person-to-service (embedding data) communications. This represents quite a departure from today since we are no longer calling a specific termination (identified by a telephone number). Rather, we are connected to a particular value point (a person, a service). Conceptually we are always connected to that value point; we just decide to do something on that existing connection. The fact that such a decision may involve setting up a path through the network(s) is irrelevant to the user, particularly so if these actions involve no cost to the user. The concept of number disappears and with it a strong asset of today's operators.

The value of contextualized personal information finds its mirror in the "sticker" communication paradigm. A single person or a machine, asks, implicitly or explicitly, to be always connected with certain information. Most of this may reside on the terminal, but a certain part can relate to the particular place the terminal is operating or to new information being generated somewhere else. Communication operates in the background, ensuring that relevant information is at one's fingertips when needed. It is more than just pushing information; it requires continuous synchronization of user profile, presence/location, and ongoing activities. This embeds concepts like mash ups of services and information, metadata, and meta-service generation. It requires value tracking and sharing. It might require shadowing (tracking data generated through that or other terminals with which that person/machine comes to interact with).

The variety of devices available for communications in any given environment, some belonging to a specific user, some shared by several users (e.g., a television) and some that might be "borrowed" for a time by someone who is not the usual owner, can be clustered to provide ambient-to-ambient communications that may be mirrored by the "cluster" paradigm. Autonomic systems will surely help in making this sort of communication possible and usual. The personal interaction point, a person will be using, will morph into a multi-window system where one could choose the specific window(s) to use for a certain communication. Similarly, at the other end, the other user will have the possibility of choosing the way to experience that particular communications. In between, there may be one or more communications links and some of these may not even be connecting the two parties, since communications may involve information that is actually available somewhere else and that is taken into play by the overall system.

This kind of communications will be, at the same time, more spontaneous (simple) to the parties involved and more complex to be executed by the communications manager. The communications manager can, in principle, reside anywhere. Surely Network operators may be the ones to propose this communications service.

Contextualized communications is going to be the norm in the future. It is a significant departure from the communications model we are all used to. We'll explore this further in the final part of this chapter, as seen from the user.

9.10 BUSINESS ECOSYSTEMS

From the previous discussion we see that the future will consist of many players loosely interconnected in the creation and exploitation of innovation. Because of this, several economists, as well as technologists, have started to wonder if the usual representation of relationships among players in a certain area can still be modeled on the bases of value chains. There is a growing consensus that value chains modeling shall be complemented by a broader view considering business ecosystems. What characterizes a value chain is the set of contractual obligations existing among the various actors. Competition may bring one actor to discontinue the relation with another actor connecting to a new one offering, at a better quality/price, the same raw product/service. Value chains tend to become more and more efficient since the competitive value of each player is to be the most efficient one in that particular place of the value chain. Innovation is pursued to increase efficiency, and, over time, in a competitive market, the value produced by efficiency at any point in the value chain tends to move to the end of the value chain so that the end customer is the one that really benefits from it. Those sustaining the cost of innovation might see increased margins for a while but the long lasting benefit is that of remaining part of the value chain because they remain competitive. Clearly, patents may lock in innovation and preserve its value to that player. This is particularly true for manufacturers, less for those offering services. Services are easier to copy, circumventing any patent.

An ecosystem is characterized by the loose relationships existing among actors. Sometimes actors do not even know each other. In a way, an ecosystem is a set of autonomic systems in the sense that each player plays his/her/its own game, trying to understand how the whole ecosystem (or the part that matters) evolves and it reshapes its behavior and interaction according to that. Innovation happens anywhere in an ecosystem and benefits basically anyone since it increases the perception of value of the ecosystem for the players. Who generates the innovation basically can keep its value since there is often a direct link to the end user, thus to the perception of that value. On the other hand, innovation is much more tumultuous and having less inertia, interest shifts more rapidly from one player to the other. Hence, there is much more pressure to innovate and this innovation is by far directly pointing to the end users rather than at something internal. The crucial point for innovators remains unchanged: how can they take the innovation to the market? The usual tactic is to piggy back on existing connections to the end market (thus exploiting advertisement, distribution, and billing). These connections are also known as "control points" because of their power in bringing innovation to the end market and controlling the flow of value to the end customers. iTunes is such a control point. Set-top boxes may be another example of control point (although today they are part of a value chain and not of an ecosystem, something that is likely to change in the future).

If we look at car manufacturing companies we see the presence of very strong value chains whose effectiveness has been tuned over the years to incredible points (just in time, no more warehousing, co-design, etc.). Around these value chains we have seen the birth of an ecosystem for add-on parts (radio, stereo, seat cover, stickers, snow ice chains). The companies producing these add-ons simply piggy back on existing car models to offer their products. There is no contractual obligation with the car manufacturers. Hence, as shown by this example, ecosystems already existed in the industrial society. What is new is the increased flexibility and openness provided by objects embedding microprocessors and software. It becomes possible to add features inside the object, not just on the outside.

The computer industry is another example, even closer to the emergence of ecosystems. Here again we have a strong value chain: from the computer components to the manufacturers, the operating systems and the world of applications. The latter may be seen as produced by independent players who take advantage of the market created by the computer industry. Some of these applications are part of the value chain (such as Microsoft Office or the similar suite produced by Apple for its computers) others are the result of investment by independent players. They choose to invest in one platform (Windows, OS X, Linux, Symbian) depending on their evaluation of the potential market. On some of these applications, more applications (sometimes called plug-ins) are developed by other players, again benefiting from the market generated by that specific application.

All applications, plug-ins, and other applications that provide ways to refine results produced by others increase the value of the ecosystem. Sometimes, this value is so high that customers are locked-in from moving to a different ecosystem. For example, moving from the ecosystem based on Windows to the one based on OS X would mean losing a number of valuable applications and this is not acceptable. Notice that this characteristic of lock-in can be found in bio-ecosystems as well. It is not the only similarity. Actually, there are many similarities between business ecosystems and bio-ecosystems resulting from a set of ground rules that apply to all complex systems where the various components are interacting on the basis of rules that are common to the ecosystem and not specific to the two parties interacting. What's more, the interaction takes different paths and leads to different results on the basis of the local status of the ecosystem (local means that part of the ecosystem that is perceived by the actors involved in the interaction at that particular time).

The future will bring many more business ecosystems to the fore as a result of the openness of objects and their flexible behavior and interaction based on microprocessors, sensors, actuators, and software (they are autonomic systems). This is not enough. One single Roomba in a living room does not create an ecosystem. What it takes is to reach a certain threshold in terms of quantity of actors. Billions of open cell phones, interacting directly or indirectly with one another, will create a huge ecosystem whose value in business terms will be enormous, well beyond the value of the sum of each cell phone.

Another way of looking at the increased value produced by independent players within an ecosystem is through the concept of mash-ups. This is basically a seed that can aggregate an ecosystem (the typical example is Google maps). Most mash-ups today are about the aggregation of information. In the future, mash-ups will comprise services as well as information. They do not cover the whole space of ecosystems, since they do not represent something like an iPod ecosystems where we can see actors producing external loudspeakers, pouches, decoration. Note however, that as objects become autonomous systems it will be progressively difficult to distinguish, from a business point of view, the atoms from the bits.

This latter observation brings us to exploration of the future of the Internet, not in terms of its physical and architectural underpinning, but in terms of value growth. Today, the Internet, the Web, consists of an endless world of information.

In these later years it has become a source of services to the point that many established actors in the services business are starting to become concerned to the point of reconsidering their business strategy. The silver bullet of a killer application sought by those services providers, like mobile television on cell phones, looks more and more unlikely to happen. Telecom operators that have prospered on the connectivity service and have progressively added other services (value-added services) in their closed and walled garden are now seeing a growing number of small players creating services. The sheer number of them is sufficient to guarantee a level of innovation that exceeds any resulting from massive efforts by Telecom operators. They are not bound to "principles." They just offer their wares at low cost leaving the broad audience of Internet surfers to try, argue, and even better their offer. The release of a service in a "Beta version" is beyond the culture of a Telecom operator. Writing, along with the service, the sentence that "no responsibility is taken on the proper functioning of the service" is unheard of in the world of Telecom operators. The Internet has brought with it the concept of "best effort" in connectivity terms, and is now creating a culture of "best effort" in the service area.

The mass market responds well to this offering. It is so wide that one can find an interested party for basically any proposition it receives. Only a few niches may be interested in acquiring and using a service, but these niches span the planet and they are sufficient to generate a return on the (usually small) investment of those who created the services.

Although connectivity is key in enabling service access and fruition, the value from the user point of view shifts from connectivity to service. Note that the trend towards the embedding of information into services is a further drive in this direction. One thing is to type on a browser the URL to access a site to get a weather forecast (the very fact that we have to "dial" an address brings communication value to the fore). It is quite a different thing to click on an icon with the symbol of the sun on the iPhone (or the iTouch) and get the forecast for those places you care about. In this latter scenario connectivity has disappeared. As a matter of fact, you may not know whether this information is the result of the click activating a connection through the network or is the result of this information being pushed to your device as you were recharging it. You may not know what kind of connectivity was used to bring the information in the terminal (one of the thousands of WiFi networks that the device can automatically and seamlessly connect to or a cellular network?) and you don't care. Communication is no longer on your radar screen.

The attempt of operators to charge for connectivity on a bit-by-bit basis is losing ground and competition will result in flat rate on mobile data as it has become the custom for fixed-line access.

Once this comes to pass, and it will, operators may not expect to see an increase of revenue from connectivity, nor from non-existing killer applications. In most Western markets, penetration has reached a point where further growth of the market is unlikely, while the progressive squeezing of margins is almost certain.

The challenge for Telecom operators, that may want to position themselves at the enabling layer of a business ecosystem, is how to transform their management infrastructure into an ecosystem management fabric, moving from a centralized, hierarchical management to a distributed multi-domain management.

9.11 INTERNET IN 2020

How is the Internet going to look in 2020 and how can we get there? There is a general consensus that the Internet will be the backbone for communications, of people, objects, and machines. This backbone uses, as physical infrastructure, a variety of communications links, both wireline and wireless. This variety will probably be hidden to most users in most places. When that becomes the case, no limits to connectivity will be perceived by the user. If that represents a general consensus, more debate exists on the way people will communicate. My personal view is that communications will mimic closely natural communications; hence it will be rooted in the specific environment where it is taking place. As in everyday communications, most of it will be a mixture of visual and aural exchange of information. Terminals will serve the purpose of annihilating distance, nothing more. Communications may take advantage of high-definition screens and projectors, where feasible 3D visualization will replace 2D and ambient hifi sound will replace mono low-quality sound.

Screens will be on many objects. We will have a number of them on our person (cell phone, watch, media player, social walk-window, glasses). EyeRes will be common. With that I mean a resolution that compares to the one of our eyes and therefore can trick us into believing it is the real thing.

In some cases, additional information exchange will be required, such as swapping of text. As happens in everyday life, when this is required, we use an additional channel. We take a piece of paper and jot notes. Similarly, we will be using the most appropriate means to swap information. It might be a nearby television screen that will be able to scan a piece of paper with the associated camera and display it on a similar screen in our correspondent living room. The use of different channels will look completely seamless to the user, as it is today picking up a piece of paper to write down a diagram, and, exactly as it happens today when we have to ask for a piece of paper or a pencil. We might have to ask for an additional communication means in the ambient.

However, a word of caution is required. We will still have the communications of today, limited to aural interaction, a cell phone between our ear and our shoulder as we look at something else and our hands are doing other things … so one thing is to say that in the 2020s we will have the possibility of a more natural communications another is to say that such a communications will have displaced what we have today. If there is going to be both, does it mean that people will be paying a premium for the extra quality delivered? My opinion is "NO." They may pay more in setting up an environment that can provide this kind of natural communications, like better screens and loudspeakers, but my bet is they won't be willing to pay anything more for the enhanced connectivity that is required. That, unfortunately, will be taken for granted.

The Internet will provide enterprises with tools to produce their goods, more effectively, often making use of components produced by other parties. In the future, we might see an evolution of production systems to include clients and users. This may take a while and may come to pass only in certain areas. Services, typically, lend themselves well to this kind of production.

Social networks are already thriving on today's Internet and will be even more important in the future. They will continue to be a marketplace and their weight on the society will tend to grow to the point that government will have to take them into serious account. It is unlikely to see social network replacing government as a new form of democracy, but they will surely affect government agendas. It might be likely though, that some political party will be created through a social network and will be able to gain election completely by campaigning on the Web.

The Internet, by 2020, has become the world infrastructure for production of many types of goods. Most enterprises, in fact, will use the Internet, not just for support to the production (orders, purchasing, inventory, relations with partners) but for creating and assembling the various product parts. Goods will most likely be embedding software, will be customized at the point of sale, and will operate part of their functions through the Internet. Additionally, we are likely to see part of the functionalities being created on the Internet by communities of users and these functionalities will be embedded in subsequent releases of the product. Enterprises will have the opportunity of monitoring the usage of their products by the clients and will be able to tweak functionalities and evolve the product based on actual usage. This evolution will most likely apply to new released products as well as to products already sold that will be updated via the Internet, in some cases, and in others will just find themselves operating differently since operation occurs by accessing various components on the Internet that are dynamically updated over time.

Innovation is the key competitive differentiator for enterprises. The price (cost) and operating cost will remain important but the overall efficiency of production and distribution chains will tend to equalize these costs across the competing enterprises. Furthermore, labor cost will tend to decrease in percentage (due to progressive automation of the production and distribution of products), will tend to be similar in different regions of the world (because of the generalized growth of wealth), and outsourcing will be even easier and more effective, letting enterprises access labor forces where it is most convenient. Technology is going to become basically accessible to all enterprises in any part of the world on an equal basis, further reducing any competitive advantage due to location. Hence, the real differentiator rests on the capacity to innovate continuously. Design, interactive design, making products fit the user ecosystem and possibly becoming seeds to attract more functionalities, will be the crucial factor for success. Those enterprises, and geographical areas that will be able to innovate, attract innovation with their product and maintain control points, will be the most successful and stable over time.

This stability is crucial for long-term success and well being of an area and can be sustained by appropriate regulation aimed at attracting and making possible, rather than defending and constraining. According to a World Bank estimate, we will have, by 2040, 438 million new people entering the work market. Ninety percent of this workforce will come from what today are developing countries. Bright innovators are, basically, evenly distributed all over the world. Education is crucial, but we are seeing a bettering of education facilities everywhere and we can expect a great decrease of differential in education skills in different parts of the world. It will hence be a crucial success factor to be able to attract these innovators. Because of the generalized bettering of welfare all over the world, individuals will have less motivation to move away from their home base. Enterprise will have to de-localize design and "soft" production to exploit innovators where ever they are.

The Internet in 2020 will be the enabling fabric for social networks, allowing communications among members and the delivery of information and services that in turn characterize that social network. It is difficult to imagine a social aggregation happening outside of the Internet at that time. Even families will use the Internet as glue for their relationships. Enterprises will exploit social networks for knowledge sharing, for knowledge management, for production, for relation with customers, and for finding marketplaces and creating them.

The Internet, once made available to anybody without restriction, is intrinsically neutral. It has no boundary and bias. Information and services are a completely different story. A political entity, national or over-national, like the European Union, will need, on the one hand, to create the right environment for stimulating the creation of services and the understanding of information. On the other hand, it will have to ensure education to all citizens to appropriate usage of the Internet, ensure protection from frauds (that are inevitably going to grow as the Internet becomes more and more used), and guarantee privacy and set limits within the culture and ethics that is acceptable to that cultural area.

As social networks grow in members and in numbers, there will be fragmentation of cultures and markets. Someone refers to this fragmentation in negative terms, comparing it to Balkanization as dispersion due to centrifugal forces no longer kept under control and leading to all sorts of problems. On the Internet, however, fragmentation may go along with aggregation since they are not mutually exclusive. Each political/cultural area needs to pay attention to these phenomena to make sure that one is not prevailing over the other (aggregation may lead to the destruction of cultural diversity and this in general is not a good thing).

9.12 COMMUNICATIONS IN 2020 (OR QUITE SOONER)

We have seen a few technological areas, forecast their evolution, and we have seen the evolution of the business framework, with the emergence of the ecosystem paradigm. However, the interesting question is: how will all this affect our way of communicating? The terminal used in most situations will be a wireless one, a cell phone of a sort, having a nice three, or more, inch screen with at least VGA resolution occupying most of the surface. Much more resolution will be offered in top-of-the-line terminals, likely using OLED technology (this provides a very bright screen visible also under sunlight with pixels of very small dimension so that high resolution can be packed onto a small surface). Some models may sport a second screen on the back, black and white, based on electronic ink technology (reflective, so that the more ambient light is available, the better the quality of vision, like paper). Actually, this terminal could be better described as a lens to kill distance and discover what may be hidden at first sight. The memory goes to the 17th century, when it became fashionable to have a magnifying lens as part of one's own dress code.

The terminal will be equipped with sensors, some of these to identify the user handling it. When we pick it up the screen will show the context we are in, usually the location that could be represented as the background image. At the center will be us, maybe our picture. If I look at it I'll see my face, if I hand it out to you, you'll see your face. This simple action, handing out the device from me to you, requires access to the identity and to the associated data. This association can happen via the network or it could be obtained by some local personal identification system (a chip embedded in the body as in the WISP initiative from Intel, or a local dingle that can only be active when in close proximity with the person it refers to).

A cell phone becomes my cell phone when I pick it up and I am authorized by the cell phone owner, otherwise it will just remain inert and useless. On the screen I will see a number of icons, some defined by me over the years and brought forward from one cell phone to the next (changing a cell phone model does not require any work on my part to transfer data, preferences and authorization, it just happens seamlessly on any cell phone I am the owner of as well as those I am authorized to use); some other icons are defined by others, as I will soon describe.

These icons are the new thing of communications in the next decade. A few examples are already visible in some of today's phones, although they will look very primitive just a few years from now. One thing that is missing is the loudspeaker. The concept is to have visual communications and to have that you must be able to look at the screen, something that would be impossible were you placing the phone over your ear. Hence, the sound will reach your ear via a "classic" Bluetooth ear plug or via more sophisticated sound beaming devices (today we have some of these but they require several loudspeakers to focus the sound in a small areas; progress in processing capacity and smart materials will make it possible to miniaturize these loudspeakers in the phone shell).

Another thing that is missing is the keypad with numbers when we want to call someone or something. The point is that all communication is contextualised (with very very few exceptions). It is like entering into a room. You see a person, or an object, and you just get close to him or it and start interacting. Now, one might wonder that placing an icon to identify whom you are calling might require an impossibly high number of icons and recognizing one might turn out to be more challenging than reading Chinese characters. Besides, what if I want to call something that I do not know, that I never called before?

The present dialing using numbers requires typing 9 to 13 digits. This lets us reach some billion terminations. On a 3.5-inch screen, like an iPhone, one can place some 20 icons. And have the same number of choices with just 7 clicks. So in terms of pure clicks we can substitute numbers with icons. But this does not solve the issue of identifying the right icon. The problem can be solved by a tree structure where every click is basically restricting the context. Suppose I want to call my friend John, who is also my partner at tennis. I can click on the "community" icon, then on "tennis" and there I'll see his face in an icon. I might also start by clicking "home" and then "agenda," "Wednesday last week" and again I'll see his icon face since we had dinner together last Wednesday at my home. You get the gist: there are many ways our brains remember things, and all of them work through association. The best interface we can think of is one based on association (and that is not the one we have been using so far by dialing numbers). This issue of identifying something I want to call is very interesting because it is not just a matter of interface but it has some deeper implication on data structuring, architecture, and it might involve several players in an ecosystem approach.

Suppose you want to call that restaurant you were taken to by a business associate last month, and, as usual, you don't remember the name of the restaurant. Well, just click on your face and click on history. Drag the arrow of time to last month and you'll start seeing a collection of memories that have been stored in your cell phone, among these that particular night at the restaurant. You did not write down the name of the restaurant because you were taken there by your business associate but your cell phone has the location tracked and stored. Now you can just click on the restaurant to view it. The location information will be sent to some directory provider (that is actually much more than a directory provider, it can be the Yellow Pages of the future; by the way, the present Yellow Pages have to start rethinking their business before they discover it is no longer there) who will push onto your screen a new context containing the restaurant icon that would allow you to get inside the restaurant, browse the menu, make a reservation, call the restaurant, create a date with several friends on that place, the transportation icon so that you know how to get there, references to comments left by other people on the place and so on. Now, it is easy to see that an interface based on context is really the interface to our thinking and its implementation requires the involvement of several data owned by different parties and located in different places, reshaped into information by some service provider that knows how "I" think, who has my profile. Part of this transformation may take place inside the terminal. Remember, it is storing 1 TB of data and it is probably the most knowledgeable entity about me. When placing questions to the outside world it may decide to mask my identity for privacy reasons.

So, personalized and contextual communications are two of the things characterizing communications in the next decade. Let's look at some more icons on our screen. One may be the shopping cart. Click on it and you'll see a number of icons replacing it. One is the "to do" list. This to do list is sharing information with your family, if that is the context you are using it, with your business associates if that's the context. Suppose it is the family one: when your wife gets the bread the "bread" item disappears from the list, if you add "light bulb" it gets on their to do list too. There is no explicit action to do that. As she is paying for the bread with her phone at the supermarket, the item disappears from anyone connected to the list. This list might also be shared with your car so that as you drive the car shares it (neutralizing your identity) with the stores on the way and will prompt you to the availability of the item in a particular store nearby. You might also elect to let the sharing to occur on a broader environment and accept ads on some of the items. That way you'll be able to see a variety of offers. Surely it does make little sense for bread but if you are interested in a digital camera or renting a flat that would be of interest.

The control of information sharing is going to be crucial both for its acceptance and for monetizing the ad value. Some third party should be involved, maybe more than one. There are so many potential actors involved and the margins are so razor thin that most of the processing should be made automatically and it should be up to the communications fabric to intermediate among the players.

Another icon shows the previous shopping sprees (also the ones of those who has decided to share it with you, such as family members). How is this information captured? Clearly any time we pay with a credit card the information is potentially traceable. However, getting it may be close to impossible, given the various databases involved, the variety of security systems, and ownership domains. Life (at least in this respect) would be simpler if one were using his/her cell phone for any transaction. That does not imply that the transaction will not be charged on a credit card of same sort or even paid cash, only that the cell phone keeps track of what is going on. It may be expected that in the future some Operators will end up developing this kind of tracer and people will be willing to use it to keep track of their life.

Another icon will show the shops in the area. This icon, once clicked, will result in several icons, one for each of the shops in the area that have pushed themselves onto your screen. Can they do that? Well, of course some technology is required, a platform to manage the information, some applications to format it in the appropriate way according to the terminal visualizing it. It requires, in addition, some agreement from the owner/holder of the cell phone to allow this kind of icons to sneak in on his screen. Surely, any click on one of those icons should bear no cost to the clicker. This possibility of entering in the menu of cell phones is a very interesting one for retailers, who will have a way to get in touch with potential customers roaming in their area.

Let's take another icon, placed on level one, the one representing my home. By clicking on it I can activate a connection to the home (call home) but probably it is more interesting to stretch the icon to fill the whole screen. In this way the home becomes my context and I will have available a variety of icons (services) through which I can interact with appliances and people who at that time are part of the home environment.

I'd like to close this part, and the chapter, with one more icon: it can have the shape of an eye since its objective is to transform the cell phone into a lens to look at information and services layered on an object. It is something that is already a reality in Japan's supermarkets or in France on billboards. A tag on the object can be read through the cell phone camera retrieving the unique identity of the object. This identity can be used to retrieve services and information associated to that object by a number of parties, the object producer or anyone else. This opens a Pandora's box to create new services since every object becomes, potentially, a point to distribute and access information and services. This is what we called before the Internet with Things.

The cell phone is likely to be the chief intermediary in our communications activities and the point of aggregation of personal information that will also be used to customize services. However, we are going to have more opportunities to communicate than the ones offered by the cell phone. The ambient we are in is a communication gateway to the world, be it our home, a hotel room, our office. Walls will display information, sensors will be present to customize the environment to our taste, and cameras will be available to bring our image to far distant places.

Any object can be overlaid by our information, in the same way as today we may want to stick a piece of paper on a box or underline a few lines of text with a crayon. Objects sent through FedEx or by bits, like digital photos, will have associate information that we have created for the receiver to listen to. Communications, in other ways, will make great use of wires (and wireless) but thanks to technology it will be ubiquitous, overcoming in some instances the needs for a classical communications infrastructure.

However it will turn out to be, of one thing I am certain: communications will be the invisible fabric connecting us and the world whenever and wherever we happen to be in a completely seamless way, connecting us so transparently, cheaply, and effortlessly that very seldom we will think about it.

These few characterisations used to describe communications in the future—invisible, transparent, seamless, cheap, effortless do require a lot of management. This management is unlikely to correspond to a single player; rather it will be the result of several management domains interacting with one another. Where the burden of management has been in the past the sole responsibility of network owners, in the future it will be shared with terminals. Actually, it will be up to terminals, enabled to access a variety of networks, to work out the best "local" strategy to get the connectivity desired. Each network management will shift its focus from being attentive to the availability and quality to make sure that such availability and quality is appreciated by the terminal that in the end will make the decision to use that particular network.

Management, thus, will become a service and at the same time a crucial competitive factor for Network operators and service providers.

9.13 REFERENCES

1. Lloyd S. 2000. Ultimate physical limits to computation. *Nature* 406(8):1047–1054.
2. Freiberger P, Swaine M. 2000. *Fire in the Valley: The Making of The Personal Computer*, Second Edition. New York: McGraw Hill.
3. Tangible Media Group. http://tangible.media.mit.edu/ (Accessed October 1, 2009).

INDEX

AAA application server, 149
Abstract syntax, 190
Access characteristics, new, 16
Access functions, 109
Access lines, 21–22, 28–29
Access Management Function (AMF), 138
Access networks, 29–30
Access-Resource Admission Control Function (A-RACF), 137
Access transport functions, 109
Accounting management (AM), 4, 111, 215, 242
  FCAPS, 140
  network-related, 142
  service-related, 143
Adaptive Dynamic Backbone (ADB), 221
Admission control
  NGN, 136
  RACS, 136
Advanced encryption standard (AES), 79
Advanced Television Systems Committee (ATSC), 60
Advertising revenues, 98
After sale phase, in the customer experience, 48
agentsInformation group, 224, 225
Aggregation/backhaul networks, 22
Aggregation layer, 37
Aggregation networks, 29
Aggregation technology, 30
American National Standards Institute (ANSI), 53
American National Standards process, 235
Analogue information, coding and transmission of, 58–62
Anmp: Ad-hoc Network Management Protocol, 213
Anmp MIB, 223–224. *See also* Management Information Base (MIB)
Anmp protocol, 220, 223
Appeals process, for the standards development process, 234
Application information table (AIT), 68
Application plane requirements, to support NG services, 120
Application Programming Interfaces (APIs), 3–4, 21, 117–118, 149
  standardized, 178
Application servers (ASs), 33, 130, 145
  examples of, 146–149
Applications functions, in Next Generation networks, 112
Apps storefront, consumer- and business-oriented, 117–118
Architectures
  functional and physical, 213–214
  information, 216–228
  logical, 214–216
  standardized, 35
  technology-neutral, 189
ATSC-PSIP program decoding, 62–63
ATSC-PSIP standard, 61–62
Authentication, in NGN, 138
Autonomic systems, 267–268

B2B realization requirements, 143–144. *See also* Business to business (B2B)
Backbone networks, 22, 31–32
Back-end, 37–39
Backhaul networks, 29
Balance of interest, in the standards development process, 234
Bandwidth, increased, 56–57
Bandwidth value, 256

Next Generation Telecommunications Networks, Services, and Management, Edited by Thomas Plevyak and Veli Sahin

Base-line privacy interface plus (BPI+), 78
Bellcore, 232
Bio-ecosystems, 272
Bit transfer rate, 259
Body-embedded communications/ computing, 116
Bound OCAP applications, 67
Bouquet Association Table (BAT), 61
Broadband Bluetooth, 104
"Broadband everywhere" strategy, 17
Broadband Forum, 123–124
Broadband multi-service, 16
Broadband PON (BPON), 103
Broadband remote access server (BRAS) functions, 22
Broadband technologies, fixed and mobile, 21
Broadband triple play, customer experience in, 47–51
Broadband, Wireless and Plain Old Telecommunications Services (BWPOTS), 253
Broadcast television, QoS for, 24–25
Business communications service, software-based, 114–115
Business Customer Premises Network (B-CPN), 10–11
Business ecosystems, 270–273
Businesses, next generation of, 126–127
Business interaction, requirements for, 196
Business management systems (BMSs), 2
Business models, new, 33–34
Business opportunities, finding, 18
Business-oriented apps storefront, 117–118
Business process framework, 159–163
Business service contracts, 195–200
Business service framework, 181
Business services, 167–170, 195–200
Business Support System (BSS) standard, 54
Business to business (B2B), 169. *See also* B2B realization requirements
Business view, 163–164, 196

CableCard, 72–73
Cable futures, 97–98
Cable industry, 53
Cable IP telephony, 84–96
CableLabs, 72, 73, 90
Cable modem CPE interface (CMCI), 79
Cable modems (CMs), 76–78, 80–81
Cable modem termination system (CMTS), 76
Cable MSOs, 55, 56, 96–97, 98
Cable operators, 98
Cable systems, 53
  functional elements of, 54
Cable telephony, 83–96
Caching, 32
Call control application server, 146–147
Call detail record (CDR), 81
Call Session Control Function (CSCF), 129, 130, 134
CapEx/OpEx, control of, 152. *See also* Operating expenses (OpEx)
Capital expenditure (CapEx), 157
  benefits of, 181
Cell phones
  data derived from, 265
  in 2020, 277, 279
Central device remote management, 33
Chief Technology Offices (CTOs), 157
China, innovation in, 257
Classifiers, 78
Cloud computing paradigm, 262, 266
Cluster heads, 217
Cluster maintenance, 220
Clusters, 220
CM transmitters, 76. *See also* Configuration Management (CM)
CMTS network side interface (CMTS-NSI), 79
Code generator, 202
Cognitive radios, 107
Collaboration agreements, 158
Combining ratio, 56, 57
Common Information Model (CIM), 165–166
Common Management Information Protocol (CMIP), 223
Common Management Information Service Element (CMISE), 232–233
Common Object Request Broker Architecture (CORBA), 233. *See also* CORBA management interfaces
Communication, to customer-facing entities, 43
Communications
  in 2020, 274, 276–280
  variety of devices available for, 270

in wireless ad hoc networks vs. Sensor networks, 211
Communications and computers (C&C) concept, xvi
Communications/computing, wearable, body-embedded, 116
Communications environment, future, 269
Communications industry, xvii
Communications management, future, 280
Communications networks, xvii
Community Antenna Television (CATV) system, 53
Computer industry, 272
Conditional access, 63
Conditional Access Table (CAT), 60
Configuration management (CM), 4, 110–111, 214, 242
FCAPS, 140
network-related, 142
service-related, 142
Configuration management database (CMDB), 166
Connectivity, 273
Connectivity Session Location Function (CLF), 138
Consumer-oriented apps storefront, 117–118
Content, focus on, 97–98
Content delivery network (CDN) architectures, 32
Content delivery network technologies, 30
Content distribution network techniques, 22
Content Encounter, 159
"Context aware" objects, 263
"Contextualization," 258
Contextualized communications, 270, 277, 278
Contextualized personal information, 270
Contract assessment, 12
Contract methodology, 168
Contract nodes, linking, 204
Contract order, 197
Contracts
benefits of, 169
linked, 198, 199
NGOSS, 167–170
SLA, 12
Contract "scaffolding," 205
Contract tooling, 203, 204
Contract variants, creating, 204
Control platform, 22
Converged/personalized/interactive multimedia services, 116–117
Convergence, 25, 101
CO-OP, 166, 167
Copper lines, 21, 28–29
CORBA management interfaces, 166. *See also* Common Object Request Broker Architecture (CORBA)
Core control, 33
Core transport functions, 109
Cost merits, IMS-related, 145
CPN network, 10–11
Craveur, Jean, xix, 15
Create connection (CRCX) message, 87–88
Cross-organization collaborations, 237
Cross-organization interactions, 238–239
Cultural islands, 9
Cultures, fragmentation of, 276
Customer care center (CCC) function, 41–42
Customer care centers (CCCs), 42, 44
Customer-centric approach, 18
Customer data function, 38
Customer equipment, 21
Customer experience
in broadband triple play, 47–51
main phases in, 47–48
Customer front-end, 36
Customer journey, stages of, 49–51
Customer Network Gateway Configuration Function (CNGCF), 138
Customer platform, 39
Customer Premise Network (CPN), 6
Customer premises devices, 113
Customer premises equipment (CPE), 26–28
additional function in, 27
Customer relationship, online-driven, 34
Customer relationship management (CRM), 21, 34, 255–256. *See also* "360° CRM"

Data analyses, statistical, 265–266
Data banks, analyzing, 266
Database repository, 200, 201
Data encryption standard (DES), 79
Data hubs, 269
Data management, 260

Data Over Cable Service Interface Specification (DOCSIS) standards, 60, 73–83. *See also* DOCSIS entries
  evolution of, 82–83
Data storage, 259–260
Data synchronization, 269
DCS-proxy (DP), 90
Deep packet inspection (DPI) feature, 122
Defense Interest Group meetings, 159
Delivery phase, in the customer experience, 48
Deployment view, 164
Desktop PCs, 2–3
Device management application server, 148
Device shift service, 8
Digital Living Network Alliance (DLNA), 127
Digital Shadow, 253
Digital storage, 259–260
Digital Storage Media—Command and Control (DSM-CC), 64–66. *See also* DSM-CC protocol
Digital subscriber line access multiplexers (DSLAMs), 22, 24
  unavailability of, 25
Digital subscriber lines (DSL), xv
Digital TV, 57–73
Digital TV Head-End, 70
Digital TV SI information, 60–61. *See also* DTV infrastructures
Digital video broadcast-handheld (DVB-H), 116–117
Digital video broadcasting (DVB) transmission standards, 57–58
Digital video recorder (DVR) functionality, 72
Direct messaging (DM). *See* DM entries
Disaster recovery planning (DRP), 45
Disjoined islands, system integration and interoperability of, 8–9
Display technology, 263–265
Distributed call signaling, 90
  architecture for, 91
Distributed Management Task Force Common Information Model (DMTF CIM), 165–166, 194–195
Distribution channels, multiplay/broadband, 49
DM notifier, 94, 95
DM server, 93, 94
DOCSIS 3.0, 77, 79, 82–83. *See also* Data Over Cable Service Interface Specification (DOCSIS) standards
DOCSIS cable modem start-up, 80–81
DOCSIS Data Link Layer, 76–79
DOCSIS IP detail records, 81–82
DOCSIS multicast operation, 80
DOCSIS Physical Layer, 74–76
DOCSIS Set-top Gateway (DSG), 54, 68, 69–70
DOCSIS versions, comparison of, 75, 76
DoD Architecture Framework (DoDAF), 191
Domains, third-party, 112–113
Domain specific languages (DSLs), 189–192, 193–194. *See also* DSL-based solution
  designing, 202–204
  linking, 204
Downloadable Conditional Access System (DCAS), 72
Downstream transmission convergence (DTC) sublayer, 74–75
DSL-based solution. *See also* Domain specific languages (DSLs)
  building, 200–205
  proposed content for, 200–201
DSM-CC protocol, 65, 68. *See also* Digital Storage Media—Command and Control (DSM-CC)
DSx messages, 77, 78
DTV infrastructures, 54. *See also* Digital TV
Due process, in the standards development process, 234
DVB-SI program decoding, 62
DVB-SI standard, 60–61
Dynamic bonding change request (DBC-REQ) mechanism, 78
Dynamic Host Configuration Protocol (DHCP), 30, 31, 80, 138
Dynamic line management (DLM), 28–29
Dynamic quality-of-service (D-QOS), 87

Early authentication encryption (EAE), 79
Eclipse, 192, 201–205
Eclipse Modeling Framework (EMF), 193
Eclipse plugin lifecycle management, 205
Eclipse tools, 193
Ecore, 193, 202

Ecosystems
business, 270–273
increasing the value of, 272
Edge functions, 109
Elementary stream (ES), 59
Element management systems (EMSs), 2, 243
Embedded MTA (E-MTA), 85
Embedded MTA (eMTA) start-up, 90–91
Embedded user equipment (E-UE), 93
EMS-OSS interfaces, 177–179
End-to-End Deployment and Interoperability Forum, 235
End-to-end PM functions, 8
End-to-end service management, 2, 9
End-to-end standards development, successful, 237
End-user functions, in Next Generation networks, 113
Energy consumption, technology evolution and, 254–255
Energy map, 215–216
Enhanced High Speed Packet Access (E-HSPA), 106
Enhanced Telecommunications Operations MAP (eTOM), 159–163, 195. *See also* eTOM entries
relationship to Infrastructure Technology Information Library, 162–163
Enhanced TV Binary Interchange Format (ETV-BIF) specification, 69
Enhanced TV (ETV)/interactive TV (iTV), 67–69
Enhanced video, in IPTV, 152
Enterprise UML model, 200
Entitlement control message (ECM), 62–63
Entitlement management message (EMM), 62–63
e-textiles, 262
Ethernet, long-haul managed, 103–104
Ethernet-based NG services, 121–122
Ethernet over Synchronous (EoS) Digital Hierarchy (SDH), 121–122
Ethernet PON (EPON), 103
Ethical issues, in technology usage, 125–126
eTOM levels, 160–162. *See also* Enhanced Telecommunications Operations MAP (eTOM)
eTOM standards, 241, 249
ETSI TISPAN standard, 136, 138, 182–183. *See also* European Telecommunications Standards Institute (ETSI); Telecommunications and Internet converged Services and Protocols for Advanced Networking (TISPAN)
EuroDOCSIS, 73, 75, 80
European Telecommunications Standards Institute (ETSI), 183. *See also* ETSI TISPAN standard
Event correlation, 10–11
Evolution Data Optimized (EVDO), 106
Evolved Social Networking Service (E-SNS), 118
Extended application information table (XAIT), 68
Extended Information Tables (EITs), 62
Extended NGN architecture, 132
Extensible Markup Language (XML), 170–174, 176
EyeRes, 274

Fargano, Michael, xix
Fault, Configuration, Accounting, Performance, and Security (FCAPS) operations, 3, 242
changes in, 4–5
requirements for, 140–141
Fault management (FM), 4, 110, 215, 242
FCAPS, 140
network-related, 142
service-related, 142–143
Faults, disaster, and overload (FDO) management, 124
Fiber accesses, 21–22
Fiber to the home (FTTH), xv, 29
Fiber to the premises (FTTP), 103
Field operation function, 41
"Find-bind and execute" paradigm, 168
Fixed access network transformation, 30–31
Fixed mobile convergence (FMC), 129. *See also* FMC service
Floor information management, 133
FMC service, IMS-based, 134. *See also* Fixed mobile convergence (FMC)
Focused Interoperability Forum, 235
Forum (Industry Forum), 234
Forum types, 245
Forward Data Channel (FDC), 64

FTTx options, 29
Fully qualified domain name (FQDN), 86
"Full routed" mode box, 31
Functional architectures, 213–214
Functional entities (FEs), 118–119
Functional frameworks, 241
Function distribution, 27

Gantt Chart, for project/program management, 238–239
Gate control, RACS, 136
Gateway functions, 109
Gateway GPRS Support Node (GGSN), 131
General network/service SDO/Forum, 239. *See also* Standards development organizations (SDOs)
General Standards Development/ Coordination Framework, 236–239, 245
General standards development process, 233–239, 248–249
Generic Framing Procedure (GFP), 121, 122
Geographical clustering, 221
Gigabit Ethernet (GEth) technologies, 30
Gigabit PON (GPON), 103
Globally Executable MHP (GEM), 67
Global MSF Interoperability (GMI), 153
Global System for Mobile (GSM) infrastructures, 19
Global telecommunications industry, 1
Global vision, 23
"Glocal" innovation, 257–259
"Glocalization," 258
Government Emergency Telecommunications Services (GETS), 5
Grand-separation, for pay-per-use service, 117
Graph-based clustering, 220–221
Graphical Modeling Framework (GMF), 193
Graphical user interface (GUI), 12. *See also* GUI editor
Guerilla, an Adaptive Management Architecture for Ad-hoc Networks, 213
Guerilla architecture, 218–219, 221, 222, 225
Guerilla MIB (GMIB), 225. *See also* Management Information Base (MIB)
GUI editor, 203. *See also* Graphical user interface (GUI)

Hand-held devices, wireless charging of, 115–116
Handtops, 3
Hard drives, 259
Hardware abstraction layer, 157–158
HD services, 58. *See also* High-definition (HD) broadcast services
HD television screens, 264
Heterogeneous information networks, xviii
HFC network maintenance, 56
HFC network upgrades, 56–57
HFC planning/inventory, 55–56
Hierarchical manager-agent communication architecture, 220
High-definition (HD) broadcast services, 65. *See also* HD entries
High-definition voice, 115
High-Speed Downlink Packet Access (HSDPA), 20
HomeGrid Forum, 124
Home networking drivers, 4
Home Networking (HNet) Forum, 123
Home network paradox, 27–28
Home networks (HNs), 26–28, 113
  changes in, 4
Home Subscriber Server (HSS) database, 183
Home Subscriber Server/Subscription Locator Function (HSS/SLF), 130
Host digital terminal (HDT), 84
Hot spot identification, 11
Humane services, NG technology-based, 125
Human machine interfaces (HMIs), 104
Hybrid fiber-coaxial (HFC) network, 55–57

Icons, in 2020, 277–279
IEEE Network Management Series, 2
Implementation view, 164, 165
IMS architecture, 33. *See also* IP Multimedia Subsystem entries
IMS-based FMC service, 134
IMS-based IPTV architecture, 152–153
IMS-based IPTV service, 134–135
IMS-based NGNs, 135

IMS platforms, 187
  advantages of, 144–153
IMS service control (ISC), 130–131
IMS services, 133–135
IMS standardization, 129
In-band (IB) channels, 64
Independent network providers, growth of, 267
India, innovation in, 257
Information, layered, 279–280
Information age, xvii
Information and communications technology (ICT), progress in, xv
Information architectures, 216–228
Information exchange, in 2020, 274
Information security solutions, 123
Information sharing, control of, 278
Information System (IS), 33–40. *See also* IS entries
Information technology (IT). *See* IT entries
Information technology SDO/Forum, 240. *See also* Standards development organizations (SDOs)
Infrastructures
  bottom-up, 258
  evolution of, 257
  fixed and mobile, 21
Infrastructure Technology Information Library (ITIL), 162–163
Innovation
  in business ecosystems, 271
  as a competitive differentiator, 275
  global-localization of, 257
Innovative offers, quick introduction of, 33–34
Innovators, as a success factor, 275
Integrated development environments (IDE), 176
Integrated receiver/decoder (IRD), 71–72
Integrated Services Digital Network (ISDN), 20–21
Intelligent Network, 268
"Intention" detectors, 265
Interface frameworks, 241
International Standards Development Organization (SDO), 235. *See also* Standards development organizations (SDOs)
Internet
  future of, 272–273
  management of, 255
  success of, 267
  in 2020, 274–276
Internet access support, transformation to triple play support, 30–31
Internet Engineering Task Force (IETF), 232–233
Internet functionalities, 275
Internet Group Management Protocol (IGMP) interaction, 24
Internet Protocol (IP), 3, 58. *See also* IP entries
  in Next Generation networks, 113–114
Internet Protocol Detail Records (IPDR), 80, 81–82, 98, 166, 167
Internet Protocol Television (IPTV), 151. *See also* IPTV entries
Internet Protocol version 4 (IPv4), 113–114, 121
Internet Protocol version 6 (IPv6), 114, 121
Internet with Things, 253
Interrogating CSCF (I-CSCF), 132
Inventory, HFC, 55–56
IP-based multimedia services, 144. *See also* Internet Protocol (IP)
IP-based NG services, 121–122
IP-based packet-switched networks, 3
IPDR Streaming Protocol (IPDR/SP), 81–82. *See also* Internet Protocol Detail Records (IPDR)
IP Multimedia Subsystem (IMS), 22, 92, 101, 118–119. *See also* IMS entries
IP Multimedia Subsystem architecture, 129–132, 182–183
  OSS interaction with, 183–185
IP Multimedia Subsystem (IMS)-based signaling, 3
IP networks, 31
IP packet loss, 26
IPTV architectures, 151–152. *See also* Internet Protocol Television (IPTV)
IPTV-GSI, 135
IPTV platform, 33
IPTV service, IMS-based, 134–135
IS architecture, for triple/quadruple/multiple play business, 35. *See also* Information System (IS)
IS equipment, 22
IS infrastructures, designing for triple and quadruple play, 24–26

Islands, disjoined, 8–9
IS/network/service platform cooperation, 39–40
IS operation functions, roles and responsibilities of, 45–46
IS tools, for service management centers, 43–44
IT infrastructure, 21
IT platforms, in triple and quadruple play contexts, 44–45
ITU-T J.83 standard, 58
ITU-T standards, 135
ITU-T Telecommunications Management Network (TMN), 155–156

J2EE for Java 5 (JEE5), 171
Jacobs, David, xx, 53
Java Reference application server (JRI), 170–171

Kawakami, Keizo, xx, 129
Kenyoshi, Kaoru, xx, 129
Key exchange mechanism, 78–79
Key performance indicators (KPIs), 11, 43
Khasnabish, Bhumip, xx–xxi, 101

lacm group, 224–225, 226
Laptops, 3
Law/regulation standards, 232
Legacy OSS, 177–178
  migration from, 182
Lifecycle management, 198–200
Link layer security, 78
Local Area Network (LAN), wireless, 208
Local loop unbundling (LLU), 48
Local operators, relationship with, 49
Local wireless drops, 269
Location application server, 148
Logical architectures, 214–216
Logical Link Control (LLC), 79
Logical management layers, 242
Lombard, Didier, 51
Long-haul managed Ethernet, 103–104
Long term evolution (LTE), 106, 264

Maintenance reduction, IMS-related, 144–145
Managed element, 186
Managed object classes, 225
Management
  within the ecosystem framework, 255–256
  of NG services, 123–124
  scope of, 240–241
  of wireless ad hoc and sensor networks, 207–230
Management functional areas, 242
Management functions, in Next Generation networks, 110–112
Management information/models, structure of, 223–225
Management Information Base (MIB), 217. *See also* Anmp MIB; Guerilla MIB (GMIB); RMON-MIB; SNMP MIBs
Management interfaces/protocols, 223
Management of Mobile Ad-hoc Networks: Information model and probe-based architecture, 213
Management SDO/Forum categories, 239–240, 249. *See also* Standards development organizations (SDOs)
Management standards
  frameworks and architecture in, 241–243, 249–250
  history of, 232–233
  principles and concepts in, 240–241
  principles in, 249
Management standards development
  challenges and trends in, 251
  key challenge for, 248
  key lessons related to, 250
  strategic framework for, 244–248, 250
Management-standards focused SDO/Forum, 240. *See also* Standards development organizations (SDOs)
Management system applications, examples of, 10–13
Manager-agent communication models, 217–223
"Manager-agent" model, 216–217
MANNA information architecture, 216, 219, 223, 225
MANNA: Management Architecture for wireless Sensor Networks, 213
MANNA object classes, 226–227
MAP messages, 76, 77. *See also* Enhanced Telecommunications Operations MAP (eTOM)
Markets, fragmentation of, 276
Mash-ups, 272

Master Guide Table (MGT), 61
Master Street Address Guide (MSAG), 56
M-Card, 72
MDSD approach, 194
Media Access Control (MAC) Sublayer, 76–78
Media control functions, 147–148
Media gateway (MG), 86
Media gateway controller (MGC), 86, 88
Media Gateway/Media Gateway Control Function (MGW/MGCF), 130
Media resource control application server, 147–148
Media terminal adapters (MTA), 84–90
Merger and acquisition (M&A) islands, 8–9
Messaging application server, 148
Metadata, 266
Meta-meta-model (M3), 190–191
Meta-model, 168, 169
Meta-Object Facility (MOF), 191, 194
Microprocessors, 261
Mission critical systems, xv
Misu, Toshiyuki, xxi, 129
Mobile Ad Hoc Networking (MANET), 106–107, 209
Mobile and Managed Peer-to-Peer (M2P2P) service, 115
Mobile Application Part (MAP). *See* MAP messages
Mobile Internet, for automotive and transportation, 117
Mobile Internet Protocol version 6 (MIPv6), 114
Mobile virtual network operators (MVNO), 96
Mobile Worldwide Inter-operability for Microwave Access (M-WiMax), 105
MOD Architecture Framework (MoDAF), 191
Model-driven architecture (MDA), 168, 193
Model-driven software design, 193–194
Modular-CMTS (M_CMTS), 83
Moral issues, in technology usage, 125–126
Motion Picture Experts Group (MPEG) standards, 58–60. *See also* MPEG entries
MPEG-2, 19–20. *See also* Motion Picture Experts Group (MPEG) standards
MPEG-2 decoder, 71
MPEG-2 Program Specific Information (PSI), 60
MPEG-2 system, 59–60
MPEG-2 transport stream, 59–60
MPEG-4, 20, 60
MTOSI APIs, 180. *See also* Multi-Technology Operations System Interface (MTOSI)
MTOSI case study, 170–175
"Multi-hop" communications, 209
Multimedia Home Platform (MHP) specification, 67
Multimedia Resource Function (MRF), 129–130
Multimedia services, converged/personalized/interactive, 116–117
Multiplay/broadband distribution channels, 49
Multi-play service offerings, defined, 48–49
Multiple-input multiple-output (MIMO) system, 106
Multiple play strategy, 18
Multiple-play technology, 52
Multi-protocol label switching (MPLS), 22, 121
Multi-service access nodes (MSANs), 22, 23, 29, 177
MultiService forum (MSF), 119–120
Multi-Technology Network Management (MTNM), 166–167
Multi-Technology Operations System Interface (MTOSI), 166–167. *See also* MTOSI entries
Mutualization, 25

National/Regional SDO, 235. *See also* Standards development organizations (SDOs)
NCS cable IP telephony, call set-up behavior of, 87–90
Network Access Configuration Function (NACF), 138
Network Address Translation (NAT), 93
Network attachment control functions, 110
Network Attachment Subsystem (NASS), 138–139
Network build, considerations for, 55
Network Control Signaling PacketCable 1.x, 85–90
Network Information Table (NIT), 61, 62

Network infrastructures
changes in, 3–4
designing for triple and quadruple play, 24–26
Networking paradigms, new, 268–270
Network interface device (NID), 83, 84
Network Interoperability Consultative Committee (NICC) standards, 168
Network/IS/service platform cooperation, 39–40
Network layer, 79–80
Network-level performance parameters, 122
Network management, 255
aspects and framework of, 212–213
importance of, 1
Network Management Forum (NMF), 233
Network management operation requirements, 141–142
Network Management Systems (NMSs), 2
Networks, value of, 16–17
Network-to-network interface (NNI), 108
Network topology, 215
Network upgrades, 56–57
New Generation Operations and Software Systems (NGOSS), 159, 183. *See also* Next Generation entries; NGOSS entries
New networking paradigms, 268–270
New services, development and introduction of, 145–149
New skills, in operations, 47
Next Generation (NG) devices, configuring, 123–124. *See also* NG entries
Next Generation management standards, 231–251
Next Generation networks (NGNs), xv–xvi, 1, 57, 108–114, 126, 155. *See also* NGN entries
applications functions in, 112
Internet Protocol in, 113–114
management functions in, 110–112
network and service management for, 139–144
positioning of SDP in, 145–146
QoS control and authentication for, 135–139
services implemented on, 150–153
Next Generation OSS architecture, 155–206
business benefit of, 179–181
Next Generation PON (NGPON), 103
Next Generation services, 114–121
billing, charging, and settlement of, 124
device configuration and management of, 123–124
faults, overloads, and disaster management of, 124
management of, 121–124
performance management of, 122–123
security management of, 123
Next Generation Society, 124–126
Next Generation technologies, 102–108
future works/trends in, 126–127
wireless, 104–107
NGN architectures, 2, 109. *See also* Next Generation networks (NGNs)
NGN legal restrictions requirements, 144
NGN management, requirements of, 139–140
NGN OSS function/information view reference model, 187
NGN OSS Service Object (NOSI) concept, 187–188
NGOSS contracts, 167–170. *See also* New Generation Operations and Software Systems (NGOSS)
NGOSS framework, 163, 165
NGOSS meta-model, 169
NGOSS principles, 188
NG services architectures, 118–120. *See also* Next Generation entries
NG technology-based humane services, 125
NG wireline devices, 103
Nodes
in wireless ad hoc networks, 209, 211
in wireless sensor networks, 210, 211
Nomadic Manager Module (NMM), 225
Nomadic managers, 217–219, 222
Non-IMS IPTV architecture, 151–152
Non-NGN IPTV architecture, 151
Normal play time (NPT), 65

Object Constraint Language (OCL), 194
Object Management Group (OMG), 189–191
Object-Oriented (OO) class, 187
OCAP middleware, 68
Off-net call, 88–90
OLED technology, 276
OMG UML specification, 191
1-to-n communication, message duplication and transmission in, 133–134

Online content, 98
"Online customer centric" vision, 36
On-net call, 87–88
OOB FDC control information, 69. *See also* Out-Of-Band (OOB) channels
Open Cable Applications Platform (OCAP), 67
Open Mobile Alliance Device Management (OMA-DM), 93–95
Open service environment (OSE), 130–131
Open-source tool environments, 201–205
Open Systems Interconnection (OSI), 225
Operating cost reduction, IMS-related, 144–145
Operating expenses (OpEx), 157. *See also* CapEx/OpEx
  benefits of, 181
  control of, 47
Operating model, 40
Operational challenge, of triple/quadruple play services, 40–47
Operational expenditure (OpEx). *See* Operating expenses (OpEx)
Operational Support System (OSS) standard, 54
Operations, new skills in, 47
Operations support systems (OSSs), 11. *See also* OSS entries
  interaction with IMS and subscriber management, 183–187
  upgrading or transforming, 47
Operations support systems architecture. *See also* Next Generation OSS architecture; OSS entries
  importance of standards to, 156–158
  information framework of, 163–165
  Next Generation, 155
  Telemanagement Forum for, 158–159
Optical fiber, 108
Optical memories, 259
Order management/delivery, 39
Organizational challenge, of triple/quadruple play, 51
Organization for the Advancement of Structured Information Standards (Oasis), 169
Orobec, Steve, xxi
Orthogonal frequency division multiplexing (OFDM), 104–105
Orthogonal frequency-division multiple access (OFDMA, O-FDMA), 106, 117
OSGi standard, 205
OSS/EMS/network interoperability reference model, 244. *See also* Operations support systems entries
OSS/J, 166, 167
OSS network management, 179
OSS standardization, 156
OSS standards, 205
OSS transition strategies, 181–182
"OSS Vision," 184–185
Out-Of-Band (OOB) channels, 64. *See also* OOB FDC control information

Packet-based transmission, 98
PacketCable 1.0, 85
PacketCable 1.5, 85
PacketCable 2.0, 91–96, 98
PacketCable 2.0 QoS interface descriptions, 97
PacketCable Application Manager (PAM), 96
PacketCable Multi Media (PCMM) specification, 78
Packet Data Gateway (PDG), 134
Packetized elementary stream (PES), 59
Packet loss rate, 26
Packet transfer, NGN, 135–136
Parlay API, 149
Parlay-X, 149
Passive optical network (PON) fiber, 103
Patent policy, 234
Pay-per-use service, grand-separation for, 117
Performance increase, 256
Performance management (PM), 4, 111, 215, 242
  FCAPS, 141
  network-related, 142
  of NG services, 122–123
  service-related, 142–143
Performance Management System (PMS) applications, 10–13
Periodic clustering algorithm, 222
Personal computers (PCs), 2–3
Personalized communications, 278
Personal Video Recorder (PVR) functionality, 72
Person-to-service (embedding data) communications, 270
Photonics technologies, 103
Physical architectures, 213–214

Physical media-dependent (PMD) sublayer, 74–75
Physical Termination Point (PTP), 167
Place shift service, 8
Plain Old Java Object (POJO), 171
Planning, HFC, 55–56
Planning and engineering systems, SMS Integration with, 13
Platform Independent Model (PIM), 193, 194
Platform Specific Model (PSM), 194
Players, increase of, 257
Plevyak, Thomas, xxi–xxii, 1
"Plug and play" equipment, 26–27
Plug-ins, 272
Point of Deployment Module/ CableCard, 72–73
Point-to-point protocol (PPP), 30
Policy and Charging Rules Function (PCRF), 96
Polymer memories, 259
Power line transmission (PLT), 27
powerUsage group, 223–224
Presence application server, 147
"Printed electronics," 261, 262
Private data, 60
Probes, 218
Processing evolution, 261–262
Process repository, 200, 201
Product, defined, 163
Profile Database Function (PDBF), 138
Program Allocation Table (PAT), 60
Program Map Table (PMT), 60
Program stream, 59
Project execution, 238–239
Project/program management Gantt Chart, 238–239
Protocol independent information modeling, 241
Provisioning function, 41
Proxy CSCF (P-CSCF), 131–132
PSTN emulation, 84. *See also* Public switched telephone network (PSTN)
PSTN simulation, 84
Public service answering points (PSAP), 56
Public switched telephone network (PSTN), 19. *See also* PSTN entries
Push Proxy Gateway, 94
Push to Talk over Cellular (PoC) service, 133–134, 150
Push to X, 150–151

QAM channel, 71. *See also* Quadrature amplitude modulation (QAM)
QoS control, in NGN, 135–136. *See also* Quality of Service (QoS)
QoS management, 2
QoS reference model, 96
QoS requirements, 6
QoS session call set-up quality parameters, 26
QoS transport quality parameters, 26
Quadrature amplitude modulation (QAM), 58, 74. *See also* QAM channel
Quadruple play. *See* Triple/quadruple play entries
Quality-of-Experience (QoE) metrics, 2
Quality of Service (QoS), 5, 11, 25–26. *See also* QoS entries
  for broadcast television, 24–25
Query View Transform (QVT) language, 194

RACS function blocks, 137. *See also* Resource and Admission Control Subsystem (RACS)
Radio Frequency Identification (RFID) technology, 263
Rating Region Table (RRT), 62
Real Time Protocol (RTP), 84
Real-time service monitoring function, 42
Real time streaming protocol (RTSP), 65
Real-time Transport Control Protocol (RTCP), 133
Regulatory changes, 5
Regulatory/legal issues, 19
Regulatory responsibilities MSO, 56
Representational state transfer (REST), 176–177
Research, sensor-related, 263
Residential gateways (RGWs), 16, 27
Resource and admission control functions, 110
Resource and Admission Control Subsystem (RACS), 136–137
Resource control, 33
Resource domain, 165, 167, 182
Resource reservation, RACS, 136
Return Data Channel (RDC), 64
RMON-MIB, 224. *See also* Management Information Base (MIB)
Roomba, as an example of an autonomic system, 267–268

Running Status Table (RST), 61
Run phase, in the customer experience, 48
Runtime code generations, 192
Runtime prototypes, 205

Sahin, Veli, xxii, 1
Sale phase, in the customer experience, 47
Saracco, Roberto, xxii–xxiii, 253
Satellite access, 21
Satellite networks, 209
S-Card, 72
SDO/Forums, 232, 234, 243. *See also* Standards development organizations (SDOs)
  mapping to general standards organization type and management standards, 246–247
SDO/Forum Types, 235–236
SDP architecture, Push to X on, 150. *See also* Service delivery platforms (SDPs)
Security management (SM), 4, 5, 111–112, 215, 242
  FCAPS, 141
  network-related, 142
  of NG services, 123
  service-related, 142–143
Self-organizing networks, 214–215
Sensor evolution, 262–263
Sensor Information Networking Architecture (SINA), 228
Sensor Management Protocol (SMP), 228
Sensor Query and Data Dissemination Protocol (SQDDP), 228
Sensor Query and Tasking Language (SQTL), 228
"Separation of concerns" principle, 203, 241
Server NG technologies, 107–108
Service, defined, 163. *See also* Services
Service and contract assurance, 12
Service and network platforms, 38
Service and resource control, 33
Service assessment, 12
Service Authentication function, 133
Service aware networks, 5–7
Service continuity, QoS parameters linked to, 25
Service control functions, 110
Service delivery, quick, 153
Service delivery platforms (SDPs), 3–4, 144. *See also* SDP architecture
  features of, 146
  roles of, 145–149
Service deployment speed, 20
Service Description Table (SDT), 61
Service development, future milestones in, 20
Service enhancement requirements, 143
Service flows, DOCSIS, 77
Service impact analysis tools, 43
Service Level Agreements (SLAs), xv, 197
Service level objectives (SLO), 46
Service maintenance function, 42–43
Service management, 43
Service management center (SMC) function, 40, 42–43
Service management centers, IS tools for, 43–44
Service Management Layer (SML), xvii, 2
Service management operation requirements, 142–143
Service management system interworking, 7
Service management systems (SMSs), 2, 5. *See also* SMS entries
  integration with planning and engineering systems, 13
Service operations center (SOC), 12
Service-Oriented Architecture (SOA), 169, 195, 183, 188
Service pilots, 47
Service platform/IS/network cooperation, 39–40
Service platforms, 22, 33
  in triple and quadruple play contexts, 44–45
Service Policy Decision Function (SPDF), 137
Service provider OSS, 165, 195–196
Service providers (SPs), 2–7. *See also* SP environment
  standards and, 156–157
Service quality analysis function, 42
Services, 269
  creation of, 276
  layered, 279–280
  Next Generation, 114–121, 126–127
Service stratum, of Next Generation networks, 110
Service user profiles, 110
Serving-call session control function (S-CSCF), 92, 130–131

Session Description Protocol (SDP), 84
Session Initiation Protocol (SIP), 90, 92, 144–145
Set-top box (STB), 60
  TV programming reception by, 71–72
Shared Information/Data (SID) model, 163–166, 167–168, 191
SID Product and Service domains, 165
Signaling gateway (SG), 86
Simple Network Management Protocol (SNMP), 80, 223
Simple Object Access Protocol (SOAP), 176. *See also* SOAP packet
Single carrier FDMA (SC-FDMA), 106
Skill centers (SkC), 41
SLA contractual performance, 12
Smart materials, 264–265
SMC service pilots, 47. *See also* Service management center entries
SMS actions, hot spot identification and, 11. *See also* Service management systems (SMSs)
SMS applications, 10–13
SNMP MIBs, 223, 225. *See also* Management Information Base (MIB)
SOAP packet, 174. *See also* Simple Object Access Protocol (SOAP)
Social networking services, 118
Social networks, in 2020, 275, 276
Society of Cable Telecommunications Engineers (SCTE), 53
  standards of, 64
"Soft Telco," 157
Software as a Service (SaaS), xvi
Software-based business communications service, 114–115
Software design, model-driven, 193–194
Solid state memories, 259
Source-specific multicast (SSM), 83
Specific network/service SDO/Forum, 239–240. *See also* Standards development organizations (SDOs)
SP environment, 181. *See also* Service providers (SPs)
Stand-alone user equipment, 93
Standardization
  ETV, 67
  among operators, 52
Standardization organizations, 134–135
Standards
  general drivers for, 232
  interest level in, 244
  OSSAI, 156–158
  real network implementation of, 177–179
  relationship between, 199
  structured approach to, 241
Standards bodies, 158–159
Standards development
  "back-game" for, 245
  "front-game" for, 244–245
  strategic human side of, 245–248, 250
Standards development guidance, 235
Standards development organizations (SDOs), 22, 233–234
Standards development process, key attributes of, 233, 234–235, 248
Standards development resources, 251
Standards drivers, 248
Standards engagement determination, strategic questions for, 244–245
Standards models, 189, 194–195
Standards work, strategic progression of, 245, 250
Statistical data analyses, 265–266
"Sticker" communication paradigm, 270
Strategic NGN management standards development, key lessons related to, 250
Strategic questions, for standards engagement determination, 244–245
Strategic standards development, 231–251
Subscriber Accounting Management Interface Specification (SAMIS), 82
Subscriber management (SuM), OSS interaction with, 185–186
Subscriber management system (SMS), 63
Subscription Management model, 165
"Supervisor/agency" model, 217
Support object classes, 225
Switched Digital Video (SDV), 65–66
  implementations of, 66
Switched digital video system, 66
Synchronous Optical NETwork (SONET), 121, 122
Sync Markup Language (SyncML), 93–94
System implementation guidelines, 238
System requirements, 237
Systems analysis, 237
Systems design, 237–238
System Time Table (STT), 62
System view, 164, 196

T1M1, 232–233
Task Assignment and Data Advertisement Protocol (TADAP), 228
Tasks, in the business service framework, 196
Technical management center (TMC) function, 40–47
Technical tool box, 21–23
Technologies
  Next Generation, 102–108
  wireline NG, 102–104
Technology evolution, 254–257, 258–259
  pace of, 256
Technology islands, 9
Technology-neutral architectures, designing, 189
Technology transformation, 125
Technology usage, ethical and moral issues in, 125–126
Technology/usage revolutions, 16
Telco Information System (IS), 33–40
Telecom Act of 1996, 72
Telecom engineers, 267
Telecommunication operators, 51. *See also* Telecom operators
Telecommunications, 126
Telecommunications and Internet converged Services and Protocols for Advanced Networking (TISPAN), 183. *See also* TISPAN entries
  goal of, 184
Telecommunications infrastructures, evolution of, 257
Telecommunications Management Network (TMN) framework, 212. *See also* TMN architecture
Telecommunications Management Network model, xvii
Telecommunications Management Network standards, 249
Telecommunications networks/services management, forecast of, 253–280
Telecommunications operations, importance of, 1
Telecom operators, 273. *See also* Telecommunication operators
  challenges facing, 17–18
  economic, service, and commercial challenges of, 18–20
  technical challenge facing, 17–18, 20–40
  triple/quadruple play for, 15–18
Telecoms Application Map (TAM), 198
TeleManagement Forum (TMF, TM Forum), 200, 240. *See also* TMF entries; TM Forum Interface Program (TIP)
  for OSS architecture, 158–159
"Telepresence," 258
Television, three-dimensional, 116
10 Gigabit Ethernet PON (10G-EPON), 103
The Open Group Architecture Framework (ToGAF), 192
Third Generation Partnership Project (3GPP) IP Multimedia Subsystem (IMS) architecture, 118. *See also* 3GPP entries
3rd Generation Partnership Project (3GPP) IP Multimedia Subsystem (IMS) specification, 92, 93
3rd Party Call Control (3PCC), 147
Third-party domains, 112–113
Three-dimensional television (3D-TV), 116
3GPP IMS model, 183, 184, 185–186, 194–195. *See also* Third Generation Partnership Project entries
3GPP NRM, 186
3GPP Website, 119
3G radio access network (RAN), 20
"360° CRM," 38
Time and Date Table (TDT), 61
Time-division multiplexing (TDM), 3, 86
Time shift service, 8
Time to market, IMS-related, 145
TIP model, 191. *See also* TM Forum Interface Program (TIP)
TIP team, 167
TISPAN model, 194–195. *See also* Telecommunications and Internet converged Services and Protocols for Advanced Networking (TISPAN)
TISPAN vision document, 188
TISPAN WG-8, 183
TMF608 data model, 166, 178, 179. *See also* TeleManagement Forum (TMF, TM Forum)
TMF NGOSS, 188
TM Forum Interface Program (TIP), 166–167. *See also* TeleManagement Forum (TMF, TM Forum); TIP entries

TMN architecture, 241–243. *See also* Telecommunications Management Network entries
Topology, of wireless ad hoc networks vs. Sensor networks, 211
topologyMaintenance group, 224
Transparency, 234
Transportation, Mobile Internet for, 117
Transport control functions, 109
Transport plane requirements, to support NG services, 120–121
Transport stratum, in next generation networks, 108–110
Transport stream, 59
Transport user profiles, 110
Tree editor, 203
Triple play services, 5–6
  true nature of, 20
Triple play support, transformation to, 30–31
Triple/quadruple/multiple play business, IS architecture for, 35
Triple/quadruple play
  designing network and is infrastructures for, 24–26
  operating IT and service platforms in, 44–45
  organizational challenge of, 51
  overall architecture supporting, 23–24
Triple/quadruple play sales, phases in, 47–48
Triple/quadruple play service offer requirements, 19–20
Triple/quadruple play services
  cable perspective on, 53–100
  management of, 15–52
  operational challenge for, 40–47
  quality problems with, 18
Triple/quadruple play tool box, 21–23
Triple shift service, 7–8
Trivial File Transfer Protocol (TFTP), 80
Tru2way, 65, 67
Trunk Gateway Control Protocol (TGCP), 86
TT Service Management Forum (itSMF), 162, 163
TV-middleware, 67

UE Provisioning specification, 93
Ulema, Mehmet, xxii, 207
Ultra Mobile Broadband (UMB), 106
UML meta-model, 190. *See also* Unified Modeling Language (UML)
UML specification, 189–192
Unbound OCAP applications, 67
Unbundled-local loop (ULL), 30
Unified Modeling Language (UML), 163, 164–165, 236. *See also* UML entries
Uniform Resource Locator (URL), 176
Universal Description, Discovery and Integration (UDDI), 168
Upstream channel descriptor (UCD), 80
User Access Authorization Function (UAAF), 138
User equipment (UE), 92–96
User-generated content (UGC), 3, 15
User interface, enhancing the quality of, 36
User profile management application server, 149
User-to-network interface (UNI), 108

Value-added service, management challenges for, 7–8
Value chains, 51
  modeling, 271
Vehicles, Mobile Internet for, 117
Video on demand (VoD), 65
Virtual Channel Table (VCT), 61
Virtual concatenation (VCAT), 121, 122
Virtualization, 107–108
"Virtual visits" service, xvi
Voice over Internet Protocol (VoIP), 5–6, 84. *See also* VoIP entries
  packet loss and, 26
Voice-over-IP (VOIP) services, 157
VoIP networks, 88
VoIP QoS KPI dependencies, 11
VoIP service quality, 25

Wavelength-division multiplexed PON (WDM-PON), 103
Wearable, body-embedded communications/computing, 116
Web 2.0 paradigm, 258
Web-based graphical user interface (GUI), 12
Web Service Definition Language (WSDL), 171, 176
Web services

scale-ability of, 170
XML-based, 174
Wi-Fi, personalized and extended, 104–105
Wi-Fi alliance, 105
Wi-Fi communications, 105
Wireless, 96–97
evolution of, 268
Wireless ad hoc networks
applications for, 209–210, 229
management aspects and framework of, 212–213
management of, 207–230
overview of, 209–210
versus wireless sensor networks, 211
Wireless cellular networks, 208
Wireless charging, of hand-held devices, 115–116
Wireless drops, 269
Wireless fixed networks, 209
Wireless Identification and Sensing Platform (WISP), 263
Wireless Mesh Networking (WMN), 107
Wireless NG technologies, 104–107
Wireless/optical networks, as a supporting infrastructure, 266
Wireless sensor networks
applications for, 210–211, 229
management aspects and framework of, 212–213
management of, 207–230
overview of, 210–211
protocol architecture for, 228
Wireline NG technologies, 102–104
Work Initiation Forum, 235
WS-* based standards, 176

xDSL techniques, 28
XMI interchange format, 192

Zachman Framework, 159–160
Zachman viewpoints, 160
ZigBee, 104